Hands-On Introduction to
LabVIEW™
for Scientists and Engineers

Hands-On Introduction to LabVIEW™ for Scientists and Engineers

Second Edition

John Essick

Reed College

New York   Oxford
OXFORD UNIVERSITY PRESS

Oxford University Press, Inc., publishes works that further
Oxford University's objective of excellence
in research, scholarship, and education.

Oxford   New York
Auckland   Cape Town   Dar es Salaam   Hong Kong   Karachi
Kuala Lumpur   Madrid   Melbourne   Mexico City   Nairobi
New Delhi   Shanghai   Taipei   Toronto

With offices in
Argentina   Austria   Brazil   Chile   Czech Republic   France   Greece
Guatemala   Hungary   Italy   Japan   Poland   Portugal   Singapore
South Korea   Switzerland   Thailand   Turkey   Ukraine   Vietnam

For titles covered by Section 112 of the US Higher Education Opportunity Act, please visit
www.oup.com/us/he for the latest information about pricing and alternate formats.

Published by Oxford University Press, Inc.
198 Madison Avenue, New York, New York 10016
http://www.oup.com

Oxford is a registered trademark of Oxford University Press

LabVIEW™ is a registered trademark of National Instruments. This book is an independent
publication. National Instruments Corporation is not affiliated with the publisher or the author
and does not authorize, sponsor, endorse, or otherwise approve this book. The information in
this book was written by John Essick and not by National Instruments.

ISBN   978-0-19-992515-5

Printing number: 9 8 7 6 5 4 3 2

Printed in the United States of America
on acid-free paper

To my wife, Katie

Contents

Preface   xiii

1. THE WHILE LOOP AND WAVEFORM CHART   **1**

1.1   LabVIEW Programming Environment   1
1.2   Sine-Wave Plot using a While Loop and Waveform Chart   2
1.3   Block Diagram Editing   3
1.4   LabVIEW Help Window   20
1.5   Front Panel Editing   22
1.6   Pop-Up Menu   27
1.7   Finishing the Program   30
1.8   Program Execution   31
1.9   Program Improvements   33
1.10   Date-Type Representations   43
1.11   Automatic Creation Feature   46
1.12   Program Storage   48
     Do It Yourself   50
     Problems   51

2. THE FOR LOOP AND WAVEFORM GRAPH   **56**

2.1   For Loop Basics   56
2.2   Sine-Wave Plot using a For Loop and Waveform Graph   57
2.3   Waveform Graph   58
2.4   Owned and Free Labels   59
2.5   Creation of Sine Wave using a For Loop   60
2.6   Cloning Block-Diagram Icons   63
2.7   Auto-Indexing Feature   65
2.8   Running the VI   68
2.9 x-Axis Calibration of the Waveform Graph   69
2.10   Sine-Wave Plot using a While Loop and Waveform Graph   75
2.11   Array Indicators and the Probe Watch Window   79
     Do It Yourself   90
     Problems   92

3. THE MATHSCRIPT NODE AND XY GRAPH   **99**

3.1	MathScript Node Basics	99
3.2	Quick MathScript Node Example: Sine-Wave Plot	102
3.3	Debugging with Error List	109
3.4	Waveform Simulator using a MathScript Node and XY Graph	111
3.5	Creating an *xy* Cluster	115
3.6	Running the VI	116
3.7	MathScript Interactive Window	117
3.8	Adding Shape Options to Waveform Simulator	121
3.9	The Enumerated Type Control	122
3.10	Finishing the Block Diagram	124
3.11	Running the VI	128
3.12	Control and Indicator Clusters	129
3.13	Creating an Icon using the Icon Editor	136
3.14	Icon Design	137
3.15	Connector Assignment	142
	Do It Yourself	146
	Problems	148

4. DATA ACQUISITION USING DAQ ASSISTANT   **152**

4.1	Data Acquisition VIs	152
4.2	Data Acquisition Hardware	153
4.3	Analog Input Modes	155
4.4	Range and Resolution	157
4.5	Sampling Frequency and the Aliasing Effect	158
4.6	Measurement & Automation Explorer (MAX)	160
4.7	Simple Analog Input Operation on a DC Voltage	164
4.8	Digital Oscilloscope	175
4.9	Analog Output	184
4.10	DC Voltage Source	185
4.11	Software-Timed Sine-Wave Generator	192
4.12	Hardware-Timed Waveform Generator	194
4.13	Placing a Custom-Made VI on a Block Diagram	197
4.14	Completing and Executing Waveform Generator (Express)	199
4.15	Modified Waveform Generator	201
	Do It Yourself	203
	Problems	204

5. DATA FILES AND CHARACTER STRINGS **211**

5.1	ASCII Text and Binary Data Files	211
5.2	Storing Data in a Spreadsheet-Formatted File	213
5.3	Storing a One-Dimensional Data Array	214
5.4	Transpose Option	217
5.5	Storing a Two-Dimensional Data Array	220
5.6	Controlling the Format of Stored Data	224
5.7	The Path Constant and Platform Portability	226
5.8	Fundamental File I/O VIs	227
5.9	Adding Text Labels to a Spreadsheet File	233
5.10	Blackslash Codes	237
	Do It Yourself	239
	Problems	243

6. SHIFT REGISTERS **252**

6.1	Shift Register Basics	252
6.2	Quick Shift Register Example: Integer Sum	255
6.3	Numerical Integration and Differentiation using Shift Registers	258
6.4	Power Function Simulator VI	258
6.5	Numerical Integration via the Trapezoidal Rule	264
6.6	Trapezoidal Rule VI using Single Shift Register	267
6.7	Convergence Property of the Trapezoidal Rule	275
6.8	Numerical Differentiation using a Multiple Shift Registers	278
6.9	Modularity and Automatic SubVI Creation	284
	Do It Yourself	289
	Problems	289

7. THE CASE STRUCTURE **296**

7.1	Case Structure Basics	296
7.2	Quick Case Structure Example: Runtime Options using Property Nodes	299
7.3	Numerical Integration using Case Structures	310
7.4	Numerical Integration via Simpson's Rule	310
7.5	Parity Determiner using a Boolean Case Structure	313
7.6	Summation of Partial Sums using a Numeric Case Structure	318
7.7	Trapezoidal Rule Contribution using a Boolean Case Structure	321
7.8	Top-Level Simpson's Rule VI	323
7.9	Comparison of the Trapezoidal Rule and Simpson's Rule	326
	Do It Yourself	329
	Problems	330

8.   DATA DEPENDENCY AND THE SEQUENCE STRUCTURE   **338**

8.1   Data Dependency and Sequence Structure Basics   338
8.2   Event Timer using a Sequence Structure   342
8.3   Event Timer using Data Dependency   349
8.4   Highlight Execution   352
      Do It Yourself   354
      Problems   355

9.   ANALYSIS VIs: CURVE FITTING   **364**

9.1   Thermistor Resistance-Temperature Data File   364
9.2   Temperature Measurement using Thermistors   367
9.3   The Linear Least-Squares Method   370
9.4   Inputting Data to a VI using a Front-Panel Control   371
9.5   Inputting Data to a VI by Reading from a Disk File   376
9.6   Slicing Up a Multi-Dimensional Array   379
9.7   Curve Fitting using the Linear Least-Squares Method   385
9.8   Residual Plot   391
      Do It Yourself   394
      Problems   397

10. ANALYSIS VIs: FAST FOURIER TRANSFORM   **405**

10.1   The Fourier Transform   405
10.2   Discrete Sampling and the Nyquist Frequency   406
10.3   The Discrete Fourier Transform   407
10.4   The Fast Fourier Transform   408
10.5   Frequency Calculator VI   409
10.6   FFT of Sinusoids   412
10.7   Applying the FFT to Various Sinusoidal Inputs   414
10.8   Magnitude of the Complex-Amplitude   417
10.9   Observing Leakage   421
10.10 Analytic Description of Leakage   427
10.11 Description of Leakage Using the Convolution Theorem   430
10.12 Windowing   434
10.13 Estimating Frequency and Amplitude   440
10.14 Aliasing   443
       Do It Yourself   444
       Problems   445

11. DATA ACQUISITION AND GENERATION USING DAQmx VIs — **451**

11.1 DAQmx VIs — 451
11.2 Simple Analog Input Operation on a DC Voltage — 453
11.3 Digital Oscilloscope — 459
11.4 Express VI Automatic Code Generation — 466
11.5 Limitation of Express VIs — 467
11.6 Improving Digital Oscilloscope using State Machine Architecture — 469
11.7 Analog Output Operations — 480
11.8 Waveform Generator — 481
Do It Yourself — 485
Problems — 485

12. PID TEMPERATURE CONTROL PROJECT — **491**

12.1 Voltage-Controlled Bidirectional Current Driver for Thermoelectric Device — 491
12.2 PID Temperature Control Algorithm — 492
12.3 PID Temperature Control System — 495

13. CONTROL OF STAND-ALONE INSTRUMENTS — **498**

13.1 Instrument Control using VISA VIs — 498
13.2 The VISA Session — 499
13.3 The IEEE 488.2 Standard — 503
13.4 Common Commands — 503
13.5 Status Reporting — 504
13.6 Device-Specific Commands — 508
13.7 Specific Hardware used in this Chapter — 509
13.8 Measurement & Automation Explorer (MAX) — 511
13.9 Simple VISA-Based Query Operation — 518
13.10 Message Termination — 522
13.11 Getting and Setting Communication Properties using a Property Node — 523
13.12 Performing a Measurement over the Interface Bus — 527
13.13 Synchronization Methods — 532
13.14 Measurement VI Based on the Serial Poll Method — 538
13.15 Measurement VI Based on the Service Request Method — 545
13.16 Creating an Instrument Driver — 551
13.17 Using the Instrument Driver to Write an Application Program — 565
Do It Yourself — 571
Problems — 572

APPENDIX I: CONSTRUCTION OF TEMPERATURE CONTROL SYSTEM — **575**
APPENDIX II: PROGRAM CROSS REFERENCE TABLE — **583**

Index — **585**

Preface

Hands-On Introduction to LabVIEW for Scientists and Engineers provides a learn-by-doing approach to acquiring the computer-based skills used daily in experimental work. This book is not a manual-like presentation of LabVIEW. Rather, *Hands-On Introduction to LabVIEW* leads its readers to mastery of LabVIEW through the process of using this powerful laboratory tool to carry out interesting and relevant projects. Readers, who are assumed to have no prior computer programming or LabVIEW background, begin writing meaningful programs in the first few pages.

Hands-On Introduction to LabVIEW can be used as a text in an instructional lab course or for self-study by individual researchers. The book is designed for flexible use so that readers can easily choose the desired depth of coverage. The first four chapters, which form the foundation appropriate for all readers, focus on the fundamentals of LabVIEW programming as well as the basics of computer-based experimentation using a National Instruments data acquisition (DAQ) device. These opening chapters can be used as the basis of a three-week introduction to LabVIEW-based data acquisition. Subsequent chapters have been written as independently as possible so that an instructor or self-learner can fill out his or her course of study as desired. Those who work through most of the text's chapters will attain an intermediate skill level in computer-based data acquisition and analysis.

The progression of topics in *Hands-On Introduction to LabVIEW* is as follows:

Chapters 1–3: Fundamentals of the LabVIEW Graphical Programming Language. Central features of LabVIEW including its control loop structures, graphing modes, mathematical functions, and text-based MathScript commands are learned in the course of writing digitized waveform simulation programs.

Chapter 4: Basic Data Acquisition: Concepts of digitized data such as resolution, sampling frequency, and aliasing are covered. Then, using LabVIEW's high-level Express VIs, programs are written that execute analog-to-digital, digital-to-analog, and digital input/output tasks on a National Instruments DAQ device. Computer-based instruments constructed include a DC voltmeter, digital oscilloscope, DC voltage source, waveform generator, and blinking LED array.

Chapters 5–8: More LabVIEW Programming Fundamentals. Implementation of data file input/output, local memory, and conditional branching in LabVIEW is investigated while writing several useful programs (e.g., spreadsheet data storage, numerical integration, and differentiation). Additionally, LabVIEW's control flow approach to computer programming is studied.

Chapters 9 and 10: Data Analysis. Proper use of LabVIEW's curve fitting and fast Fourier transform functions is investigated. Using Express VIs to control a DAQ device, two computer-based instruments—a digital thermometer and a spectrum analyzer—are constructed.

Chapter 11: Intermediate-Level Data Acquisition. Programs are written to carry out analog-to-digital, digital-to-analog, and digital counter tasks on a DAQ device using the conventions of DAQmx. This lower-level approach (in comparison to the high-level Express VIs) allows utilization of the full available range of DAQ device features. A DC voltmeter, DC voltage source, waveform generator, and frequency meter are constructed, as well as a sophisticated digital oscilloscope based on the state machine architecture.

Chapter 12: Temperature Control Project. A wide range of the LabVIEW skills acquired throughout the book are used to construct a Proportional-Integral-Derivative (PID) temperature control system. Appendix I gives a design for the hardware required for this project.

Chapter 13: Control of Stand-Alone Instruments. Using LabVIEW's VISA communication driver, control of a stand-alone instrument over the General Purpose Interface Bus (GPIB) as well as the Universal Serial Bus (USB) is studied. An Agilent 34410A Multimeter is used to demonstrate the central concepts of interface bus communication between a PC and stand-alone instrument.

Key features of *Hands-On Introduction to LabVIEW* include its emphasis on real-world problem solving, its early introduction and routine use of data acquisition hardware, its "Do It Yourself" projects at the end of each chapter, and its healthy offering of back-of-the-chapter homework problems.

Real-World Problem Solving: Chapter topics and exercises provide examples of how commonly encountered problems are solved by scientists and engineers in the lab. LabVIEW features, along with relevant mathematical background, are introduced in the course of solving these problems. The "best practice" strategies presented (such as modularity and data dependency) equip readers to optimize their use of LabVIEW.

Data Acquisition Usage Throughout: LabVIEW's Express VIs allow exercises involving DAQ hardware to appear early and then routinely in *Hands-On Introduction to LabVIEW*. Express VIs package common measurement tasks into a single graphical icon and so allow the user to write a program with minimal effort. Of particular note, following the book's first three software-only chapters that teach the fundamentals of the LabVIEW programming language, data acquisition using a DAQ device is covered in Chapter 4. For a professor or self-learner who wishes to devote only three weeks (or so) to instruction in computer-based data acquisition, Chapters 1 through 4 will provide the needed instructional materials. For those planning a more comprehensive study of LabVIEW, the Express VIs allow construction of a computer-based digital thermometer and spectrum analyzer in Chapters 9 and 10, respectively. In Chapter 11, the control of a DAQ device via the more advanced programming DAQmx icons is covered. In contrast to the Express VIs, the DAQmx icons enable a user to utilize the full available range of the DAQ-device features. In Chapter 12, readers use a DAQ device to precisely control the temperature of an aluminum block, and in Chapter 13, data are acquired remotely from a stand-alone instrument using the GPIB and/or USB interface bus.

Do It Yourself Projects: To allow a reader to gauge his or her understanding of the presented material, each chapter of *Hands-On Introduction to LabVIEW* concludes with a Do It Yourself project. Each of these projects poses an interesting problem and (loosely) directs the reader in applying the chapter's material to find a solution. In some chapters, this project involves writing a program that functions as a stopwatch (Chapter 1) or determines a person's reaction time (Chapter 8); in other chapters the reader constructs a computer-based instrument including a digital thermometer (Chapter 9), a spectrum analyzer (Chapter 10), and a frequency meter (Chapter 11).

Back-of-the-Chapter Homework Problems: A selection of homework-style problems is included at the end of each chapter so that interested readers can further develop their LabVIEW-based skills. In some of these problems, readers test their understanding by applying the chapter topics to new applications (e.g., Bode magnitude plot); in others, readers use programs written within the chapter to explore important experimental issues (e.g., frequency resolution of a fast Fourier transform). Finally, a number of problems introduce readers to features of LabVIEW relevant to, but not included in, the chapter's text (e.g., data storage in binary format).

IMPROVEMENTS TO THE SECOND EDITION

This new edition includes the following improvements:

- All chapters are fully updated to the latest version of LabVIEW. Functionality of the new Probe Watch Window (Chapter 2) and Icon Editor (Chapter 3) is explained.
- MathScript Node's online help and automated data-type formatting features are covered (Chapter 3).
- Low-cost DAQ hardware commonly used in instructional laboratories and self-learning is highlighted, including the USB-6009, myDAQ, PCI-6251, and ELVIS II DAQ devices (Chapter 4).
- "Quick Example" sections at the chapter beginnings give concise introductions to the MathScript Node, Shift Register, and Case Structure (Chapters 3, 6, and 7).
- Property Node usage is included in earlier chapter text (Chapter 7).
- Both GPIB and USB control of stand-alone instrumentation is implemented using a late-model Agilent 34410A Digital Multimeter; presentation is compatible with the older Agilent 34401A multimeter (Chapter 13).
- As an aid to instructors designing their courses, a table is provided that cites sections where an earlier-written program is needed as a subroutine (Appendix II).
- Solutions to even-numbered back-of-the-chapter problems are available to all at www.oup.com/us/essick; for instructors who adopt this book for a course, a password-protected link to the solution set for every problem is available from Oxford University Press.

Hands-On Introduction to LabVIEW is compatible with both the Full Development System and the Student Edition of LabVIEW. An instructor might consider having students purchase personal copies of the low-cost Student Edition software (the Student Edition can now be purchased by itself at a very affordable price; that is, it is no longer necessary to buy an expensive bundled book/software package). With their own LabVIEW software, students can perform non-hardware-related chapter sections and/or back-of-the-chapter problems as homework on their own computers.

To aid readers in creating their LabVIEW programs, the following conventions are used throughout the book: **Bold** text designates the features such as graphical icons, palettes, pull-down menus, and menu selections that are to be manipulated in the course of constructing a program. The descriptive names that label controls, indicators, custom-made icons, programs, disk files, and directories (or folders) are given the straight font. *Italic* text highlights character strings that the programmer must enter using the keyboard and also signals the first-time use of important terms and concepts.

Any suggestions or corrections are gladly welcomed and can be sent to John Essick, Reed College, 3203 SE Woodstock Boulevard, Portland, OR 97202, USA, or jessick@reed.edu.

Updates, answers to frequently asked questions, and ancillary materials for *Hands-On Introduction to LabVIEW* are available at http://academic.reed.edu/physics/faculty/essick.

Additionally, solutions to the even-numbered back-of-the-chapter problems can be downloaded at www.oup.com/us/essick. Instructors who adopt this book for a course can obtain a password-protected link to the solution set for every problem from Oxford University Press.

For their advice and assistance in preparing this revision of *Hands-On Introduction to LabVIEW*, I thank John Challice, Caroline DiTullio, Claire Sullivan, and Dan Pepper of Oxford University Press and Mark Walters and Adam Foster of National Instruments. For their helpful comments and suggestions, I express my appreciation to the reviewers:

Geoffrey Brooks, Florida State University

Mark Budnik, Valparaiso University

Shannon Ciston, University of New Haven

Juan I. Collar, University of Chicago

James Hetrick, University of Michigan–Dearborn

Eric Landahl, DePaul University

Jed Marquart, Ohio Northern University

Casey Miller, University of South Florida

David Roach, Mott Community College

as well as the reviewers of the first edition:

William DeGraffenreid, California State University, Sacramento

Marty Johnston, University of St. Thomas

David Loker, Penn State Erie

Hakeem Oluseyi, Florida Institute of Technology

David Pellett, University of California, Davis

Ian Robinson, University of Illinois at Urbana–Champaign

Perry Tompkins, Samford University

Finally, to my family: Thank you for your love and support while I worked on this project.

John Essick
Portland, Oregon

CHAPTER 1

The While Loop and Waveform Chart

1.1 LABVIEW PROGRAMMING ENVIRONMENT

Welcome to the world of *LabVIEW*, an innovative graphical programming system designed to facilitate computer-controlled data acquisition and analysis. In this world, you—the LabVIEW user—will be operating in a programming environment that is different from that offered by most other programming systems (for example, the C and BASIC computer languages). Rather than creating programs by writing lines of text-based statements, in LabVIEW you will code programming ideas by selecting and then properly patterning a collection of graphical icons.

A LabVIEW program consists of two windows, the *front panel* and the *block diagram*. Once you have completed a program, the front panel appears as the face of a laboratory instrument with your own design of knobs, switches, meters, graphs, and/or strip charts. The front panel is the program's user interface; that is, it facilitates the interaction of supplying inputs to and observing outputs from the program as it runs.

The block diagram is the actual LabVIEW programming code. Here reside the graphical images that you have appropriately selected from LabVIEW's well-stocked libraries of icons. Each icon represents a block of underlying executable code that performs a particular useful function. Your programming task is to make the proper connections between these icons using a process called *wiring*, so that data flow amongst the graphical images to accomplish a desired purpose. Because the icon libraries are designed specifically with the needs of scientists and engineers in mind, LabVIEW enables you—the modern-day experimentalist—to write programs that perform all of the laboratory tasks required for your state-of-the-art research, including instrument control, data acquisition, data analysis, data presentation, and data storage.

To begin developing your skill in LabVIEW graphical programming, in this first chapter I will guide you through the steps of creating a LabVIEW program. Together, we will write a program that systematically increments the argument of the sine function and plots the resulting sine wave as time progresses. In the process of writing this program, you will learn about the *While Loop*, one of LabVIEW's four fundamental program control structures, as well as one of LabVIEW's three commonly used graphing modes, the *Waveform Chart*.

1.2   SINE-WAVE PLOT USING A WHILE LOOP AND WAVEFORM CHART

Find the LabVIEW program (which may have a desktop icon or may be in a folder named National Instruments, LabVIEW, or LabVIEW Student Edition, depending on your particular computing system) and launch it by double-clicking on its icon or name.

After a few moments, a **Getting Started** window will appear. Place your mouse cursor over the **Blank VI** selection and then click. An untitled front panel will appear in the foreground of your screen with a slightly offset block diagram in the background. In addition, a menu bar for an array of pull-down menus that contain LabVIEW program editing items will appear at the top of the window.

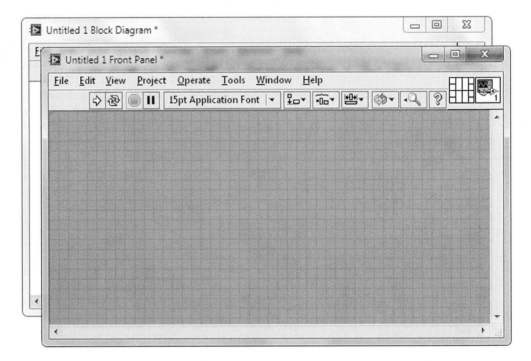

There are three ways to toggle the block diagram between the background and foreground:

- Select **Show Block Diagram** (when it is in the background) or **Show Front Panel** (when the block diagram is in the foreground) from the **Window** pull-down menu.
- Click your mouse cursor on a visible region of the window (block diagram or front panel) that is currently in the background.
- Use the keyboard shortcut by typing *<Ctrl+E>*.

Practice toggling the block diagram between the foreground and background using each of the three available methods.

1.3 BLOCK DIAGRAM EDITING

To begin programming, situate the blank block diagram in the foreground. The objects to be placed on this block diagram as you write your LabVIEW program are found in a repository called the *Functions Palette*. If a Functions Palette is not already visible, activate one by selecting **Functions Palette** from the **View** menu.

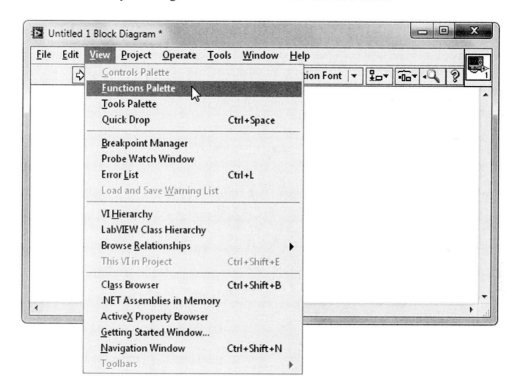

A "floating" Functions Palette will appear on your screen, which can be placed in a convenient spot by "clicking and dragging" on its title bar.

The Functions Palette is an organized collection of *categories* (such as *Programming*, *Express*, and *Mathematics*), where each category contains a palette of related programming objects. The Functions Palette may be configured in a multitude of ways, and the particular configuration that you see will depend on choices made by past users (if any) of your LabVIEW system.

For our work, configure the appearance of the Functions Palette as follows:

- Click on the **Customize** button 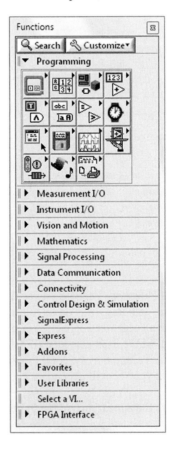 near the upper right corner of the Functions Palette (for older LabVIEW versions, it is called the **View** button). In the menu that appears, select **Change Visible Palettes…** In the dialog window that follows, click on **Select All**, and then press the **OK** button.
- Click on the **Customize** button again. This time, in the menu, click on **Options…** In the dialog window that appears, select **Controls/Functions Palettes**, and then, under **Formatting**, choose **Category (Standard)** for the **Palette** option. Finally, press the **OK** button.

The Functions Palette is now configured. A particular category's palette can be toggled between visible ("open") and hidden ("closed") by clicking on its name. Open the **Programming** palette and close all others, so that your Functions Palette appears as follows (the categories listed on your Functions Palette may differ slightly from this illustration if your system has some optional LabVIEW add-ons).

Introduce yourself to the contents of the Programming palette. By passing the mouse cursor over each button in this palette, the name of the button will appear in a small box called a *tip strip*. An example of a tip strip is shown below.

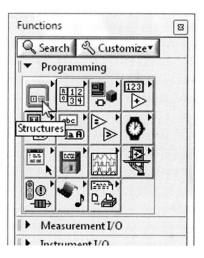

Now place the cursor over the **Structures** button and click. The **Structures** sub-palette will then appear as shown below. To close this subpalette and return to the Programming palette, simply click on **Programming**.

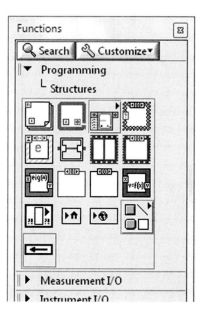

Move the cursor over the objects in the Structures subpalette and watch their names appear in a tip strip. Select the **While Loop** structure from the subpalette by placing the cursor over this object (which looks like a gray arrow bent into the shape of a square) and then clicking. Once you've made this selection, the cursor will appear as a miniature

of the While Loop structure 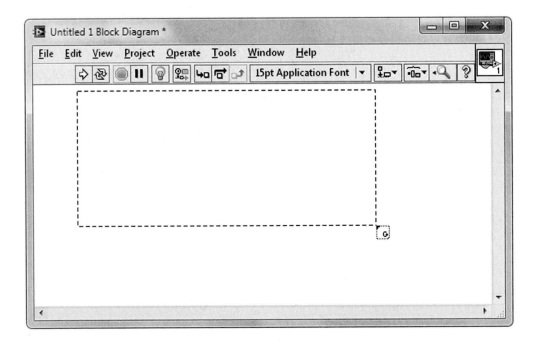 when it is placed within the block diagram window. To

place a While Loop within the diagram, click where you want the upper left corner of the loop to be. Then, while holding down the mouse button, drag the cursor to define the size of your loop.

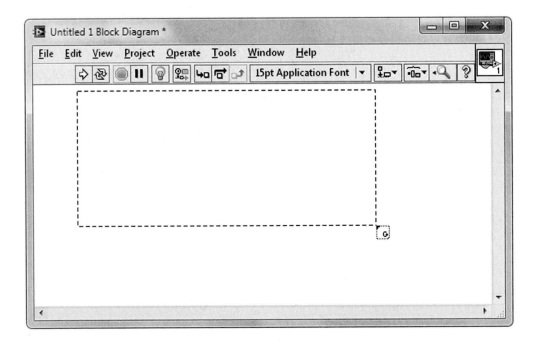

When you release the mouse button, the While Loop will appear as shown here.

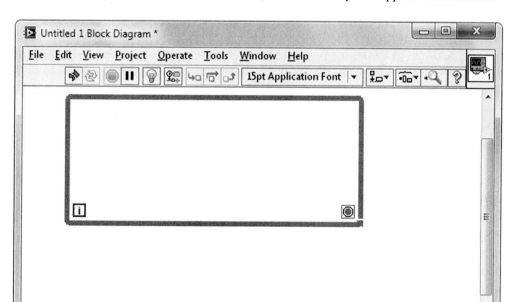

If you are dissatisfied with the dimensions of your While Loop, this can be remedied using the *Positioning Tool* ⬉, whose job is to select, move, and resize objects. The ⬉ is one of the several available LabVIEW editing tools displayed in the *Tools Palette*, which may or may not be visible at the moment.

Activate the Tools Palette by selecting **Tools Palette** in the **View** pull-down menu.

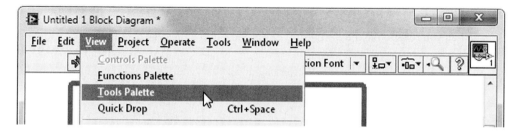

The "floating" Tools Palette will appear as shown next and, just like the Functions Palette, can be relocated to a convenient location by "clicking and dragging" its title bar.

The Positioning Tool may already be selected on your Tools Palette, in the manner shown above. If not, use the mouse cursor to select it now. A short description of each of the ten tools can be obtained from the tip strip that appears when placing the cursor over each tool's button.

The Positioning Tool is used to resize your While Loop by the following procedure. Place the ↖ at one of the loop's corners. At the corner, the ↖ will transform into a *resizing handle* ↘. Click and drag this cursor to redefine the dimensions of your While Loop. When you release the mouse button, the While Loop of desired size will appear.

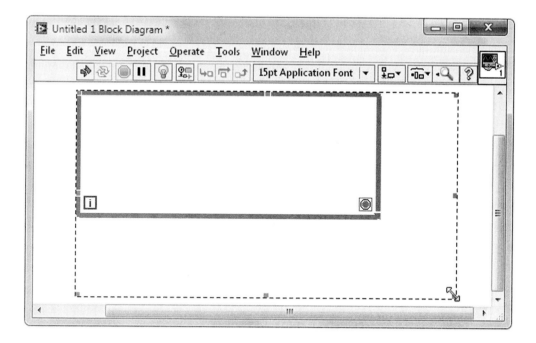

The While Loop structure is used to control repetitive operations. By default, it will repeatedly execute the subprogram written within its borders (called the *subdiagram*) until a specified Boolean value is no longer FALSE. Thus, this structure is equivalent to the following code:

```
Do
        Execute subprogram (which sets condition)
While condition is FALSE
```

Within the While Loop, you will find the *iteration terminal* as well as the *conditional terminal*, which is set to its **Stop if True** state by default. At the end of each loop iteration, LabVIEW checks the value of . If it is FALSE, the value of is incremented by one, and the loop begins another execution; if TRUE, the loop ceases execution. The initial value of (during the first iteration of the While loop) is zero. Thus, in a While Loop where is continuously FALSE until becoming TRUE during the tenth iteration, the loop will execute exactly ten times and the final value of will be nine. In this chapter, we will cause a While Loop to execute numerous iterations by connecting

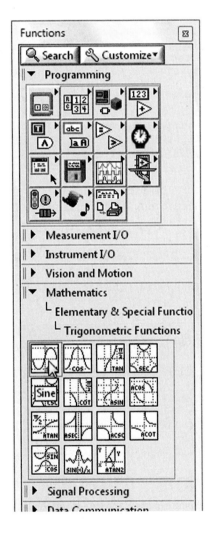 to a Boolean control terminal that is usually FALSE.

Now, let's write a program that will generate and plot a sine wave. With a While Loop already on your block diagram, select the **Sine** icon from the Functions Palette. To find this icon, first open the Mathematics palette by clicking on the **Mathematics** category. Then, select the **Elementary & Special Functions** subpalette, and then finally the **Trigonometric Functions** sub-subpalette. From now on, such a sequence of choices will be indicated as follows: **Functions>>Mathematics>>Elementary & Special Functions>>Trigonometric Functions**.

Once you've selected the **Sine** icon from the Functions Palette by clicking the mouse, place the mouse cursor at the location within your block diagram at which you wish the icon to reside.

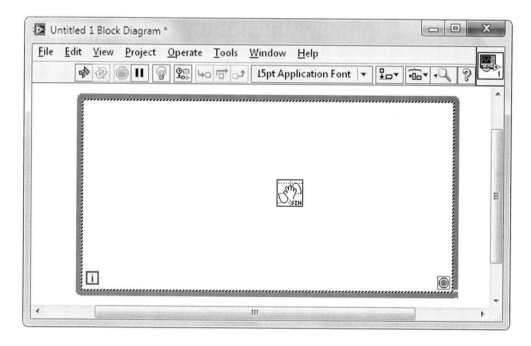

Then click the mouse and—Whoomp!—there the icon is.

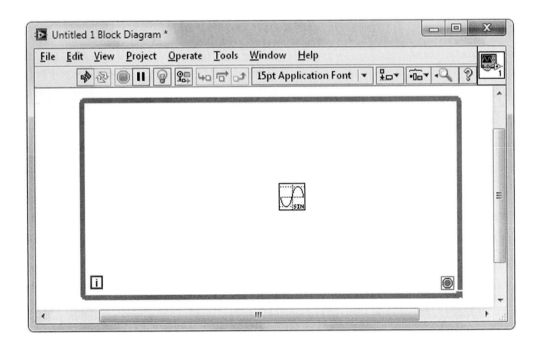

If you subsequently decide that you'd like to move this icon to some other location, you can do so through use of the Positioning Tool. With the ⬚, click on the **Sine** icon.

When thus selected, the icon will become highlighted by a moving, dashed border called a *marquee*. With the mouse button depressed, drag the highlighted icon to the newly desired location within your While Loop. When properly placed, release the mouse button and move the cursor to an empty spot on the block diagram. Then click the mouse to deselect the icon.

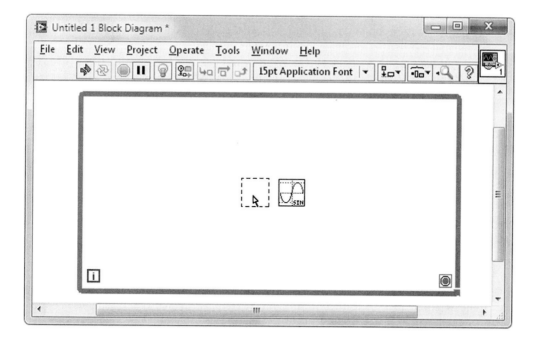

Once an object is highlighted with a marquee, there are a couple of other handy movement techniques. If you hold down the *<Shift>* key and then drag the object, LabVIEW will only allow purely horizontal or vertical motion. Also, you can move the selected object in small, precise increments by pressing the keyboard's *<Arrow>* keys, rather than dragging with the mouse. Try moving the **Sine** icon using both of these techniques.

The function of the **Sine** icon is described in the following *Context Help Window:*

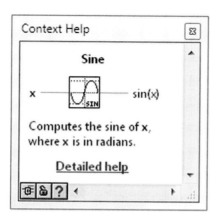

In a Context Help Window, the icon's input connections are shown on the left and output connections on the right. Thus, we see that the **Sine** icon accepts an argument **x** in radians and returns the value for **sin(x)** at its output.

You may see this Context Help Window for yourself by selecting **Show Context Help** from the **Help** pull-down menu. In our shorthand convention, this menu selection is **Help>>Show Context Help**.

After the Context Help Window appears, place it in a convenient position using the

⬉. Then place the ⬉ over the **Sine** icon to view its description.

When you have no further need for the Context Help Window, it can be toggled off in the **Help** menu by selecting **Show Context Help**. As an alternative to mouse control, try toggling the Context Help Window on and off using the following keyboard shortcut: <Ctrl+H>.

Now let's write a program that generates a sequence of points that follows the sine function. First, we need to configure the While Loop so that it will repetitively perform the operation defined within its borders. One (crude) method of accomplishing this goal is the following: Select a **False Constant** from **Functions>>Programming>>Boolean**.

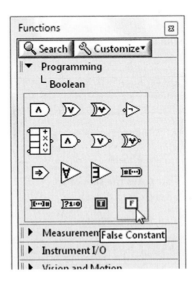

Place the **False Constant** ⬚ (in older LabVIEW versions, this icon appears as ⬚) near the While Loop's conditional terminal ⬚. If you place ⬚ too close to ⬚, LabVIEW's *automatic wiring* feature will automatically connect these two icons with a green wire. In the future, you may want to take advantage of this effort-saving feature. At present, however, we are trying to gain experience in wiring two icons together manually, so if automatic wiring occurs, select **Undo Create** in the **Edit** pull-down menu. LabVIEW's **Undo** function will revert your block diagram to its state prior to the most recent editing action (i.e., the addition of the False Constant). You can then place a new ⬚ on your block diagram a bit farther from ⬚ so that autowiring does not occur.

As an aside, remember the **Edit>>Undo** editing trick, whose keyboard shortcut is *<Ctrl+Z>*. It's the easy way to erase editing mistakes when they occur.

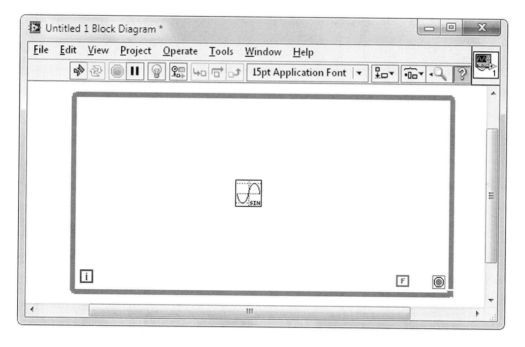

We will now connect the **False Constant** to ⬚ using the *Wiring Tool* ⬚. Go to the Tools Palette and select the Wiring Tool by clicking on it.

Position the ◈ over the **False Constant** icon, so that ⨍ begins to blink. Then click the mouse. You have now tacked down one end of a wire to ⨍.

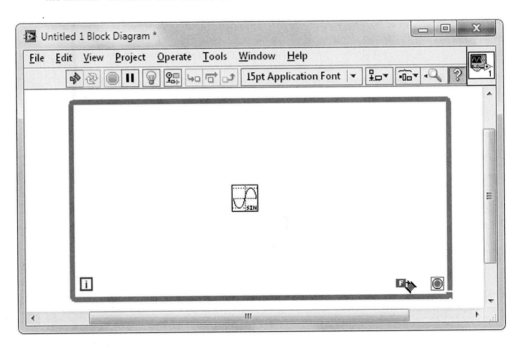

Smoothly move the Wiring Tool to the right until it is over the conditional terminal and the ◉ is blinking. Click the mouse.

If all goes well, you will see a dotted green wire connecting the 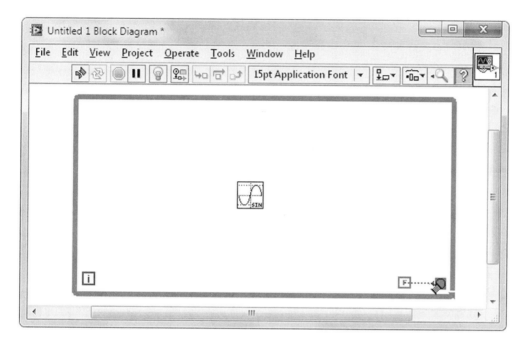 F and ⊚ icons. Such a green wire is LabVIEW's method of indicating that Boolean data are to be passed between two programming objects, in this particular case, from the False Constant F to the conditional terminal ⊚.

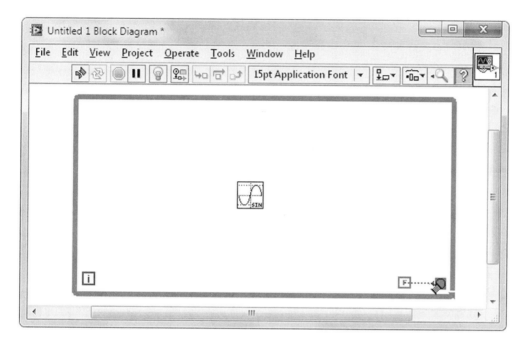

Here are two features of wiring, which are demonstrated in the next illustration:

First, while using the 🖉, if you make a mistake and try to form an improper connection between two objects, the resulting *broken wire* will appear as a black dashed line with a ✖ through it. Such a mistake can be erased by selecting **Edit >>Remove Broken Wires** or, more simply, using the keyboard shortcut: *<Ctrl+B>*. Alternately, you can highlight the broken wire using the Positioning Tool and then erase it by pressing the *<Delete>* key on your keyboard.

Second, while using the 🖉, if at some location you wish to make a right-angle bend in the wire, click the mouse when you arrive at that location, and then move the 🖉 off in the newly desired (perpendicular) direction.

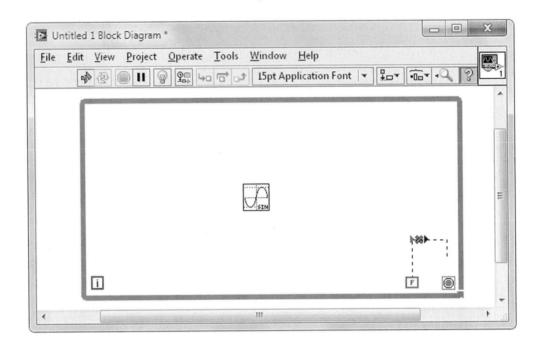

Our accomplishment so far is the creation of a While Loop that iterates continually. To see why this is so, remember that the conditional terminal is set to its default state **Stop if True**. Thus, during each and every loop iteration, a Boolean FALSE value will flow through the wire from [F] to [⊙], setting the conditional terminal to FALSE. When LabVIEW checks the value of [⊙] at the end of the iteration, the While Loop will be instructed to re-execute (rather than stop).

We will now use the While Loop's iteration terminal [i] as a source of ever-increasing argument **x** for our sine function. Use the ⬢ to position [i] in the neighborhood of the **Sine** icon, and then wire the [i] to the Sine's argument **x** input.

Here is a method of producing proper wiring. First, activate the Context Help Window by the keyboard short cut *<Ctrl+H>*, and then position the Context Help Window in a convenient, out-of-the-way place. Now, select the Wiring Tool from the Tools Palette, position the ◆ over [i] so that this icon blinks, and click to tack down one end of a wire to [i]. Move the ◆ smoothly over to the terminal for the Sine's argument **x** input, using the Context Help Window as your guide. When the ❖ is positioned correctly, the terminal for the **x** input will begin to blink both on the block diagram and in the Context Help Window. To erase any further doubt regarding the identity of the terminal to which you are wiring, the terminal's name is supplied by a tip strip. Click the mouse to complete the wiring. As an additional wiring aid, *whiskers* appear from each input and output as the Wiring Tool closely approaches the icon. These whiskers are especially helpful when wiring to icons with a multitude of inputs and outputs.

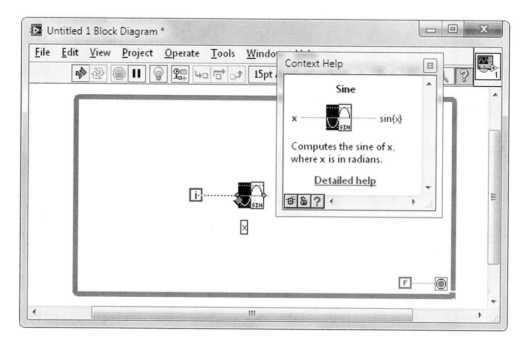

If done correctly, a colorful wire will connect ⓘ to the **x** input, indicating that numerical data are to be passed between these two objects. Remember, if you make a mistake, the illegal wiring will result in a black dashed line that can be most easily erased by the **Remove Broken Wires** command's keyboard shortcut: *<Ctrl+B>*.

In LabVIEW, the color of a wire carrying numerical data indicates the manner in which the number is represented. Blue wires and icons denote integers, which come in one-, two-, four-, and eight-byte varieties, while orange indicates single-precision (four-byte) and double-precision (eight-byte) floating-point numbers. We will soon discover how to control the exact formatting of a particular numerical value. For now, simply note that ⓘ and wires emanating from it are blue, indicating integer values. However, a red *coercion dot* appears where this blue wire connects to the Sine icon input, denoting that the Sine icon is automatically converting this integer input into the floating-point format it requires for its argument **x** input.

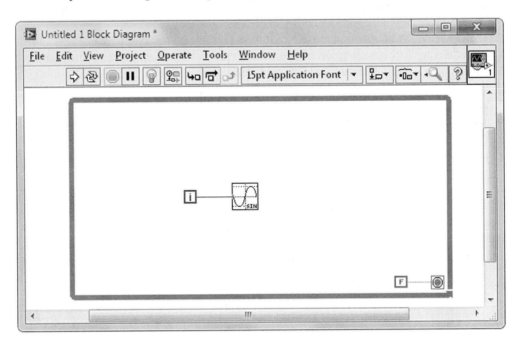

1.4 LABVIEW HELP WINDOW

To obtain more detailed information about the inner workings of a particular icon, the online reference resource *LabVIEW Help* can be accessed by clicking on the blue

Detailed Help hypertext or the Question Mark 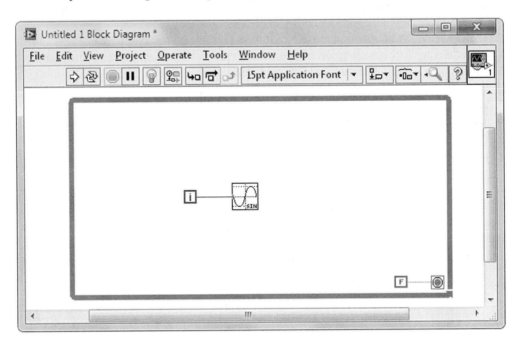 at the bottom of the icon's Context

Help Window. Alternately, you can select **Help>>LabVIEW Help...** and then search for the LabVIEW Help Window describing your icon of interest.

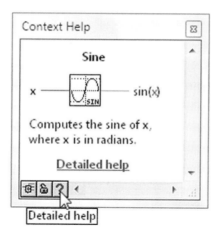

The LabVIEW Help Window for the Sine icon is shown next.

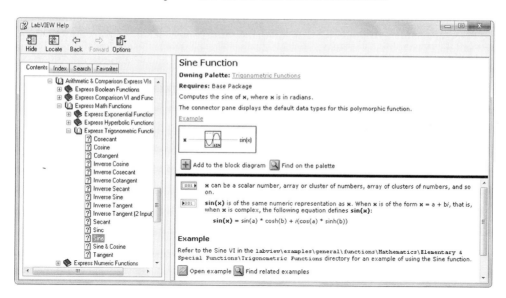

This window informs us that the default data type for the **x** input (called its *representation*) is [DBL], that is, a double-precision floating-point number. However, we also are told that Sine is a *polymorphic function*. The definition of polymorphic can be found by clicking on the **Index** tab, typing the keyword **polymorphic**, and then selecting the subtopic **functions**.

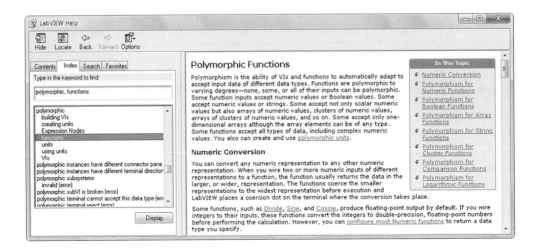

Reading the relevant sections of this window, we find that a polymorphic function can adapt (from its default settings) to accept a variety of data types at its input. In the case of a numeric trigonometric function like **Sine** (whose default input representation is

[DBL]), the output of the icon will always have the same representation as the input. So,

for example, a single-precision floating-point input will yield a single-precision floating-point output. The only exception to this rule is for an integer input, in which case the output is a double-precision floating-point number.

In our work ahead, we will routinely use Context Help Windows and only occasionally need the more detailed information available in LabVIEW Help Windows. Hence, to save words, I will refer to a Context Help Window simply as a "Help Window," while I will call a LabVIEW Help Window by its full name, that is, a "LabVIEW Help Window."

1.5 FRONT PANEL EDITING

Now, let's instruct LabVIEW to plot the sine function as each While Loop iteration generates a new value. A graph falls under the generic category of user interface, which is the domain of the front panel. Switch to the front panel using the keyboard shortcut *<Ctrl+E>*.

A user typically wishes to interact with a program by supplying inputs and observing outputs. In LabVIEW, these operations are facilitated by a wide choice of knobs, switches, meters, and graphs found in the *Controls Palette*. If it is not already visible, activate the Controls Palette by selecting **Controls Palette** in the **View** pull-down menu. Once activated, the Controls Palette will appear whenever the front panel is brought to the foreground. It will disappear and be replaced by the Functions Palette when the block diagram is subsequently toggled into the foreground. If only a few categories are visible in your Controls Palette, click the **Customize** button near its upper right corner, and then

select **Change Visible Palettes ... >>Select All**. Also choose **Customize>>Options ... >>
Controls/Functions Palettes>>Formatting>>Palette>>Category (Standard).**

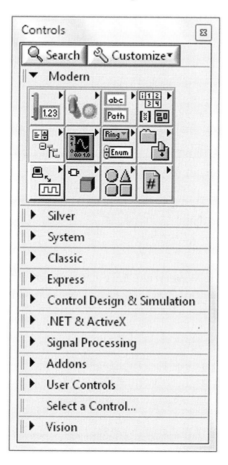

While front-panel editing, those with a later version of LabVIEW will have the choice of obtaining needed programming objects (called *controls*) from either the **Modern** or the **Silver** palettes. The analogous controls found in these two palettes are functionally the same; they differ only in appearance. The **Modern** controls have been available in many versions of LabVIEW, while the **Silver** controls were recently introduced to commemorate the 25th anniversary of LabVIEW and have a contemporary look with rounded edges and sleek shadowing. Because not all readers have the latest LabVIEW update, I will use the **Modern** controls throughout the book. You may, however, enjoy using the equivalent **Silver** controls instead if you have them available.

In **Controls>>Modern>>Graph**, select a **Waveform Chart**. The Waveform Chart, one of three commonly used graphical modes in LabVIEW, behaves like a laboratory strip chart, producing a real-time plot as each new data point is generated. In contrast, the **Waveform Graph** and **XY Graph** (which we will study later) display a full array of data that were produced at an earlier time.

After you place the **Waveform Chart** on the front panel with a single mouse click, if you don't click the mouse further, a highlighted region will appear above the upper left corner of the plot containing the default text Waveform Chart.

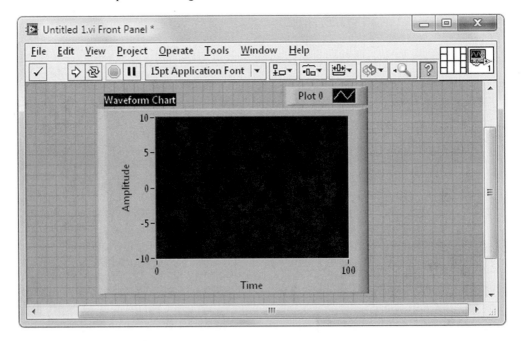

This highlighted region is the Chart's *Label.* You can use the keyboard to give the plot a descriptive name, but let's just keep the default **Waveform Chart** for this program. Secure this label by either clicking on the Enter button ✓ at the upper left end of the front panel, pressing *<Enter>* on the Numeric Keypad of your keyboard, or simply clicking the mouse cursor on an empty region of the front panel. In the future, if you accidentally click the mouse so that the label loses its highlighting before you have a chance to enter a name, the label region can be re-highlighted using the *Labeling Tool* [A] found in the Tools Palette.

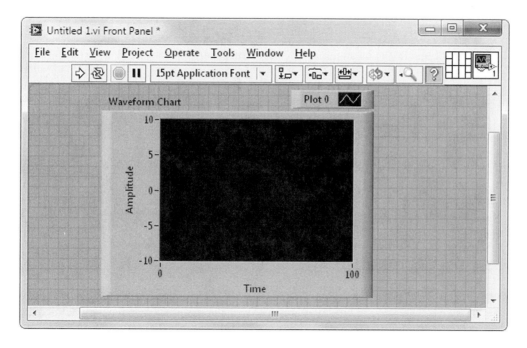

In addition to the chart region and the Label, the Waveform Chart includes the *Plot Legend* Plot 0 . As we will discover shortly, the Plot Legend allows control over the plot style. Through it, one can choose such plot characteristics as whether data will be plotted as points or as interpolated lines and the shape, if any, of the data points.

Let's explore how to reposition and resize the Waveform Chart. First, using the ◄, highlight the entire Waveform Chart (including the chart region, Label, and Plot Legend) with a marquee by placing the ◄ over the chart region and clicking. With the mouse button depressed, drag the highlighted object to the newly desired position. Once properly placed, release the mouse button and move the cursor to an empty spot on the block diagram. Then click the mouse to deselect the object. In addition, the Label and Plot Legend can each be moved independently. To demonstrate this feature, highlight just the Plot Legend with a marquee by placing the ⬉ directly over this object and clicking. Then you will be able to drag this single object to a convenient location. Finally, to resize the chart region, position the ⬉ at one of the Waveform Chart's corners. At the corner, the ⬉ will transform into a resizing handle ⬊. Click and drag this cursor to redefine the dimensions of your chart region. Experiment with this resizing for a few moments.

You will find that you can resize both the actual chart region as well as its background frame.

A more important adjustment is the manner in which the axes should be scaled. In a moment, we will be plotting sine-wave values on the Waveform Chart's y-axis, so this axis must be prepared to chart data in the range of -1.0 to $+1.0$. Note that the default setting for the Waveform Chart's y-axis data range is -10.0 to $+10.0$.

It is the job of the *Operating Tool* 🖑 to change values appearing on both the front panel and the block diagram. Select the 🖑 on the Tools Palette. Then, change the y-axis data range from its default setting through the following procedure: First, use the 🖑 to highlight the upper limit of the y-axis data range. Once highlighted, enter the newly desired value, which in our case is *1.0*, and then click on ✓ (or click the mouse on an open area of the front panel). Then, in a similar manner, define the lower limit of the y-axis data range to be *–1.0*. LabVIEW will automatically redefine the intermediate y-axis labeling.

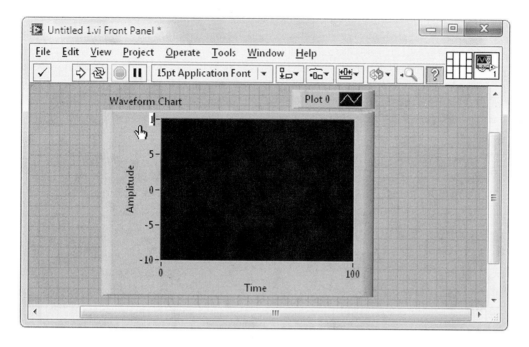

To save the redefined *y*-axis labeling scheme, simply select **Edit>>Make Current Values Default**.

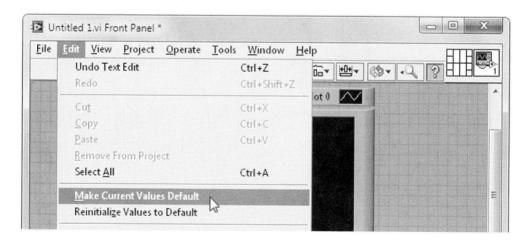

1.6 POP-UP MENU

You are about to enter a hidden world and ascend to a higher level as a LabVIEW programmer. You might wish to pause and reflect upon your life prior to the enlightenment that you are about to attain. Here is the deep secret: Almost every LabVIEW object has its own associated *pop-up menu* that is hidden from the uninitiated. By gaining access to a pop-up menu, you, the programmer, are empowered to control the functioning of the associated object. How does one gain access to a pop-up menu? By performing the simple operation of "*popping up*" on the object. To pop up on an object, place the mouse cursor directly over it, and then click the right mouse button ("right-click"). You can also pop up on an object using the *Pop-Up Tool* (also called the *Object Shortcut Menu Tool*) found in the Tools Palette.

Pop up on the chart region of the **Waveform Chart**. As a first example of controlling the features of this object using its pop-up menu, toggle the Label on and off by selecting **Visible Items>>Label**. You will find that when a feature such as Label is activated ("on"), a check mark appears by its name in the pop-up menu.

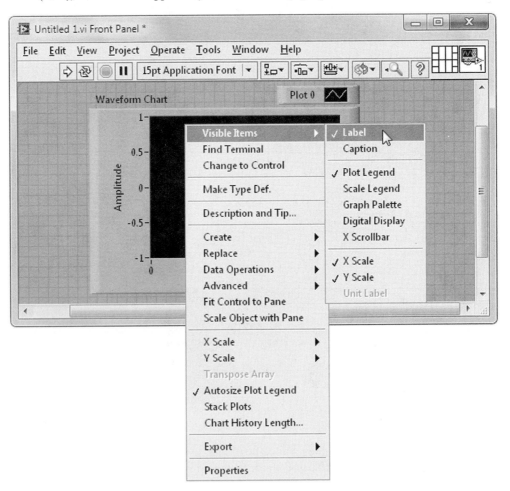

Now, let's use the Waveform Chart's pop-up menu to assist us in appropriately scaling its *y*-axis. A few moments ago, we manually chose proper scaling for this axis using the Operating Tool. A simpler method to accomplish this same goal is via activation of the Waveform Chart's autoscaling feature: Pop up on the chart region of the Waveform Chart and inspect the **Y Scale>>Autoscale Y** option. You will find it checked, meaning it is already activated. By default, the autoscaling on the Waveform Chart's *y*-axis (but not its *x*-axis) is activated.

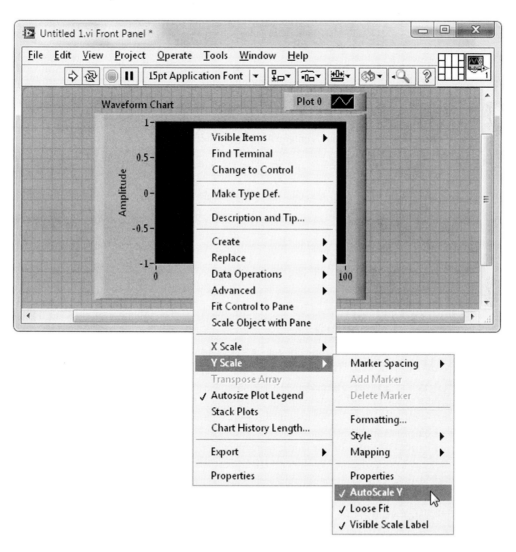

Finally, let's use the pop-up menu to activate one of the Waveform Chart's as-yet-unseen features. Pop up on the chart region and select **Visible Items>>Scale Legend**. The *Scale Legend*, which allows access to several useful functions that determine the scaling and labeling of the *x*- and *y*-axes, will then appear below the Waveform Chart.

Here, we see that, by default, the *x*- and *y*-axes are labeled **Time** and **Amplitude**, respectively. We, of course, will be plotting *sin(x)* vs. *x*, so these default labels are inappropriate choices for our plot. Use the ⍇ to highlight **Time**, replace it with **x (radians)**, and press *<Enter>*. Similarly, replace **Amplitude** with **sin(x)**.

Additionally, the Scale Legend provides a simple method for activating autoscaling of axes. Simply use the ⍇ to click on the **Scale Lock** button of the desired axis. When the Scale Lock is open 🔓, autoscaling is OFF. When it is closed 🔒, autoscaling is continually ON. If you just want to autoscale, say, the *x*-axis once (i.e., not continuously), click on the *Autoscale* ⍐ button. The small green indicator in the Autoscale button lights when autoscaling is activated.

1.7 FINISHING THE PROGRAM

With one last editing step, you'll be ready to run your first LabVIEW program! Use the keyboard shortcut *<Ctrl+E>* to return to your block diagram.

There you will find the Waveform Chart's *icon terminal* ▦. This icon terminal is the Waveform Chart's block-diagram portal, that is, it accepts the block-diagram data to be plotted on the front-panel Waveform Chart. The icon terminal contains a small picture of plotted data (indicating it is associated with the Waveform Chart) and has an orange border with text **DBL** denoting that the data input to it should be double-precision floating-point numbers. It is further identified by the **Waveform Chart** label that you entered on the front panel. By popping up on the icon terminal and selecting **View as Icon**, the terminal will morph into its *data-type terminal* ▦ guise, whose smaller size can be helpful if you need to conserve block-diagram real estate. To toggle back to the icon terminal form, simply select **View as Icon** in the pop-up menu. Throughout this book, icon terminals will always be used in block-diagram illustrations. However, feel free to use data-type terminals on your diagrams, if you wish.

Using the ↖, place the Waveform Chart's icon terminal near the **sin(x)** output of the

Sine icon, and then use the ◈ to connect the **sin(x)** output to ▨. Remember that the Help Window can be activated with <*Ctrl+H*> to aid in this wiring operation.

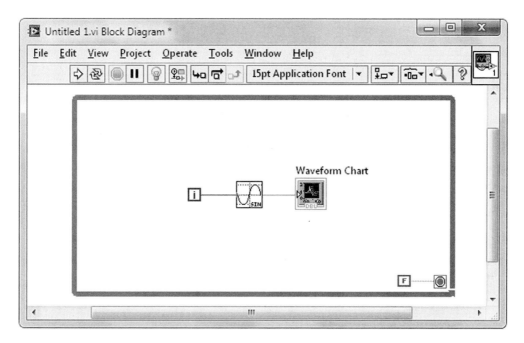

1.8 PROGRAM EXECUTION

Now return to the front panel. You are ready to run your program, with the help of the *toolbar* shown below.

The leftmost button on the toolbar is the **Run** button 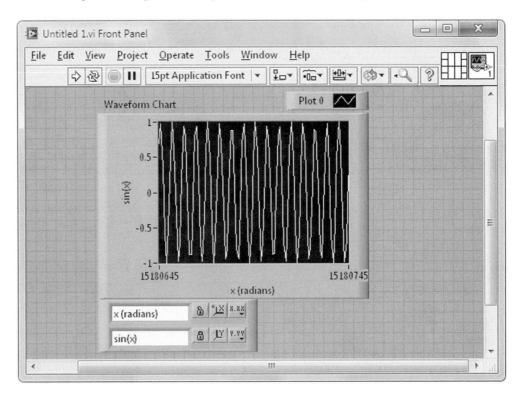. To start your program, simply click on the . As your program executes, you will observe a rather jagged-looking sine wave produced on your Waveform Chart in strip chart fashion.

After watching the sine wave move across the Waveform Chart for awhile, you may start to wonder about a few things. For example, the Waveform Chart's block-diagram icon is supplied only with *y*-axis (sine-wave) values, so how is the Waveform Chart producing the associated *x*-axis values? What determines the speed at which the sine wave appears to move? And, now that the program is running, how do we turn it off?

Let's answer the first question first. Concisely stated, the *x*-axis denotes the *count indices* of the plotted data. That is, the Waveform Chart keeps track of the number of data values it has received and associates a count index with each datum. For the *i*th data value, the plot displays an (x, y) point, where $x=i$ and $y=$ the actual data value. In its default setting, the Waveform Chart's *x*-axis values are the indices of the last 101 *y*-axis (in our case, sine-wave) values supplied to its block-diagram icon.

In regard to the second question, the speed at which the sine wave moves across the chart region is determined by the calculational time delay necessary to produce each new data point. So this speed simply reflects the time it takes for each iteration of your block diagram's While Loop. Because we are allowing this program to run freely, this time per iteration is determined by the speed of your computer's processor, which (most likely) is very fast. On my computer, the x-axis values are on the order of 15 million after ten seconds of runtime. Thus, the time per iteration is approximately $10 \text{ s}/15 \times 10^{6} \approx 7 \times 10^{-7} \text{ s} = 0.7 \ \mu\text{s}$.

Finally, how can the program be turned off? On your block diagram, a False Boolean value is constantly being read by the While Loop's conditional terminal at the end of each loop iteration, so your program will run forever. Because there is no way provided within the program to halt execution, your only recourse at the moment is to click on the **Abort Execution** button in the toolbar. Halt your program by clicking on the **Abort Execution** button.

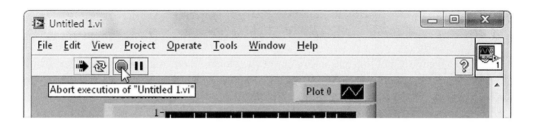

Now that you have used the **Abort Execution** button, let me mention that it is bad practice to do so. This button, at the moment it is activated, causes your computer to immediately cease operation of your program. For the present program, such an action is probably no big concern. However, in more sophisticated programs, pressing **Abort Execution** at the wrong time could halt execution while reading data from a file or during communication with data acquisition devices connected to your computer. These situations can lead to data corruption and other undesirable consequences. Thus, it is always best to code a built-in stopping mechanism into your program.

1.9 PROGRAM IMPROVEMENTS

Based on the preceding observations, let's upgrade your program in the following three ways: (1) furnish a front-panel button that, when pressed, allows the program to complete its current While Loop iteration and then halt its operation, (2) provide a control over the speed at which the While Loop iterates, and (3) improve the resolution of the sine wave being produced.

1.9.1 Front-Panel Switch

LabVIEW provides a multitude of front-panel switches that can be used to bring your programs to a graceful stop. Halt your program so that you may edit its front panel. In **Controls>>Modern>>Boolean**, select a **Stop Button**.

Using the ⬉, put this button in a convenient location on your front panel. The default Boolean value for the Stop Button is FALSE.

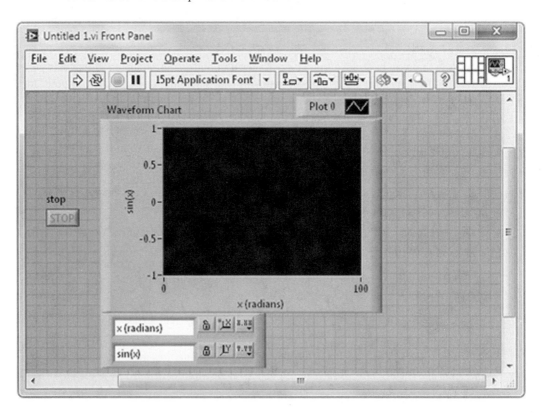

There are six modes in which a LabVIEW Boolean switch can behave in response to being pressed. These modes are listed in the switch's pop-up menu under the **Mechanical Action** option. Pop up on the **Stop Button** and select **Mechanical Action.** You will find that this switch is set to **Latch When Released** by default. In this mode, the user places the Operating Tool over the **Stop Button** switch and depresses the mouse button. Then, at the later time when the user releases the mouse button, the switch changes from its default value to the opposite Boolean value. The switch retains ("latches") this new value until the program reads it once, at which time the switch returns to its default setting. You can find detailed explanations of each mode by selecting **Help>>LabVIEW Help…** and then searching for **Mechanical Action**.

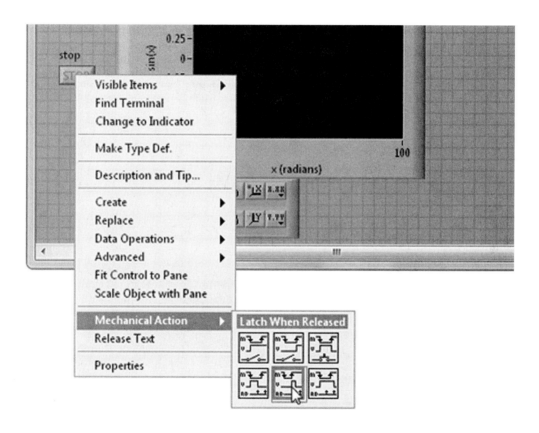

Now toggle to the block diagram. There, you will find the icon terminal 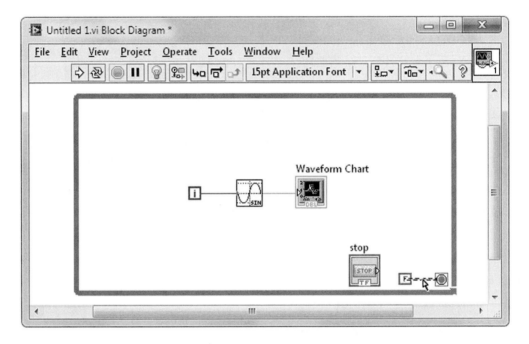 that feeds the value of the front panel's **Stop Button** to the block diagram. Our plan is to con-

nect to the While Loop's conditional terminal. To accomplish this, select the wire connecting the **False Constant** to the conditional terminal by clicking on it with the . Once it becomes highlighted with a marquee, erase this wire by pressing *<Delete>* on your keyboard.

In a similar way, delete the **False Constant** . Now drag the Stop Button's icon terminal near the and wire these two objects together. Because is FALSE by default, the While Loop will continually iterate when the program is started. At some later time, almost assuredly during some intermediate portion of the While Loop cycle, the Stop Button will be pressed and then released, thus changing to the TRUE state.

Because the While Loop only checks the value of at the end of an iteration, the Loop will fully execute its calculations during that final iteration, before ceasing operation.

Once the Stop Button's TRUE value is read, it will return to its default value of FALSE so the program is ready for the next time it is run.

1.9.2 Controlling Iteration Rate

While we are on the block diagram, let's also provide a method for controlling the rate at which the While Loop iterates. Through an icon called **Wait (ms)**, LabVIEW provides a means of delaying subsequent program operations for a specified time period. In **Wait (ms)**'s Help Window shown next, we see that by specifying a numerical constant at the icon's input, the program will wait for that specified number of milliseconds.

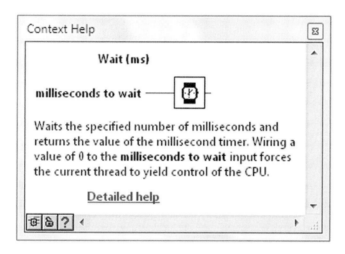

In **Functions>>Programming>>Timing**, select the **Wait (ms)** icon and position it within the While Loop. In the Help Window above, the blue wire emanating from the **milliseconds to wait** input indicates that this terminal should be wired to an integer value. Select a **Numeric Constant** from **Functions>>Programming>>Numeric**. When the **Numeric Constant** is first selected, its interior is highlighted and ready for you to enter a number. If you click the mouse, this highlighting will disappear. To restore the highlighting, use the ⌲. Type the integer *100* into the **Numeric Constant**, press *<Enter>* (or click on a blank region of the block diagram), and then wire this icon to the input of **Wait (ms)**.

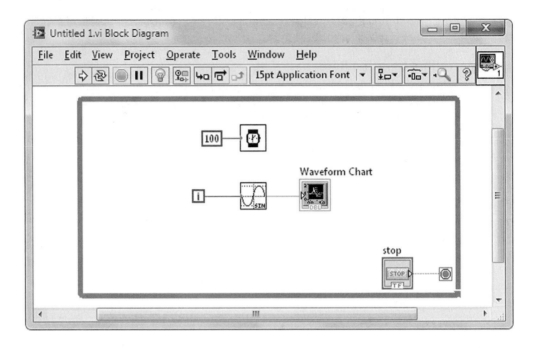

The block diagram above causes one to ponder an obvious question. During each loop iteration, does the wait precede or succeed the sine-wave plot? The surprising answer is that LabVIEW effectively processes these two operations in parallel. That is, the wait and the sine-wave plot begin at the same time and the loop iteration concludes when both of these operations are complete. Thus, the time for an iteration is determined by whichever of the two is the slowest. Earlier we found that when only the sine-wave plotting was within the While Loop, the loop iterated much faster than ten times a second. Thus, by including a delay of 100 milliseconds, this wait operation should provide the "rate-limiting step" that determines the time per iteration.

Now return to the front panel and run your new, improved program by pressing ⇨. Because of the **Wait (ms)** icon, you will find the sine wave is now generated in an easy-to-observe manner. Use the front panel's **Stop Button** to stop your program gracefully. After halting the program, if you would like to clear your plot, pop up on the Waveform Chart and select **Data Operations>>Clear Chart** (this pop-up menu is altered while the program runs, but the "runtime" menu also allows the chart to be cleared).

1.9.3 Improving Sine-Wave Resolution

To see why our sine wave has a jagged appearance, pop up on the black region of the Waveform Chart's **Plot Legend** `Plot 0 ◌`. Under **Interpolation** we discover that, by default, the Waveform Chart presents data in a "connect-the-dots" fashion. That is, a straight line is drawn from each data point to its neighbor and the entire collection of these lines represents the waveform. Also, the default **Point Style** is **None**, which means that no symbol is placed at each data point. However, we can make the actual data points visible by selecting, for example, a large **Solid Dot** for the **Point Style**. While you're at it, you might like to spend a few minutes exploring the effect of selecting the various **Point Style**, **Line Style**, **Interpolation**, and **Color** options available.

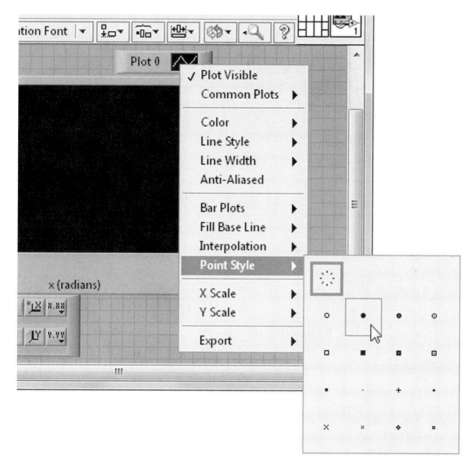

With the data points now visible, we see that each cycle of the sine wave is only represented by a few sampled locations. It is this sparse sampling that leads to the jagged waveform appearance when lines are interpolated between adjacent data points.

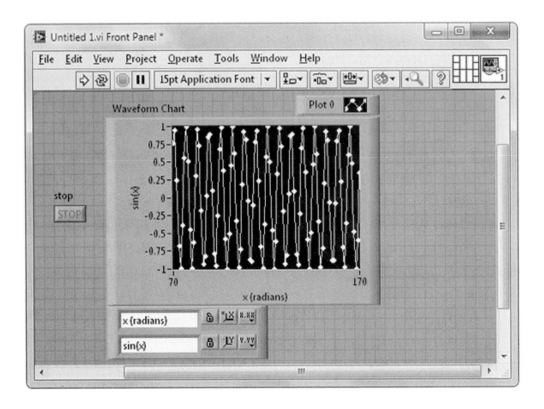

The reason for this paucity of data per cycle is, of course, the result of our programming decision to take the sine function argument **x** to be the value of the While Loop's iteration terminal . Because increments in steps of one and the sine function completes each new cycle every time **x** increases by $2\pi \approx 6$, only about six locations are sampled during each sine-wave cycle.

To sample the sine function with higher resolution, let's take **x** to be one-fifth of , rather than simply . Then it will take five times more iterations of the While Loop for **x** to increase by 2π, resulting in five time more sampled points per cycle. To accomplish this feat, select the **Divide** icon from **Functions>>Programming>>Numeric** and place it on your block diagram.

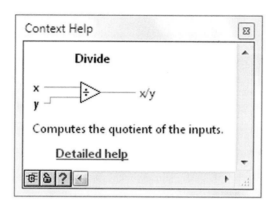

Then, also from **Functions>>Programming>>Numeric**, obtain a **Numeric Constant**. When you first position the **Numeric Constant** on your diagram (before the next mouse click), it is a blue-bordered rectangular box with its interior highlighted, ready to receive the numerical value of your choosing (remember that if the mouse is clicked accidentally, the interior of the **Numeric Constant** can be highlighted again using the

✋). Enter the number *5.0*. The Numeric Constant's border will change from blue to orange, indicating that it contains a floating-point (as opposed to integer) number.

If instead you had programmed the **Numeric Constant** with the number *5* (rather than *5.0*), the icon would remain blue, indicating that it contains an integer. On your diagram, although you programmed the **Numeric Constant** with *5.0*, it may appear as

an orange 5 rather than an orange 5.0 because an option called **Hide tailing zeros** is

activated. To deactivate this option, pop up on the **Numeric Constant** and select **Display Format...** In the dialog window that appears, choose **Default editing mode**, and then make **Digits** and **Precision Type** equal to **1** and **Digits of precision**, respectively, uncheck the **Hide trailing zeros** box, and finally press **OK**. In this book, I will always take the trouble to uncheck **Hide trailing zeros** so that it is apparent when a **Numeric Constant** contains a floating-point number. On your block diagrams, you may or may not wish to go to this trouble. The icon's blue or orange border will tell you whether it is programmed with an integer or floating-point number, respectively.

Position the **Divide** and 5.0 icons conveniently and wire them so that one-fifth of

i is input to the Sine's **x** input. Note the red coercion dot at the iteration terminal's con-

nection to the **Divide**. This dot denotes that the integer from i is converted to a floating-point number. Thus, the Divide icon performs a floating-point divide operation.

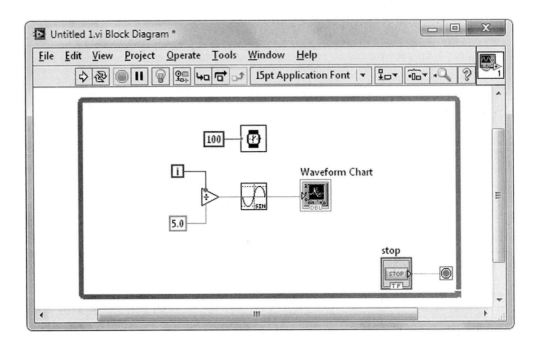

Return to the front panel. There is one thing we still need to fix—the *x*-axis calibration. If we make no further changes, the quantity plotted on the *x*-axis will be the count index associated with each sine-wave value on the *y*-axis. Remembering that we have modified our program so that the sine function argument for a particular data point is now the count index value *i* divided by 5, we see that the count index is no longer equal to the radian argument **x**, and so the wrong quantity is being plotted on the *x*-axis. However, this axis can be calibrated as **x** in radians by simply multiplying each count index by 0.2. Such calibration by a multiplicative constant is a common need when using the Waveform Chart, and thus LabVIEW provides a mechanism for implementing it. Pop up on the **Waveform Chart** and select **Properties**. In the **Chart Properties** dialog window that appears, click on the **Scales** tab. Then under the **x (radians) (X-Axis)** menu, set **Multiplier** equal to *0.2* and then press the **OK** button.

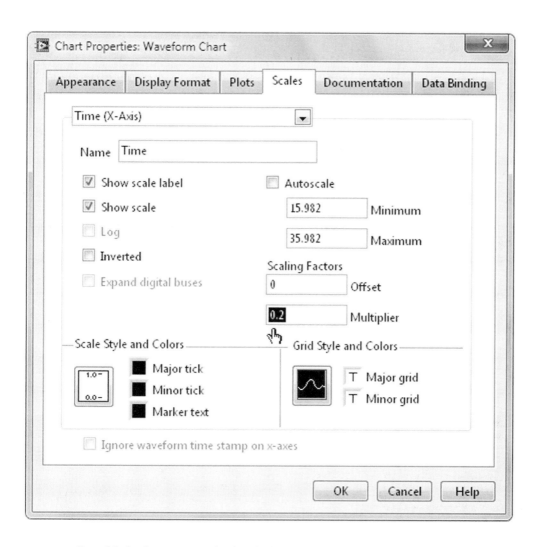

Run this final program and enjoy the beautifully realized sine wave being generated. While your program is running, imagine writing the mountains of code necessary to replicate this real-time sine-wave plot, complete with the attractive user interface, using a text-based language such as C. I think you'll agree that the graphical-based LabVIEW programming language is simple, yet very powerful.

1.10 DATE-TYPE REPRESENTATIONS

There's one last perplexing detail to resolve before our block diagram is complete. You may have noticed that a coercion dot appears at the input of the **Wait (ms)** icon, indicating a numeric-format mismatch, although we have seemingly wired the requisite integer-formatted number to this input. This puzzle can be solved by consulting the

LabVIEW Help Window, which can be accessed by clicking on the 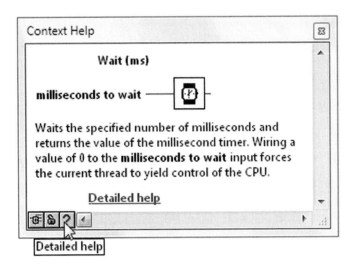 in the **Wait (ms)** Help Window.

The LabVIEW Help Window for **Wait (ms)** is shown below.

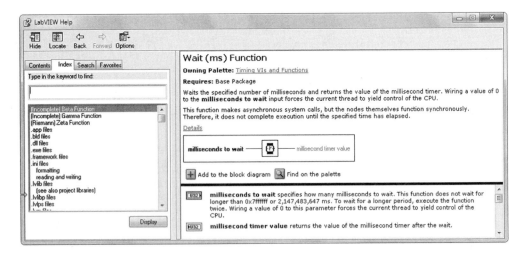

Here, we find that the **milliseconds to wait** input is configured to accept an unsigned four-byte (or 32-bit) integer, a numeric data type whose shorthand name is **U32**. This type of integer is always positive and can range in value from 0 to

$(2^{32} - 1) = 4,294,967,295$. The unsigned integer data type is in contrast to the signed integer, which sacrifices one of its bits for use as a plus or a minus sign. For example, the four-byte signed integer, a format called **I32**, can range from -2^{31} to $+(2^{31} - 1)$.

To discover the data type of the integer on your block diagram, pop up on $\boxed{100}$, and then select **Representation** from the pop-up menu. You will find that this integer is of type **I32**, the default integer data type for a **Numeric Constant**.

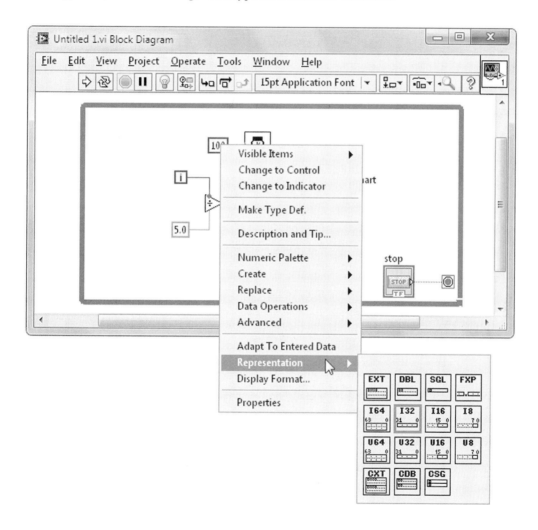

Change the data type of this number to an unsigned integer by selecting **U32** from the **Representation** palette. You will then find that the coercion dot disappears. Another mystery solved!

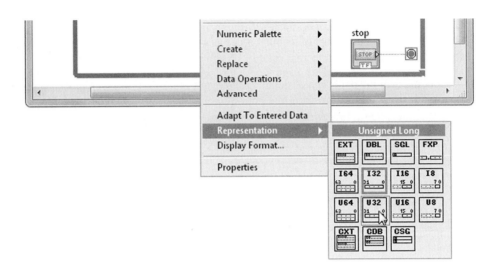

1.11 AUTOMATIC CREATION FEATURE

Now that you understand some of the subtleties involved in wiring up constants to icon terminals, here's a time-saving shortcut I think you'll appreciate. First, delete the Numeric Constant ⌜100⌝ and its wire from your block diagram. Next, place the ◆ over the **milliseconds to wait** input of **Wait (ms)**, and then pop up and select **Create>>Constant** from the menu that appears.

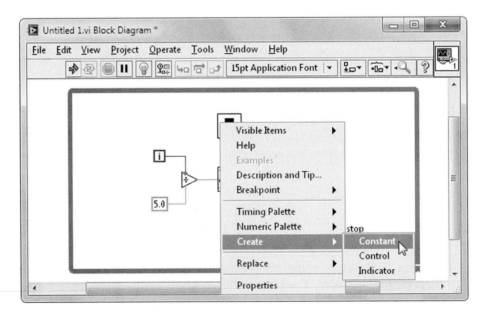

Like magic, an already-wired **Numeric Constant** of the correct data type (in this case, **U32**) will appear, with its interior highlighted.

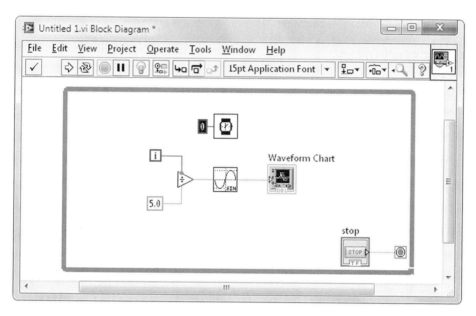

All you need to do is type in the desired integer value of *100* from the keyboard, press *<Enter>*, and you're finished.

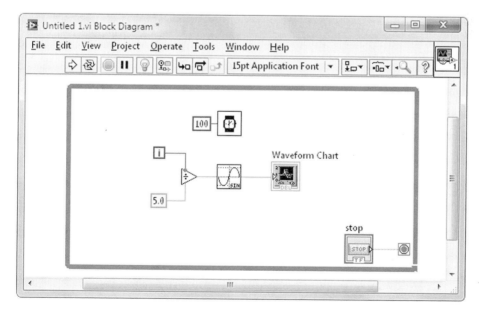

Automatic creation of "terminal-appropriate" objects is a universal feature of LabVIEW icons. Use of these time-saving "creation" options, available in the pop-up menus at the input and output terminals of all block diagram icons, will speed up your program development time.

1.12 PROGRAM STORAGE

LabVIEW programs simulate the function of laboratory instruments. As such, these programs are given the name *virtual instruments* (VIs). Your While Loop-based VI behaves as a strip chart plotting a sine wave, so let's save this VI under the descriptive name **Sine Wave Chart (While Loop)**.

In the **File** pull-down menu, select **Save**, which will activate the **Name the VI** dialog window shown next. Because you will be writing and saving many VIs in your study of LabVIEW, let's first create a folder in which to store this work. Use the **Save in:** box to navigate to a desired storage place on your computing system (e.g., **Desktop**, **Documents** folder, flash drive), and then click on the **Create New Folder** button.

In the highlighted box that appears next to the folder icon, entitle the new folder **YourName**, and then click **Open** to open this folder.

The VI you have written is associated with the first chapter of this book. Thus, within the **YourName** folder, create a subfolder named **Chapter 1** and save the VI there as follows: With **YourName** open, click on the **Create New Folder** button. In the highlighted box that appears next to the folder icon, name the new folder **Chapter 1**, and click **Open** to open this folder. Finally in the **File name:** box, type **Sine Wave Chart (While Loop)**, and then press **OK**. Your VI will be saved in the **Chapter 1** folder, which is within the **YourName** folder (a location we will denote as **YourName\Chapter 1**) with the extension **.vi** appended to its name.

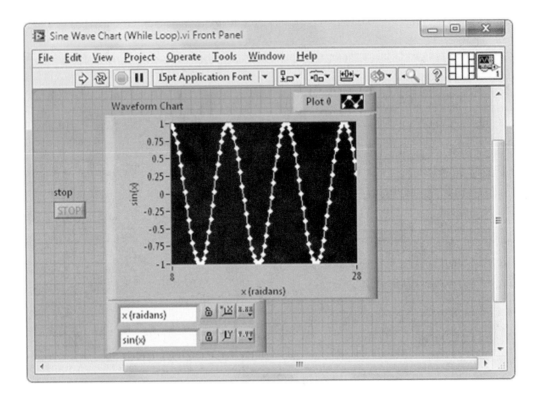

DO IT YOURSELF

Write a VI named **Stopwatch** that functions as a stopwatch with a precision of 0.01 second. Design **Stopwatch** so that it continuously displays the elapsed time from when its **Run** button is pressed until its **Stop Button** is pressed. The front panel of **Stopwatch** should appear as shown below. Here, the elapsed time is displayed on a **Numeric Indicator** (found in **Controls>>Modern>>Numeric**) labeled **Elapsed Time (second)**. Choose the Stop Button's **Mechanical Action** appropriately.

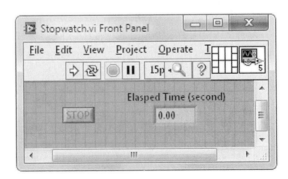

PROBLEMS

1. Write a VI called **Greater Than Ten** that lights a **Round LED** indicator (found in **Controls>>Modern>>Boolean**) on its front panel when the iteration terminal 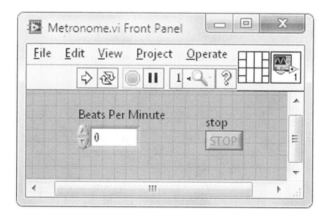 of a While Loop on its block diagram has a value greater than 10. You may find the following icons useful: **Select** and **Greater?** in **Functions>>Programming>> Comparison**.

2. Write a VI called **Single Sine Cycle** that stops automatically after the While Loop on its block diagram has iterated exactly 1000 times. As the While Loop performs these 1000 iterations, make a Waveform Chart plot one cycle of a sine wave (where the final data point is at the equivalent point on the sine wave as the initial point). You may find the following icons useful: **Equal?** in **Functions>>Programming>> Comparison** and **Pi Multiplied By 2** in **Functions>>Programming>>Numeric>> Math & Science Constants**.

3. On the block diagram of **Sine Wave Chart (While Loop)**, pop up on While Loop's conditional terminal and select its **Continue if True** mode. Then, modify the program as needed so that it executes in the original manner (i.e., charts the sine wave until the user presses the **Stop Button**).

4. Write a VI called **Metronome**, whose front panel includes a **Numeric Control** (found in **Controls>>Modern>>Numeric**) labeled **Beats Per Minute** as shown below.

When run, construct **Metronome** so that it produces N beeps every minute until the **Stop Button** is pressed, where N is the value entered in the **Beats Per Minute** control and a beep is a 264-Hz sound wave of 100-ms duration. To create a beep, use **Beep.vi**, which is found in **Functions>>Programming>>Graphics & Sound**, with its **use system alert?** input set to FALSE.

5. The odd integers are given by $2i+1$, where $i+0, 1, 2, \ldots$ Write a program called **Odd Integer Search** that answers the following question: What is the smallest odd integer

that is divisible by 3 and also, when cubed, yields a value greater than 4000? Display the answer to this question in a front-panel **Numeric Indicator**. You may find icons in the following palettes useful: **Functions>>Programming>>Numeric** (e.g., **Quotient & Remainder**), **Functions>>Programming>>Comparison**, **Functions>> Programming>>Boolean**, and **Functions>>Mathematics>>Elementary & Special Functions>>Exponential Functions**.

6. Construct a VI named EvenOdd whose front panel has two **Round LED** indicators and a **Stop Button** (all found in **Controls>>Modern>>Boolean**). Label one indicator Even and the other Odd as shown below. Place a While Loop on the block diagram, which iterates once every 0.5 seconds, and construct a subdiagram within it so that the Even indicator is lit (and Odd is unlit) during iterations for which the value of the iteration terminal 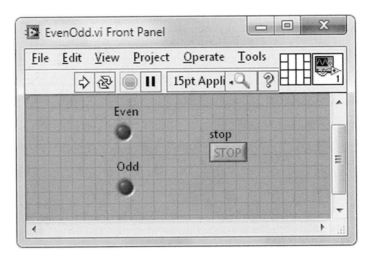 is even, while the Odd indicator is lit (and Even is unlit) during iteration for which the value of is odd. The following icons may be useful: **Select** and **Equal To 0?** in **Functions>>Programming>>Comparison** as well as **Quotient & Remainder** in **Functions>>Programming>>Numeric**.

7. Explore the impact of While Loop execution on CPU usage. Open the **Windows Task Manager** by right-clicking an empty area on the taskbar at the bottom of your monitor and then selecting **Start Task Manager** or by pressing <*Ctrl+Shift+Esc*>. You will find that the **Task Manager** displays CPU usage (expressed as a percentage of your computer's maximum value) at the bottom of its window.

   (a) Open Sine Wave Chart (While Loop), and run this VI with **Wait (ms)** programmed to produce one While Loop iteration every 100 ms. What is the approximate increase in CPU usage that results when the program is running?

(b) Next, program **Wait (ms)** to produce one While Loop iteration every 1 ms, and then run the VI. What is the approximate increase in CPU usage that results when the program is now running?

(c) Finally, delete **Wait (ms)** from the VI's block diagram so that the While Loop will iterate as fast as possible on your computing system (e.g., millions of iterations per second), and then run the program. What is the approximate increase in CPU usage that results when the program is now run? [Note that such "memory hogging" should be avoided unless your program is executing a high-speed task (i.e., not simply updating a plot as in this case).]

8. Write a program called **Iterations Until Integer Equals Five**, which randomly generates an integer in the range from 1 to 10 and iterates this random process until the integer equals 5. The number of iterations required until the randomly generated integer equals 5 is then displayed on the front panel in a **Numeric Indicator** (found in **Functions>>Programming>>Numeric**), which is labeled **Required Iterations** as shown below.

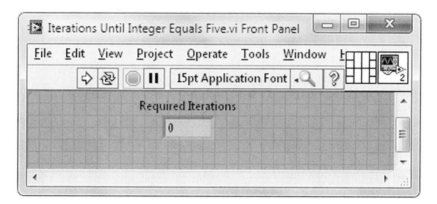

To create a random integer in the range from 1 to 10, use **Random Number (0–1)** to generate a random floating-point number in the range from 0 up to (but not including) 1, multiply this number by 10, and then use **Round Toward + Infinity** to round the floating-point number to the next highest integer (note **Round Toward + Infinity** rounds x.000 to x). All of these icons are found in **Functions>>Programming>>Numeric**. Think carefully about how to determine the value of **Required Iterations**. Also, you will need to use the **Equal?** icon, which is found in **Functions>>Programming>>Comparison**.

Knowing that the integers are created randomly (i.e., equal probability for producing integers 1 through 10 each iteration), what do you expect the value of **Required Iterations** to be on average? Run **Iterations Until Integer Equals Five** 20 times and record the value of **Required Iterations** resulting from each run. Is the average of these 20 values close to the value that you expected?

9. A characteristic of the Waveform Chart can be controlled within a program using a **Property Node**.

    (a) In this chapter, you set a Waveform Chart's **X-Axis Multiplier** equal to *0.2* using the **Properties** dialog window. Alternately, this property of the Waveform Chart can be set on the block diagram via a Property Node as follows: with **Sine Wave Chart (While Loop)** open, pop up on the Waveform Chart's icon terminal and select **Create>>Property Node>>X Scale>>Offset and Multiplier>>Multiplier**, and then place the resulting Property Node on the block diagram. This Property Node is created as an indicator, hence the small outward-directed black arrow at its right side. Change this icon to a control by popping up on its lower section and selecting **Change To Write**. The black arrow will now be inward directed on its left side. Using **Create>>Constant**, wire a value of *0.2* to the **XScale.Multiplier** property control as shown below.

    Run **Sine Wave Chart (While Loop)** and assure yourself that the Property Node is indeed setting the **X-Axis Multiplier** to 0.2.

    (b) Program **Sine Wave Chart (While Loop)** so that its Waveform Chart is cleared at the start of each run. The appropriate Property Node is made by selecting **Create>>Property Node>>History Data**. After changing this Property Node to a control by selecting **Change to Write** in its pop-up menu, **Create>>Constant** will produce the needed input called an **Empty Array** as shown below.

    To clear the Chart at the start of each program execution (i.e., immediately after the Run button is pressed), should this Property Node be placed inside or outside of the While Loop? Run **Sine Wave Chart (While Loop)** several successive times and demonstrate that the Property Node performs as intended.

    (c) Change the background color **Sine Wave Chart (While Loop)**'s Waveform Chart during runtime as follows: place a Property Node within the While Loop using **Create>>Property Node>>Plot Area>>Colors>>BG Color**, and then change this Property Node to a control by selecting **Change To Write** in its pop-up menu. On the front panel, place a **Framed Color Box**, found in **Controls>>Modern>>Numeric**, and then on the block diagram wire its terminal to the Property Node. Run **Sine Wave Chart (While Loop)** and demonstrate that the Waveform Chart's background color can be controlled using the Operating Tool.

CHAPTER 2

The For Loop and Waveform Graph

For the remainder of the book, only the code to be written on block diagrams will be illustrated, rather than the entire block-diagram windows. Front-panel windows, however, will be shown in their entirety. Also, the **Modern** Palette will be used for front-panel controls. If you prefer to use the **Silver** Palette, select **File>>VI Properties>>Category>>Editor Options>>Control Style for Create Control/Indicator>>Silver Style**, so that **Silver** (rather than **Modern**) controls are produced when you implement LabVIEW's automatic creation feature.

2.1 FOR-LOOP BASICS

The LabVIEW programming language provides two possible loop structures to control repetitive operations in a program. In the previous chapter, we explored the While Loop, which (by default) executes the subdiagram within its borders until the Boolean value wired to its conditional terminal is TRUE. That is, the While Loop executes until a specified condition is no longer FALSE. Now we will focus our attention on the *For Loop*, LabVIEW's other available loop structure. In contrast to the While Loop, which is controlled by the value of a specified condition, the For Loop simply executes the subdiagram within its borders a specified number of times. This loop structure is found in **Functions>> Programming>>Structures** and appears on your block diagram as shown below.

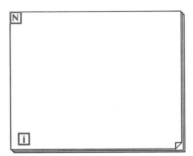

In this graphical structure, your "to-be-iterated" subprogram is written in the currently blank region within the borders. Once this subdiagram is written, the value of the *count terminal* ▣ determines the total number of times N that the For Loop will iterate. The value of the count terminal is set by wiring a **Numeric Constant** (located outside the loop) to the ▣. The For Loop's other internal icon is familiar from your work in the previous chapter. As in the While Loop, the iteration terminal ▣ contains the current number of completed loop iterations. The value of ▣ is 0 during the first loop iteration, 1 during the second, ..., and $N - 1$ during the last iteration. Thus, the For Loop is equivalent to the following text-based code:

> For $i = 0$ to $N - 1$
> Execute Subprogram

In this chapter, you will write a sine-wave plotting program based on a For Loop. In the process of writing this VI, you will explore LabVIEW's method of storing numerical data in the form of an array and learn to operate a *Waveform Graph*, the second in LabVIEW's triad of commonly used graphing options. Additionally, you will further hone your LabVIEW editing skills.

2.2 SINE-WAVE PLOT USING A FOR LOOP AND WAVEFORM GRAPH

Create a fresh VI by selecting **Blank VI** in the **Getting Started** window or, if a program is already open, **File>>New VI**.

Here's a shortcut for activating the Controls Palette, if it isn't already visible: Pop up (right-click the mouse button) on the blank front panel and a Controls Palette will appear. This palette is temporary, in that it will disappear as soon as you release the mouse button. However, you may secure the palette by placing the mouse cursor over the thumbtack located in the upper left corner of its window and then releasing the mouse button. This "tacked-down" Controls Palette can then be placed in a convenient location by "clicking and dragging" on its title bar. By the way, a similar "thumbtacking" procedure can be used to make any frequently used subpalette of the Controls Palette or Functions Palette continually visible.

Place a **Waveform Graph** on the front panel from **Controls>>Modern>>Graph**.

Secure its default label **Waveform Graph** by pressing <*Enter*> and use the ⬨ to position the entire graph nicely on the front panel. Pop up on the **Waveform Graph** and select **Visible Items>>Scale Legend**. You will be creating a program to plot *sin(x)* vs. *x*, so (with the help of the 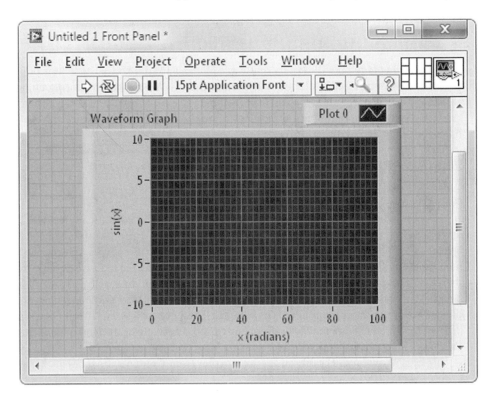) use the Scale Legend to label the *x*- and *y*-axes as **x (radians)** and **sin(x)**, respectively.

By default, autoscaling is activated on both the *x*-axis and the *y*-axis for a Waveform Graph. You can verify this fact yourself by noting that the **Scale Lock** buttons for both axes are closed 🔒. Keep autoscaling activated on both axes by changing nothing and then toggle off (i.e., hide) the Scale Legend by popping up on the Graph and selecting **Visible Items>>Scale Legend**.

If, as in this present case, labeling the axes is the only required task, rather than using the Scale Legend, you can simply highlight each axis label using the Labeling Tool 🅰 from the Tools Palette and type in the desired text directly. Try this time-saving tip.

2.3 WAVEFORM GRAPH

What is the Waveform Graph and how does it differ from the Waveform Chart? In our previous work, we saw that a Waveform Chart acts as a real-time data plotter. As each new data point is generated on the block diagram, this single numerical value is passed to the Waveform Chart's terminal and then immediately displayed in the front-panel plot. The new data point is appended to the already existing data plot, so you can view the current value in the context of the previous values.

In contrast to the Waveform Chart's interactive method of plotting data, the Waveform Graph (as well as the yet unstudied XY Graph) accepts an entire block of data that has been generated previously. The Waveform Graph's terminal accepts this data block as a one-dimensional (1D) array of N numerical elements with the order of the elements indexed by integers in the range of 0 to $N-1$. For example, a 10-element array would have the following form.

Index	0	1	2	3	4	5	6	7	8	9
Array	1.20	1.30	1.40	1.50	1.60	1.70	1.80	1.90	2.00	2.10

The Waveform Graph then assumes that the numerical value of each element is the y-value of a data point to be plotted. The x-value of each plotted point is taken to be the index of that point within the 1D array. That is, the x-value of the array's ith element is i and the y-value is the numerical value of the ith element. Thus, the Waveform Graph implicitly assumes that data points are evenly spaced along the x-axis, a situation that is often realized in practice. For example, the time-varying nature of an analog signal is commonly represented by a set of discrete data points that were digitized ("sampled") during a sequence of equally spaced time intervals. You will find later that the XY Graph allows one to plot the more general case of an array of unequally spaced data points (as well as multivalued functions such as circular shapes).

Switch to the block diagram. There you will find the Waveform Graph's terminal, in its icon terminal guise, with its label located nearby. Note that the icon terminal contains a distinctive small picture of plotted data (indicating it is associated with the Waveform Graph) and has an orange border with text **DBL**, denoting that the data input to it should be double-precision floating-point numbers. If you pop up on this icon terminal and select **View As Icon**, the terminal will change its data-type terminal form, where we note that the DBL data-type specifier is enclosed within a pair of square brackets. The square brackets indicate that the Waveform Graph terminal should be wired to accept a numerical array, rather than just a single number as in the case of the Waveform Chart discussed above. To toggle back to the icon terminal manifestation, pop up and select **View As Icon**.

2.4 OWNED AND FREE LABELS

Let me digress for a moment and discuss LabVIEW labels. The Waveform Graph's label currently on your block diagram (which contains the text **Waveform Graph**) is what is known as an *owned label*. An owned label belongs to a particular object (in this case, the Waveform Graph's terminal) and moves with that object. To demonstrate this fact, use the

↖ to highlight the terminal, and then drag it to a new location on the block diagram. Note that the label accompanies the icon to its new location. Deselect the icon by clicking on an empty spot of the block diagram, and then click on top of the label itself. A marquee will then appear solely around the label, and you will be able to drag just the label to some newly desired location relative to its owning icon. An owned label can be made visible or invisible using the **Visible Items>>Label** option in the object's pop-up menu. Experiment a bit on the block diagram until you understand the functioning of an owned label.

An alternate form of LabVIEW annotation is called the *free label*. Free labels are not attached to any object and can be created, moved, or disposed of independently. They can be used as explanatory comments on your front panels and block diagrams. To create a free label, select the Labeling Tool ⟦A⟧ from the Tools Palette and click on the desired location for the label. A small, bordered box appears, ready to accept text input. After you type the text message, press *<Enter>*, and then the free label is complete. Try producing a free label on your block diagram. After you succeed in creating a free label, click on it with the ↖ and delete it.

2.5 CREATION OF SINE WAVE USING A FOR LOOP

Now select a **For Loop** from **Functions>>Programming>>Structures** and add it to your block diagram (the Functions Palette can be activated by popping up on a blank region of the block diagram). If you are dissatisfied with your initial choice of dimensions, you can place the ↖ at a corner (or at the midpoint of a side) of the For Loop, where it will morph into the resizing handle and allow you to reshape the loop borders.

Our programming strategy is this: use the For Loop to create a one-dimensional array of data points that represent a few cycles of the sine function, and then pass this array to the Waveform Graph's icon terminal for plotting on the front panel. Because we desire data to be passed to the Waveform Graph's terminal only after the For Loop has completed its final iteration, we must place the terminal in a region outside the For Loop's boundary. Use the ↖ to build the following block diagram.

Waveform Graph

Now let's write the code for creating the sine-wave values within the For Loop. To achieve this goal, obtain a **Sine** icon from **Functions>>Mathematics>>Elementary & Special Functions>>Trigonometric Functions** and place it inside the For Loop.

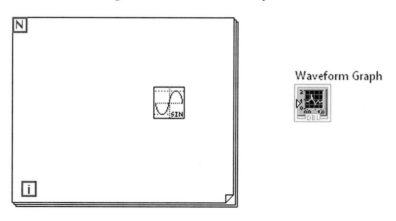

Waveform Graph

Now use **Divide**, from **Functions>>Programming>>Numeric**, to write the sine-wave creation code as follows. After placing **Divide** on your diagram, pop up on its bottom input and select **Create>>Constant** to automatically create the **Numeric Constant**  (if the constant appears as an orange , you can change it to  using **Display Format…** in the pop-up menu, if desired). Next, wire  to the top input of **Divide**, and then wire Divide's output to the **x** input of Sine. If you reverse the order in which  and  are wired to **Divide**, the **Numeric Constant** will be autocreated as an integer rather than a floating-point number (which can be corrected by selecting **Representation>>DBL** in the Numeric Constant's pop-up menu).

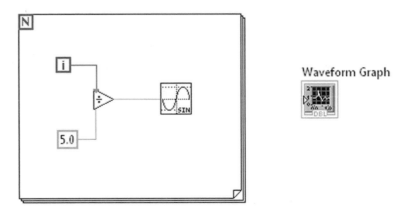

Waveform Graph

Ready for another editing shortcut? When coding a block diagram, you will find that the most commonly used tools are ♦ and ♦. Rather than having to go to the Tools Palette each time to switch back and forth between these tools, simply press the *<Spacebar>* on your keyboard. With each depression of the *<Spacebar>* key, the mouse cursor will toggle back and forth between the Positioning and Wiring Tools. Alternatively, by clicking on the **Automatic Tool Selection** button near the top of the Tools Palette, its green LED indicator will illuminate, denoting that the *automatic tool selection* option is activated. Then, when you move the cursor to a position on the block diagram (or front panel), LabVIEW will automatically select the tool that is most appropriate for that location. When this option is activated (indicated by the illuminated green LED), it can be toggled off by clicking on the **Automatic Tool Selection** button. Finally, after you have toggled off Automatic Tool Selection, your keyboard's *<Tab>* key can be used to manually select tools. As you press *<Tab>* repeatedly, the mouse cursor will cycle from ♦ to ♦ to A to ♦ in the Tools Palette.

Now we need to tell the For Loop the number of iterations *N* we would like it to perform. We know from our experience of writing similar block-diagram code in the last chapter that about 30 loop iterations will be required to generate one cycle of the sine function on this block diagram. Let's set *N* = 100, so that we generate (a little more than) three sine-wave cycles. To accomplish this setting, wire a **Numeric Constant** (defined to be *100*) to the N. You already know two methods for acquiring the required **Numeric Constant** icon—the "hard" way, by obtaining it directly from

Functions>>Programming>>Numeric, and the "easy" way, by simply popping up on the and selecting **Create>>Constant**. Be good to yourself—use the "easy" way.

2.6 CLONING BLOCK-DIAGRAM ICONS

Here's one final editing option that you can add to your bag of tricks. It's called *cloning* and can be used in your present situation to produce the required Numeric Constant (to be wired to the) because an equivalent icon already exists on the block diagram. To clone the existing **Numeric Constant** on your block diagram, place the Positioning Tool over this icon.

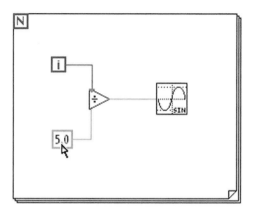
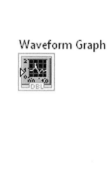

Waveform Graph

Now click the mouse button while depressing *<Ctrl>* on your keyboard. Then, by moving the mouse while still holding down the button, you will drag a copy of the icon with you, while the original stays in place. Once you arrive at the newly desired location, release the mouse button and the cloned icon will reside there (if you place within the autowiring range of the , a wire will automatically connect the two).

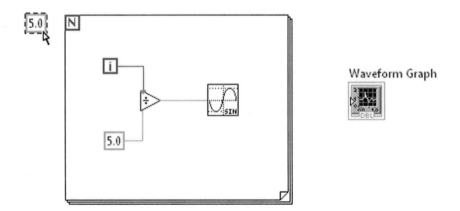

Use the Operating Tool to change the value of this new **Numeric Constant** to *100* and then wire it to the 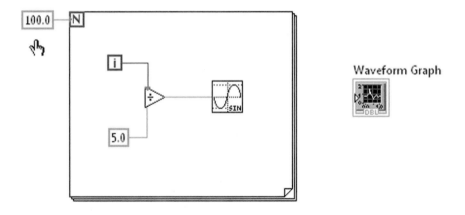.

The For Loop's count terminal is configured to accept the numeric format **I32** (signed four-byte integer, also called *long integer*). Depending on how its **Display Format...** option is configured, the Numeric Constant may or may not automatically adjust its data type to **I32**. If this adjustment is not made automatically, a red coercion dot, indicating a data-type mismatch, will appear as shown in the above diagram. This coercion dot can then be erased by popping up on the **Numeric Constant** and selecting **Representation>> I32**, as shown next.

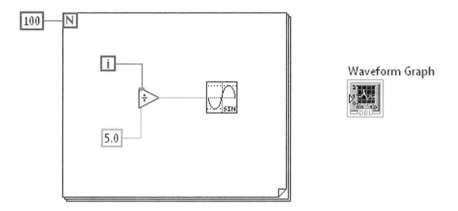

Waveform Graph

2.7 AUTO-INDEXING FEATURE

Finally, we need to store the sequence of sine-function values produced by the For Loop in a 100-element array so that we can pass this array to the Waveform Graph's terminal for plotting. Does it sound like there's some hard programming ahead? Surprisingly, there's not! Array storage of the sequence of values generated by the multiple iterations of a loop structure is such a commonly required operation that LabVIEW provides it as a built-in optional feature of both the For Loop and the While Loop. Perhaps it's easiest if we first implement this feature and then decipher what we've done.

Using ✎, simply wire the **sin(x)** output of the **Sine** icon to the Waveform Graph's terminal. You're finished!

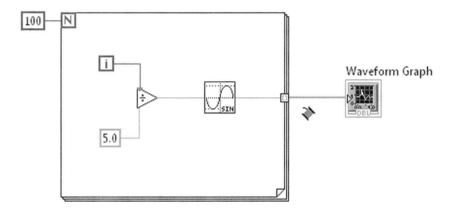

Waveform Graph

Here, we've accomplished our goal by implementing the *auto-indexing* capability of the For Loop. With this feature, as the For Loop creates a new value with each loop iteration, this new value is indexed and appended to an array at the loop boundary. After the loop completes its final iteration, the array is passed out of the loop to the Waveform Graph's terminal. Note that, in the region within the For Loop, the wire emanating from the Sine output is thin, denoting the fact that this icon produces just a single numerical value with each loop iteration. However, when this wire passes through the black-bordered square □ at the loop's boundary (called a *tunnel*), the wire becomes much thicker. This thick wire is LabVIEW's way of denoting a one-dimensional array of data values.

Let's explore this auto-indexing feature a little further. First, note the pair of square brackets contained within the tunnel icon □. Such a bracket pair is LabVIEW code for an array. Thus, these brackets are the graphical indictor that an array is being created at the tunnel as the For Loop iterates (i.e., auto-indexing is activated).

Place the Positioning Tool over the tunnel at the For Loop's boundary.

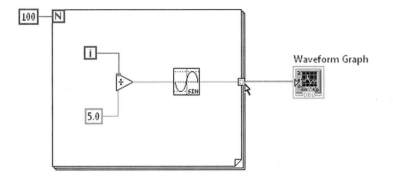

Then, pop up on this tunnel to reveal its pop-up menu.

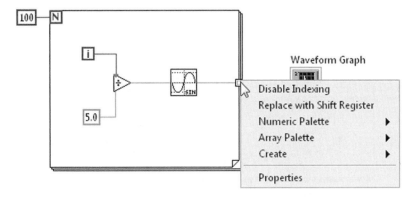

This menu offers you the ability of toggling the loop's auto-indexing feature on and off. As we surmised earlier, a For Loop's default setting is **Enable Indexing**, so the menu now offers the option of toggling this feature off with the selection **Disable Indexing**. Toggle the auto-indexing off by selecting **Disable Indexing**.

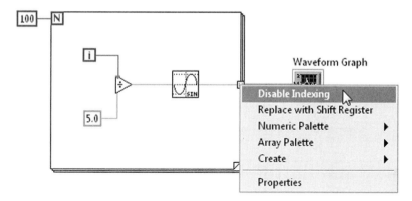

Now, rather than accumulating all the loop's iterated values into an array, the tunnel will only pass out a single value, the value of the sine function determined during the last loop iteration. To reflect this functional change for the tunnel, the bracket pair is no longer present within its icon, replaced instead by a solid color. As shown below, with the auto-indexing disabled, a broken wire now appears connected to the Waveform Graph's terminal because this icon expects to receive an entire array, not just a single numerical value.

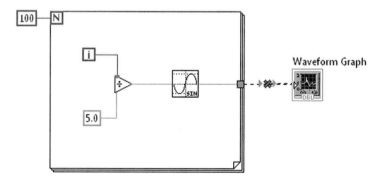

To restore the proper programming, pop up on the tunnel and toggle the auto-indexing back on by selecting **Enable Indexing**.

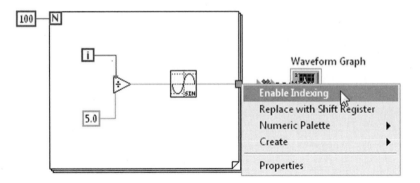

2.8 RUNNING THE VI

Let's see if your program works. Return to the front panel. As mentioned previously, autoscaling is activated by default on both axes of a Waveform Graph (you can, of course, turn off this feature in the Waveform Graph's pop-up menu or in the Scale Legend). Before running the VI, it is a good idea to pause for a moment and anticipate how the axes will be autoscaled when the program is functioning properly. The code is designed to plot a sine wave, which is discretely sampled at 100 points. Thus, the y-axis values should range from -1.0 to $+1.0$. The x-axis value of each sine wave value is taken to be its index within the one-dimensional array of data points. Because LabVIEW ascribes an index value of 0 to the first array element, the index of our last data is 99. Thus, the x-axis values range from 0 to 99. Perhaps the scaling will be more pleasing if LabVIEW's autoscaling feature chooses 100 as the x-axis upper limit (which is the default value for the Waveform Graph).

OK, run your program. Hopefully it will produce a plot as shown next, where the axes are scaled as we anticipated.

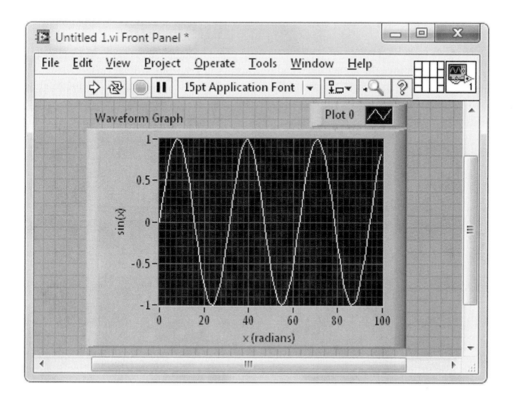

2.9 *X*-AXIS CALIBRATION OF THE WAVEFORM GRAPH

Pretty nice program, eh? There is one thing, however, we need to fix—the *x*-axis calibration. As is, the number labels on this axis reflect the integer indexing derived from the array storage of your data, while the text label indicates that this axis is intended to plot the radian argument of the sine function being graphed on the *y*-axis. Remembering that our program takes the sine function argument for a particular data point to be the array index value *i* divided by 5, we see that the *x*-axis can be calibrated in radians by simply multiplying each number label by 0.2. Alternately, we can re-express this fact by saying that, by default, the Waveform Graph assumes the *x*-axis spacing between adjacent data points Δx is 1. Calibration is then accomplished by defining Δx to be 0.2.

Such calibration by a multiplicative constant is a common need when using the Waveform Graph, and therefore LabVIEW provides mechanisms for this procedure. In the previous chapter, we accomplished *x*-axis calibration of a Waveform Chart by defining a Multiplier in the **Chart Properties** dialog window. This method can be implemented on the Waveform Graph also. Here, we will introduce an alternative calibration method, which appears explicitly on the block diagram. This calibration mechanism involves a new LabVIEW process: bundling together several objects into a *cluster*. The resulting cluster is not a mathematical object, but rather a LabVIEW convenience. By grouping a number of objects together into a single cluster, one is able to transport a large quantity of data across a block diagram using a single cluster wire. Thus, clustering greatly simplifies the appearance and readability of your block diagrams.

Return to the block diagram, place the mouse cursor over the Waveform Graph's terminal, and activate its Help Window.

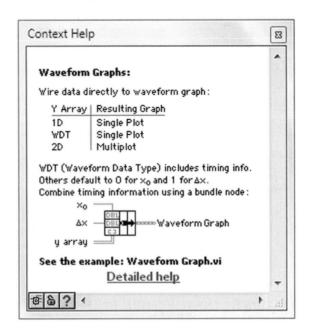

This Help Window, which is divided into a top and bottom portion, appears above. In the top portion, we find that a Waveform Graph with default data-point spacing $\Delta x = 1$ is created by simply connecting an array wire to the Waveform Graph terminal. This is the manner in which we've used the Waveform Graph thus far. We produced a single plot by feeding a one-dimensional data array to the Waveform Graph's terminal. Note that two (or more) waveforms can be displayed simultaneously on a Waveform Graph by connecting a two-dimensional (2D) array to the terminal. The bottom portion of the Help Window indicates that, when given an appropriate cluster, the Waveform Graph produces a plot on the front panel with its x-axis properly calibrated. The appropriate cluster is formed by bundling together the following sequence of objects: a value for the x-axis origin x_0, a value for the x-axis spacing $\Delta x = 1$, and the array of numerical data to be used as the y-axis values of the waveform's points.

Here's how to include the appropriate cluster in your code. First, remove the array wire connected to the Waveform Graph terminal. Using the Positioning Tool, click on this wire to produce a marquee.

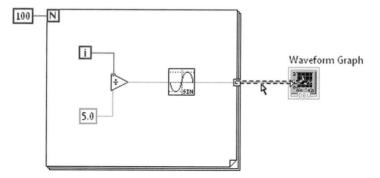

Then press your keyboard's *<Delete>* key to remove this wire.

Select **Bundle** from **Functions>>Programming>>Cluster, Class, & Variant**. The **Bundle** icon is used to collect several input objects into a cluster.

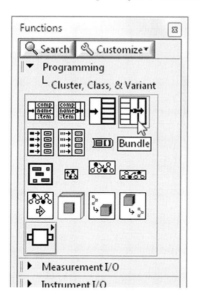

Then place the **Bundle** icon on your block diagram. You will find that its leftmost section consists of two inputs by default. We require three inputs, so you must resize the icon.

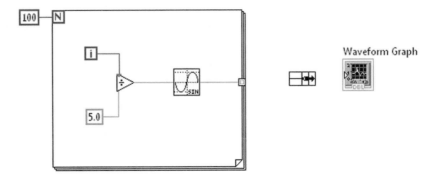

To resize the **Bundle** icon so that it has three inputs, place the ⬚ at the icon's bottom center (or top center), so that the tool morphs into a resizing handle.

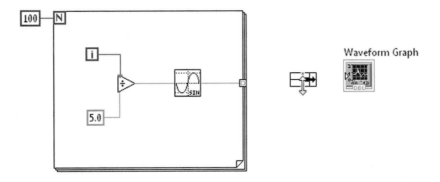

Then drag the handle until an additional input appears.

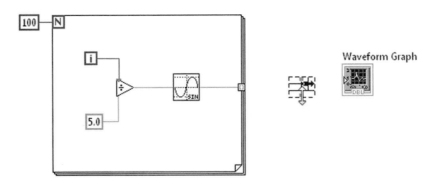

Secure this choice of inputs by releasing the mouse button. Use the Positioning Tool to position the Bundle and Waveform Graph terminal as shown next.

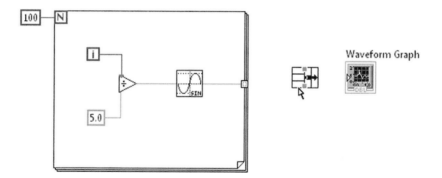

Wire the For Loop's tunnel to the Bundle's bottom input. Once this connection is made, square brackets will appear within this input, the graphical specifier for an array data type.

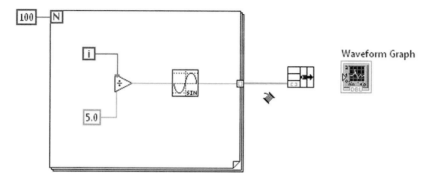

Now wire a **Numeric Constant** of *0.0* and *0.2* to the Bundle's top and middle inputs to define x_0 and $\Delta x = 1$, respectively. Note that unbracketed **DBL**s appear within these inputs, indicating that each consists of a solitary (i.e., non-array) double-precision floating-point number.

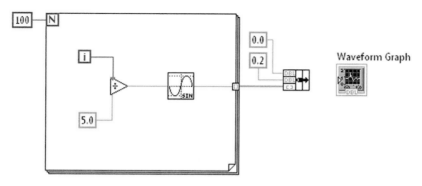

The **Bundle** icon receives these three inputs on its left side, bundles them together, and outputs the resulting cluster at its rightmost section. Wire this cluster output to the Waveform Graph's terminal.

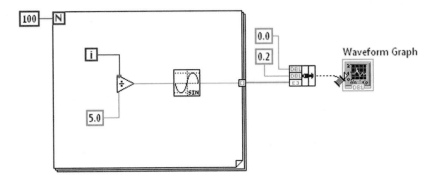

You will find that as soon as this connection is made, the icon terminal's data-type specifier morphs from a floating-point numeric specifier into a cluster specifier . In addition, the wire connecting the Bundle output to the Waveform Graph's terminal has a pink banded appearance, denoting cluster data.

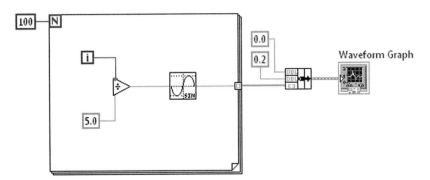

Now return to the front panel and run your final program. Note the x-axis is calibrated in radians.

Using **File>>Save**, create a folder named **Chapter 2** within the **YourName** folder, and then save this VI under the name **Sine Wave Graph (For Loop)** in the **YourName\ Chapter 2** folder.

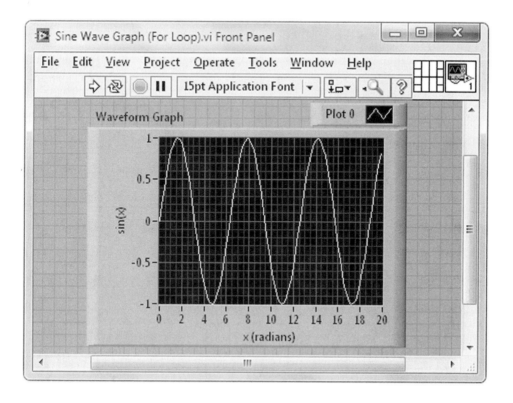

2.10 SINE-WAVE PLOT USING A WHILE LOOP AND WAVEFORM GRAPH

Let's try accomplishing the above goal of plotting a sine wave on a **Waveform Graph**, but this time using a While Loop rather than a For Loop to generate the data array. Select **File>>New VI**, and then place a **Waveform Graph** (from **Controls>>Modern>>Graph**) with its default label **Waveform Graph** on the fresh front panel. Autoscaling will be enabled on both axes by default. Using the (or the Scale Legend), label the x- and y-axes as x (radians) and sin(x), respectively.

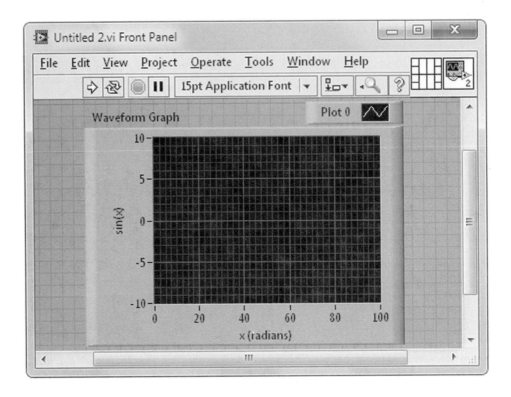

Switch to the block diagram and write the following code.

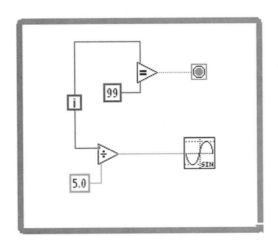

Waveform Graph

The **Equal?** icon is found in **Functions>>Programming>>Comparison**. Its function is described by the Help Window shown next. Remember, you can always access assistance from such Help Windows while wiring by toggling on **Help>>Show Context Help** or, more simply, through the keyboard shortcut <*Ctrl+H*>.

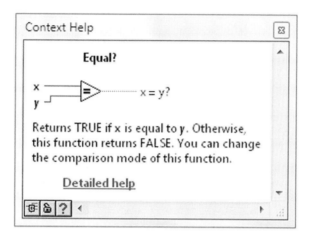

The While Loop on your block diagram will generate 100 values of the sine function before ceasing execution. Using your knowledge of the detailed operation of a While Loop, can you explain why this is so? In particular, explain why the proper value for the **Numeric Constant** in the comparison statement is *99* (and not *100*).

You now need to store the 100 sine-function values in a one-dimensional array and pass this block of data to the Waveform Graph's terminal for plotting on the front panel. Making use of the auto-indexing feature of the loop structure, we need simply to wire the **sin(x)** output of the Sine icon to the Waveform Graph terminal. Perform this wiring as shown here.

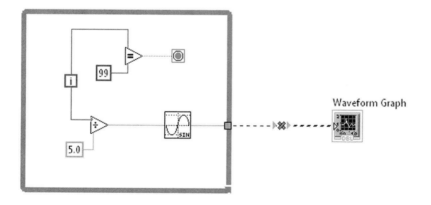

You will note that this seemingly simple connection produces a broken wire. To find out why, pop up on the tunnel at the While Loop's border.

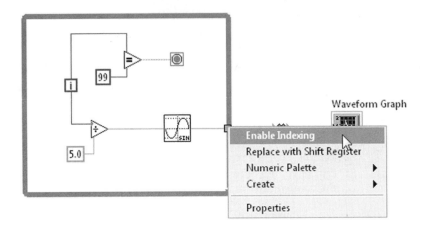

In the tunnel's pop-up menu, you will be presented the option of enabling the loop's auto-indexing feature. Thus, we see that, in contrast to a For Loop, the auto-indexing feature of a While Loop is toggled off by default. Turn the auto-indexing on by selecting **Enable Indexing**. Also include a **Bundle** icon to scale the *x*-axis in radians, as shown below.

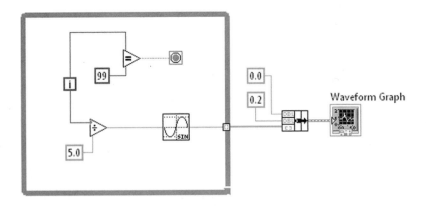

Your program is now complete. Return to the front panel and run it. You might try using the keyboard shortcut for the **Run** command: <*Ctrl+R*>.

2.11 ARRAY INDICATORS AND THE PROBE WATCH WINDOW

Are you unconvinced that there are actually 100 elements in the array you produced? You can easily check this fact by inspecting the array using an *Array Indicator*. An Array Indictor is constructed as follows. Select an **Array** shell from **Controls>>Modern>> Array, Matrix & Cluster**. As described in the next paragraph, the **Array** shell can be made to operate in either of two modes: to input an array from the front panel to the block diagram or to output an array from the block diagram to the front panel.

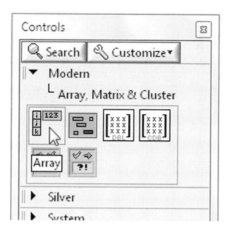

Place the **Array** shell on your front panel, and then enter the label **Array**. You will find that the shell consists of an *index display* on the left and a large blank region on the right. By filling the blank region with an appropriate object (called the *element display*), one may choose whether the **Array** shell will function as an input or output device. In the present situation, we wish to construct an Array Indicator that outputs an array to the front panel, so we will choose our element display to be a **Numeric Indicator**. If instead we wished to construct an *Array Control* that inputs an array to the block diagram, we would use a **Numeric Control** for the element display.

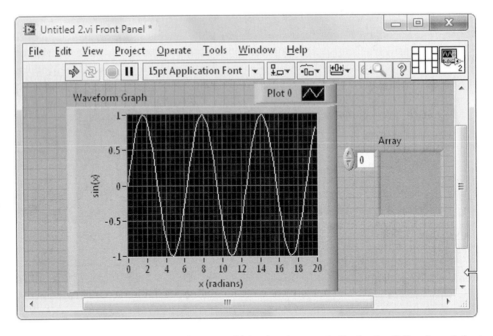

To place a **Numeric Indicator** within the **Array** shell, do the following. Select a **Numeric Indicator** from **Controls>>Modern>>Numeric**.

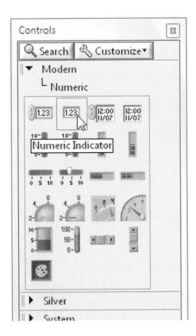

Then place the cursor in the large blank region of the **Array** shell.

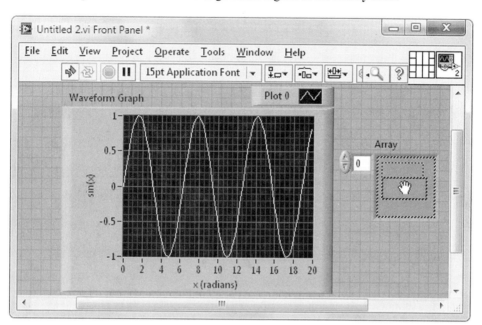

When you click the mouse button, the Array Indicator will be complete, as shown below.

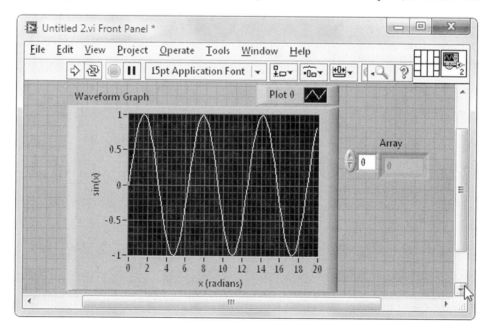

Now go to the block diagram and place the Array Indicator's terminal in a convenient location. In this program, the While Loop creates a 1D array of data. Using the Wiring Tool, connect any point on the thick wire representing this one-dimensional array to the Array Indicator's terminal.

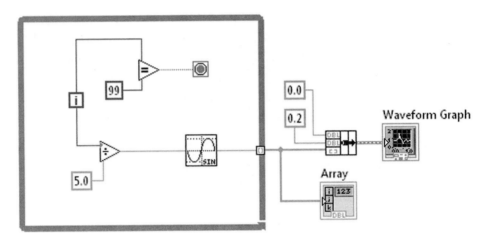

Waveform Graph

Array

Return to the front panel and run your program. After the VI completes execution, you can inspect the array generated via the Array Indicator (if you need to resize it, use

the). Using the , look at various elements in the array. To scan the array, you may increment or decrement the index number in steps of one by repeatedly clicking on the index display's up/down arrows. Alternately, to view a particular array element, you can

use the to highlight the index display, then enter the desired element's index value

from the keyboard, and press <Enter>, or click on .

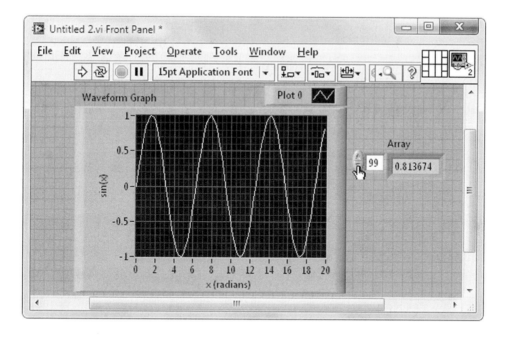

Through such exploration, you should find that your array consists of 100 elements, with indices running from 0 to 99.

LabVIEW's *Probe Watch Window* provides another method for checking the contents of your array. This handy feature allows you to examine the contents of any wire on your block diagram. To inspect the elements of sine-wave array using the Probe Watch Window, place the mouse cursor on the wire carrying these data.

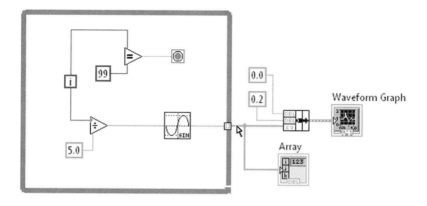

Then pop up on this wire and select **Probe**.

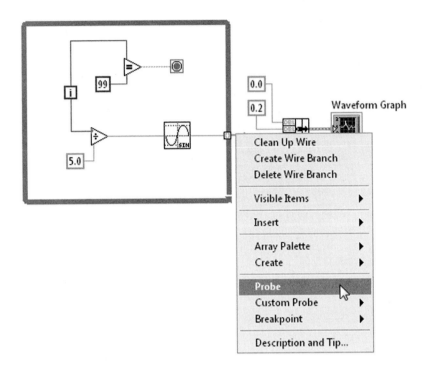

A Probe Watch Window will appear containing a probe for the selected wire, along with an associated numbered rectangle on the block diagram marking the wire location being probed. The Probe Watch Window can also be activated by clicking on the wire of interest with the *Probe Tool* found in the Tools Palette.

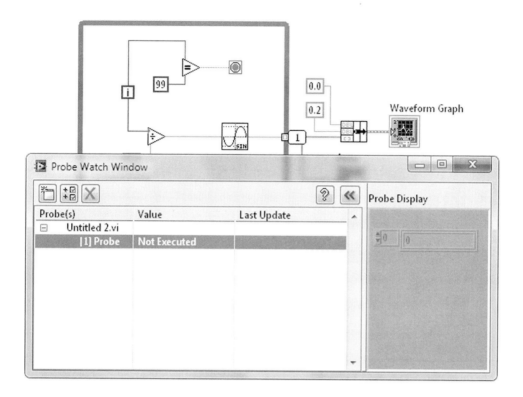

If you'd like to probe another wire as well, for example, the wire carrying the value of the iteration terminal, simply pop up on that wire and select **Probe** (or else click on the wire with), and a probe for that wire will be added to the Probe Watch Window.

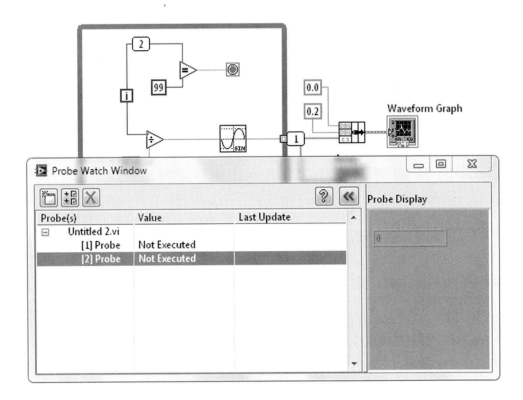

Run your program. Then, within the Probe Watch Window, click on **[1] Probe** to select the probe associated with the wire carrying the elements of sine-wave array. In the Probe Display region, that probe's Array Indicator will be made available to inspect the elements of the sine-wave array. In a similar way, you can inspect the other probe that, during runtime, reports the current value of the iteration terminal and, when the VI has completed its run, gives the final iteration terminal value.

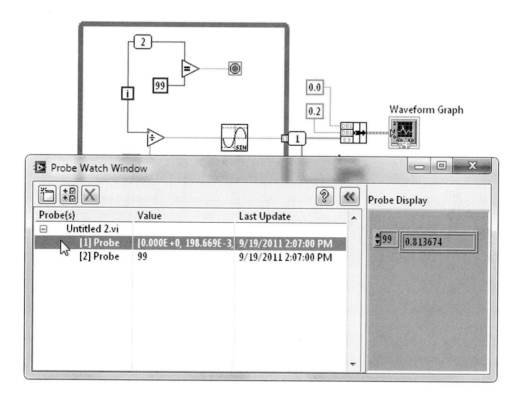

Within the Probe Watch Window, pop up on, for example, **[1] Probe** and explore the options available in a probe's pop-up menu such as deleting the probe or opening its own stand-alone window. These options are also available in the Probe Watch Window's toolbar.

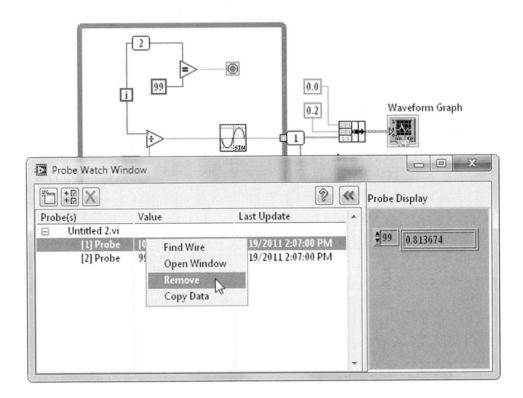

Close the Probe Watch Window when you're done and it, along with its associated numbered rectangles on the block diagram, will disappear. The Probe Watch Window is an invaluable tool for debugging LabVIEW programs that execute, but are producing questionable or unexpected results.

Finally, you might try producing an Array Indicator the "painless way." For example, pop up on the thick orange wire (or the tunnel) that carries the sine-wave array output from the While Loop, and then select **Create>>Indicator**.

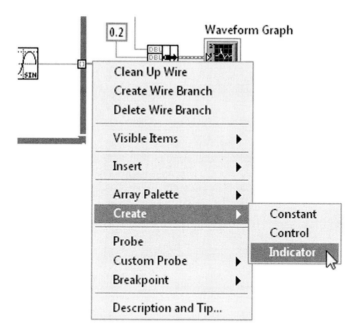

Now check the front panel. You will find an automatically produced Array Indicator there. Run your program and discover whether it works as expected.

Using **File>>Save**, save this VI in **YourName\Chapter 2** (that is, the **Chapter 2** folder within the **YourName** folder) under the name **Sine Wave Graph (While Loop)**.

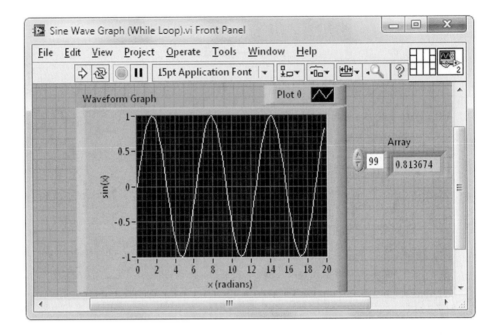

DO IT YOURSELF

A flipped coin can land in one of two states—*heads* or *tails*. Write a VI called **Four Flipping Coins** that simulates simultaneously flipping four coins and determines how many of these coins land in the heads state.

(a) First, build the front panel of **Four Flipping Coins** as shown here, where the state of each coin is represented by a **Round LED** Boolean indicator, which is found in **Controls>>Modern>>Boolean**. A lit (unlit) LED corresponds to heads (tails). Also include a **Numeric Indicator** labeled **Number of Heads** that gives the number of coins in the heads state.

On the block diagram, model the flip of each coin using a **Random Number (0–1)** icon, which is found in **Functions>>Programming>>Numeric**. When run, if this icon's **number (0 to 1)** output is greater than or equal to 0.5, assign the corresponding coin to the heads state; otherwise, that coin is in the tails state. Tally the results of all four coins to determine the number of coins in the heads state and output this integer to the **Number of Heads** front-panel indicator. Note that there are five possible outcomes—0, 1, 2, 3, and 4 heads. In writing the block diagram, you may find the following icons useful: **Greater or Equal?** in **Functions>>Programming>>Comparison, Boolean To (0,1)** in **Functions>> Programming>>Boolean,** and **Compound Arithmetic** in **Functions>> Programming>>Numeric.**

Run **Four Flipping Coins** to verify that it is working correctly.

(b) The present code on your block diagram simulates one *trial* of simultaneously flipping the four coins. Now simulate repeating such a trial over and over again. Place this code within a **For Loop** so that you can execute N trials, where N is given by the value in a front-panel **Numeric Control** labeled **Total Number of Trials**. Store the value of **Number of Heads** from each trial in an array using the For Loop's auto-indexing feature. Then, after the N trials are complete, plot a histogram on a **Waveform Graph** that displays the number of trials whose outcome was 0, 1, 2, 3, and 4 heads. To create a histogram, use the **Histogram. vi** icon found in **Functions>>Mathematics>>Probability & Statistics** with its **intervals** input wired to an integer 5 (because there are five possible outcomes for **Number of Heads**). The front panel of **Four Flipping Coins** will then appear as shown next.

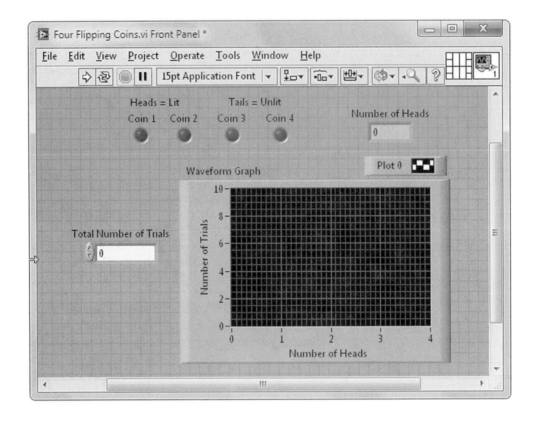

Run **Four Flipping Coins** with several choices of **Total Number of Trials**. In the Waveform Graph's **Plot Legend**, you may wish to select a **Bar Plots** option in its pop-up menu. From statistical theory, it is expected that when the number of trials N is large, the number of trials with an outcome of zero or four heads should equal $N\left(\frac{1}{2}\right)^4 = \frac{1}{16}N$ and the number of trials with an outcome of two heads should be $N\frac{4!}{2!2!}\left(\frac{1}{2}\right)^4 = \frac{3}{8}N$. Above what (approximate) value of N are your results consistent with this expectation?

PROBLEMS

1. I have agreed to hire you for a 30-day month. I pay you one penny on the first day, two pennies on the second day, and continue to double your daily pay on each subsequent day up to (and including) day 30. How many total dollars do you earn by the end of the month? Write a For Loop–based program called **Month's Total Pay** that determines the answer to this question. You may find **Add Array**

Elements in **Functions>>Programming>>Numeric** useful, along with the icons in the **Functions>>Programming>>Mathematics>>Elementary & Special Functions>>Exponential Functions** subpalette.

2. A finite geometric series of n terms is given by $y_n = 1 + x + x^2 + x^3 + \ldots + x^{n-1}$, where $|x| < 1$. It is well known that when $n = \infty$, $y_\infty = \dfrac{1}{1-x}$.

    (a) Write a For Loop–based program called **Finite Geometric Series** that, given n and x as inputs via front-panel **Numeric Controls**, determines y_n, calculates its percentage deviation from y_∞, and displays these two results in **Numeric Indicators**. You may find **Add Array Elements** in **Functions>> Programming>>Numeric** as well as icons in the **Functions>>Programming>> Mathematics>>Elementary & Special Functions>>Exponential Functions** subpalette useful.
    (b) Use your VI to answer the following question: If $x = 0.5$ and one wishes y_n to deviate from $y_\infty = 2.000\ldots$ by less than 0.1%, what is the minimum required value of n?

3. Write a program called **Chirp**, which outputs a quick succession of N sound waves, each with a unique frequency f and duration of 2 ms. Make the frequencies of the first through the last sound wave obey the following pattern: 1000 Hz, 1010 Hz, 1020 Hz, $\ldots$, 2990 Hz, 3000 Hz. To create each sound wave, use **Beep.vi**, which is found in **Functions>>Programming>>Graphics and Sound**, with its use **system alert?** input set to FALSE.

4. A unity-amplitude square wave $y(t)$ can be approximated by the following sum of three sine waves:

$$y(t) \approx \frac{4}{\pi} \left\{ \sin[x] + \frac{1}{3}\sin[3x] + \frac{1}{5}\sin[5x] \right\}$$

where x is in radians. Write a VI named **Sum of Three Sines** that evaluates this sum at 300 equally spaced x-values in the range from $x = 0$ to $x = 6\pi$ (i.e., for three cycles of the first sine function in the sum) and then plots the resulting y vs. x on a Waveform Graph, whose x-axis is calibrated in radians. You may find **Compound Arithmetic** in **Functions>>Programming>>Numeric** as well as the icons in the **Functions>>Programming>Numeric>>Math & Scientific Constants** subpalette useful. When **Sum of Three Sines** is run, it should produce the front-panel plot shown next.

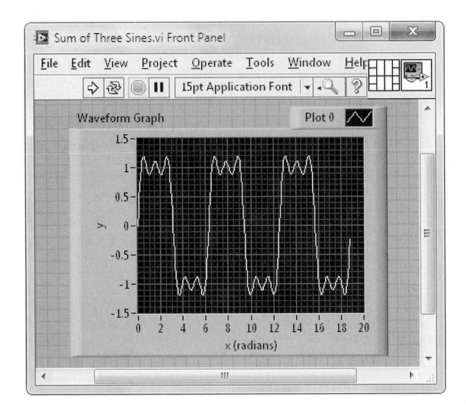

5. Write a VI called **Wait Test** that performs the following task N times in a row: Execute the **Wait (ms)** icon with **milliseconds to wait** equal to *1*, and then store the resultant value of **millisecond timer value** (which is the number of milliseconds that have elapsed since your computer was turned on). After this sequence of N operations, make your program plot the resultant N- element array of **millisecond timer value** values. Also, display this array in an Array Indicator called **Array** (this indicator can be resized using the ⬉). The front panel of **Wait Test** should appear as shown.

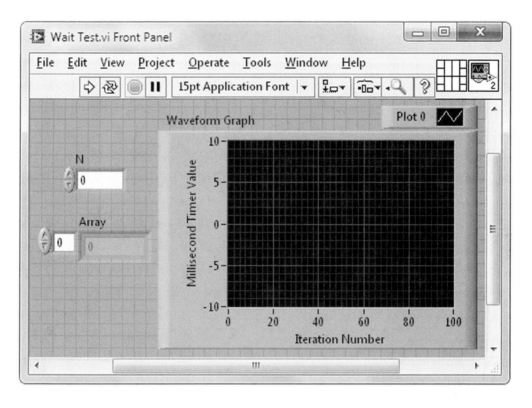

(a) When **Wait Test** is run, a straight-line plot should result. What is the expected slope for this straight line?

(b) Setting N equal to, say, 20, run the program several times under different operating conditions (e.g., several other programs besides LabVIEW running, holding down a key on the keyboard, moving the mouse, etc.). Does the resulting plot yield a straight line with the expected slope most of the time?

(c) You may observe anomalies in the plot caused by your processor attending to other business than running **Wait Test**. Let's call such an event a *lapse*. If you do observe some lapses, using **Array**, determine the typical length (in milliseconds) of an observed lapse. Is the frequency with which lapses occur dependent on prevailing operating conditions? If so, what conditions seem to enhance the occurrence of a lapse?

6. It's moving day and you are dragging a heavy dresser at constant velocity across the horizontal floor of your new apartment as shown in Figure 2.1. The force you are applying to the dresser has magnitude F and it is directed at angle θ above the

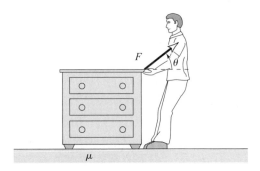

FIG. 2.1   Problem 6

horizontal. The weight of the dresser is W and the coefficient of kinetic friction between it and the floor is μ.

Newton's laws can be used to show that

$$F = \frac{\mu W}{\cos\theta + \mu\sin\theta}$$

or, expressing the force as a fraction of the dresser's weight (i.e., $F' = F/W$),

$$F' = \frac{\mu}{\cos\theta + \mu\sin\theta}$$

(a) Write a For Loop–based program called **Dragging Dresser** that, given μ from a **Numeric Control** label **Coefficient of Kinetic Friction**, calculates F' from $\theta =$ 0 to $\theta = 90°$ in $1°$ increments and then displays the resulting array of F' values on a Waveform Graph as well as in an Array Indicator labeled **Force**.

(b) Run **Dragging Dresser** with $\mu = 0.1$, $\mu = 0.4$, and $\mu = 1.0$ and determine the angle θ_{min} at which F' is smallest in each case (i.e., θ_{min} is the angle at which the minimum magnitude force is required to drag the dresser at constant velocity for the given μ).

7. (a) Write a program called **Mean of Random Array**, which generates an array of 100 random double-precision floating-point numbers in the range of 0 to 1 and then determines the mean of this array of numbers. Display the mean in a front-panel indicator called **Mean** with its **Digits of precision** set to *3*. You will find the following two icons helpful: **Random Number (0–1)** in **Functions>>Programming>>Numeric** and **Mean.vi** in **Functions>> Mathematics>>Probability & Statistics**.

(b) It is expected that the nominal value of **Mean** should be 0.500. Run **Mean of Random Array** N times in a row, where $N = 10$. Because of the random manner in which the array is generated each time, you will find that most (say, 7 out of 10) of the **Mean** values fall within a range given by $0.500 \pm \Delta x$. What value do you observe for Δx? What then is the observed *percentage fluctuation of the mean* in this data set, where percentage fluctuation is defined as $\dfrac{\Delta x}{0.500} \times 100\%$?

(c) Automate this process. Place a front-panel control called **Number of Times**, which will allow a user to enter a value for N. When run, the VI will determine the mean of a 100-element array of random numbers N times and save these mean values in an N-element array. The fluctuation of the mean can then be defined as the standard deviation σ of this array of mean values. Use **Std Deviation and Variance.vi** (in **Functions>>Mathematics>>Probability & Statistics**) to determine σ and display it in a front-panel indicator labeled **Fluctuation**. Save your work, and then run the VI with **Number of Times** set to *10*. Is the value of **Fluctuation** consistent with the value you determined for Δx in part (b)?

8. Problem 8 in Chapter 1 explains how to code the following task: Randomly generate an integer in the range from 1 to 10, and iterate this random process until the integer equals 5. The number of iterations required until the randomly generated integer equals 5 is output as a quantity called **Required Iterations**.

   Place the code that carries out the above-described task as a subdiagram within a For Loop so that, when executed, the For Loop will repeat the task N times and store the resulting N values of **Required Iterations** in an array. Then, use **Mean.vi** (found in **Functions>>Mathematics>>Probability & Statistics**) to calculate the average of the N values of **Required Iterations** and display this average in a front-panel **Numeric Indicator** labeled **Average Number of Required Iterations**. Take $N = 1,000,000$. Save this VI under the name **Iterations Until Integer Equals Five (For Loop)**.

   Run **Iterations Until Integer Equals Five (For Loop)**. Is the value you obtain for **Average Number of Required Iterations** equal to the value that you expected?

9. The *waveform* data type provides an alternate method for producing a Waveform Graph with its x-axis calibrated properly, where it is assumed that the x-axis quantity is *time*. To investigate the use of this data type, write a program called **Sine Wave Graph (Waveform)**, which plots the function $y = \sin(2\pi t)$ from time $t = 0$ to $t = 0.99$ in increments of $\Delta t = 0.01$. The block diagram for this VI is shown next. Here, **Build Waveform** (found in **Functions>>Programming>> Waveform**) combines an initial time **t0** (called a *time stamp*), the sampling time increment **dt**, and the array of y-values. **Build Waveform** must be resized to expose its three inputs. The time stamp is created using **Get Date/Time In Seconds** (found in **Functions>>Programming>>Timing**) and **Pi Multiplied by 2** is found in **Functions>>Programming>>Numeric>>Math & Scientific Constants**.

Run **Sine Wave Graph** (Waveform) and verify that it executes as expected.

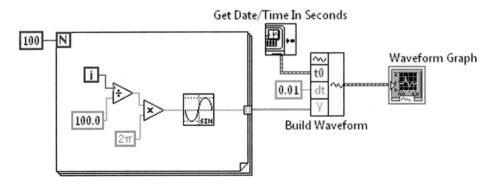

CHAPTER 3

The MathScript Node and XY Graph

> In this chapter, you will gain experience with a LabVIEW feature called the **MathScript Node.** If the MathScript Node does not appear in your **Functions>>Programming>>Structures** subpalette, re-run your LabVIEW installation disk and when the window appears in which you choose the programs to be installed, select "MathScript RT Module." If your installation package does not include the MathScript RT Module, a free 30-day trial version of the MathScript RT Module can be downloaded on the National Instruments web site (http://www.ni.com).

3.1 MATHSCRIPT NODE BASICS

In previous chapters, we have seen that the graphical nature of the LabVIEW programming language greatly simplifies the coding of desired computer operations. However, in the case of encoding mathematical relations, graphical programming is much more cumbersome than text-based languages. For this reason, LabVIEW offers the *MathScript Node*, a resizable box that you can use to enter text-based mathematical formulas directly on the block diagram. To demonstrate the utility of the MathScript Node, consider the relatively simple equation $y = 3x^2 + 2x + 1$. If you code this equation using LabVIEW arithmetic icons from **Functions>>Programming>>Numeric**, the subdiagram will appear as below.

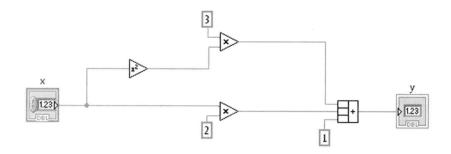

Alternately, you can program the same equation using a MathScript Node, which is found in **Functions>>Programming>>Structures**. The resulting subdiagram is as shown in the following illustration. As you can see, the MathScript Node is much easier to write and to read.

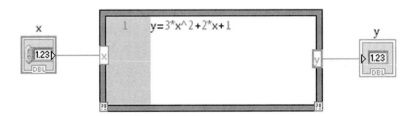

An alternate (but less powerful) text-based structure called the *Formula Node* is also available in **Functions>>Programming>>Structures**. The MathScript Node and Formula Node are similar in that both can be used to carry out common mathematical operations, including evaluating trigonometric and logarithmic functions as well as implementing Boolean logic, comparisons, and conditional branching. Here the parity between the two structures ends, however. Where the Formula Node is capable of performing only these relatively simple tasks, we will see that the MathScript Node functions as a high-level computing language (called *MathScript*) with wide-ranging functionality in digital signal processing and data analysis.

To illustrate the power of MathScript-based programming, consider coding the problem we solved in the previous chapter: Evaluate $\sin(x)$ at 100 evenly spaced x-values in the range from $x=0$ to $x=99/5=19.8$. Using graphical programming solely, we found that the solution to this problem is as shown next.

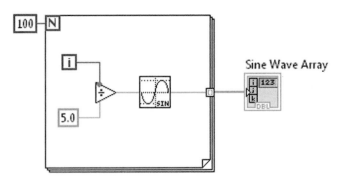

The MathScript language contains the text-based sine function $\sin(x)$, so we can simply transliterate the above code using the MathScript Node as follows.

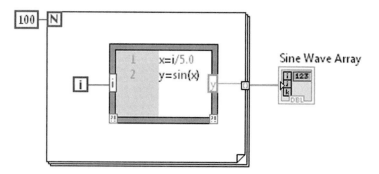

The above program will indeed execute as desired, but it fails to take advantage of the full power of the MathScript language. Namely, in MathScript, the command $x = start:step:stop$ generates a 1D array of x-values, beginning at $x = start$ and ending at $x = stop$, with the x-values evenly spaced by intervals $\Delta x = step$. Also, given x as an array of N elements, the command $y = \sin(x)$ will evaluate the sine function at each of the N x-values and return the results in an N-element array called y. Thus, when using a MathScript Node to solve our stated problem, a graphical For Loop is unneeded. The following diagram is sufficient.

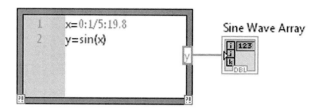

If you would like to explore some of the available MathScript functions, select **Help>>LabVIEW Help...** When the LabVIEW Help Window opens, click on the **Index** tab, and then search for *MathScript*. Under the MathScript heading, you will find many topics of possible interest, including *Classes of Functions*, which provides an organized listing of the (over 800) available functions, along with the proper syntax for implementing them.

In this chapter, you will rewrite the **Sine Wave Graph** VI using a MathScript Node. You will then use this subdiagram to supply data to an *XY Graph*, the most general of LabVIEW's three commonly used graphing options. This new program will become the basis for **Waveform Simulator**, a VI you will use to simulate discretely sampled data in your future work with file input/output, waveform generation, and fast Fourier transform–based programs. Finally, you will learn how to create our own custom-made icon that can then be used as a *subVI* in higher-level "calling" programs.

3.2 QUICK MATHSCRIPT NODE EXAMPLE: SINE-WAVE PLOT

To introduce ourselves to programming the MathScript Node, let's first write the program just discussed (i.e., a VI that evaluates sin (x) at 100 evenly spaced x-values in the range from x=0 to x=19.8). Open a new VI, and then place a **Waveform Graph** on its front panel. Using the (or, alternately, Waveform Graph's **Scale Legend**), label the x- and y-axes as **x (radians)** and **sin(x)**, respectively. By default, autoscaling will be activated on both axes.

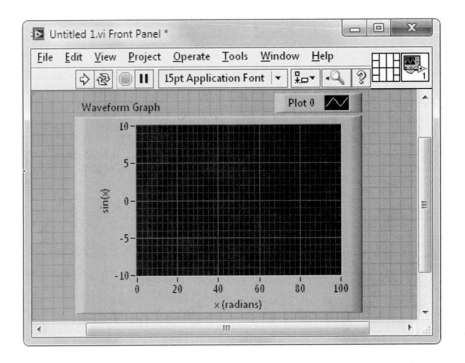

Switch to the block diagram and place a **MathScript Node** on it (found in **Functions>>Programming>>Structures**).

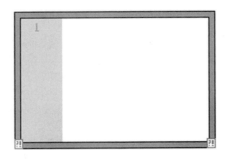

Waveform Graph

We'll now program our MathScript Node with text-based commands and equations, which will generate the desired sine-wave data. As you write coded commands and equations inside the MathScript Node, you can be as creative as you like in naming necessary variables within the limitation of the following rules: A valid variable name must begin with a letter and, after that, be composed of only letters, digits, or underscores. Names are case sensitive (that is, the capitalized version of a particular letter is distinct from its lowercase rendering) and, if desired, can be very long (up to a maximum of about 60 characters). Long variable names, of course, have the disadvantage of occupying considerable diagram space. For our present work, we require two variables, one to denote the argument of the sine function and the other to represent the sine-wave value. Let's conserve valuable diagram real estate by giving these two variables uncreative short names—how about x and y, respectively.

Using the ✋ or $\boxed{A}$, click on the interior of the MathScript Node and enter the following two lines of text-based code:

$$x = 0 : 1/5 : 19.8$$
$$y = \sin(x)$$

This code will create the two desired 1D arrays x and $y = \sin(x)$. After typing it, place the cursor outside the MathScript Node and then click the mouse to secure your entry.

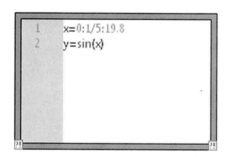

Waveform Graph

Next, we need to output y, the 1D array of sine-wave values, from the MathScript Node for plotting. To create this output, first place the ↖ at a location on the right border of the MathScript Node.

Then pop up on this border and select **Add Output>>y** from the MathScript Node's pop-up menu.

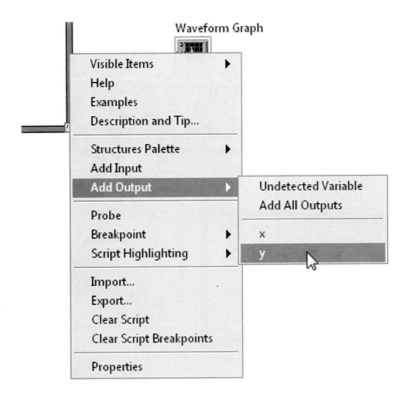

An automatically labeled indicator for this *y* output will appear in the form of a small rectangular box perched on the MathScript Node's border (for earlier versions of LabVIEW, the *y* label must be added manually using the 🖑).

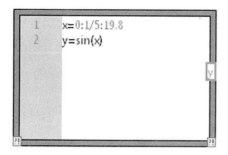

Waveform Graph

A variety of data types are possible for output terminals at the border of a MathScript Node (e.g., real and complex scalars, real and complex arrays). By default, LabVIEW automatically assigns the data type for such terminals based on how the associated variables are used inside MathScript Node. In making its automated data-type assignment, LabVIEW will account for all possible values of the output variable. For example, the variable $z = \sqrt{x}$ would be assigned a complex (as opposed to real) data type, just in case its argument x is ever negative. However, if in your use of $z = \sqrt{x}$ you know that x will always be positive, you might wish to override LabVIEW's automatic assignment to avoid the complication of working with complex data types.

On your block diagram, note that the y indicator is orange; thus, LabVIEW has correctly surmised that y is a floating-point numeric. To see LabVIEW's full assignment, pop up on the output y and select **Choose Data Type**. In the pop-up menu, you will find that **Auto Select Type** is selected by default (denoted by the check mark) and that LabVIEW's automated choice (marked by an asterisk) is **DBL 1D**. That is, LabVIEW has selected a one-dimensional array of double-precision floating-point numerics, a fine choice to represent the variable $y = \sin(x)$.

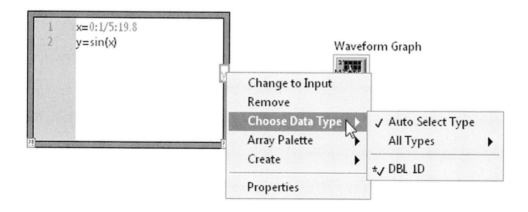

To see all the possible data-type choices for this output terminal, pop up on it and select **Choose Data Type>>All Types**. If you had reason to override LabVIEW's data-type choice for *y*, you could carry that action out in this pop-up menu.

You can also check the data type of *y* using Context Help. By activating Context Help via *<Ctrl+H>* and then placing the mouse cursor over the variable *y* within the MathScript Node, a Help Window will appear. As shown below, this window reports that the data type of *y* is **1D DBL**.

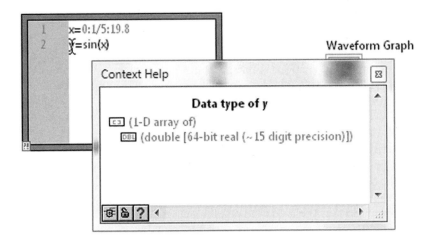

In addition, Context Help can be used to obtain useful information about other items within the MathScript Node. For example, by placing the mouse cursor over *sin* (*x*), the Help Window describing this MathScript function will be activated as shown next.

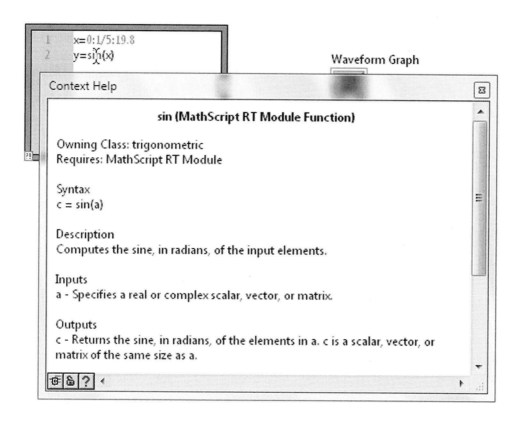

Finally, complete the block diagram as shown below. **Bundle** is found in **Functions>> Programming>>Cluster, Class, & Variant**. The **y array** indicator can be created with **Create>>Indicator** after popping up on the thick array wire. If you get a broken wire when wiring the *y* output, pop up on this object and make sure that the correct data type has been selected and also that it is indeed an output (that is, it wasn't accidentally created as an input).

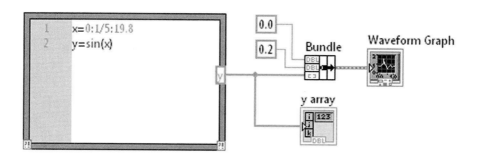

Return to the front panel. Using **File>>Save**, create a folder named **Chapter 3** within the **YourName** folder, and then save this VI under the name **Sine Wave Graph (MathScript Node)** in **YourName\Chapter 3**.

Now run your program. It should produce a few beautiful cycles of the sine function, as shown. You may then view the actual values of the sine function using the **y array** indicator.

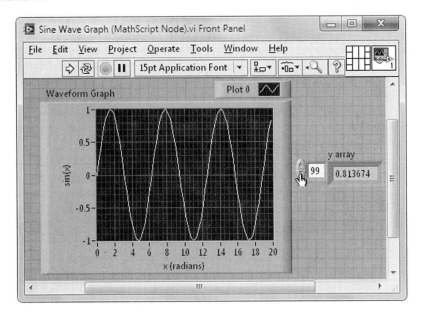

If interested, you can view the arrays produced by this VI using the MathScript Probe. To activate this tool, go to the block diagram, pop up anywhere within the MathScript Node, and select **Probe**.

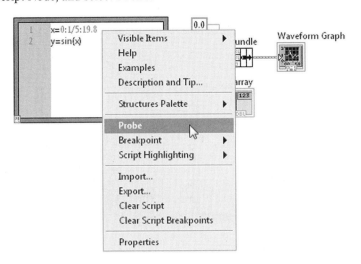

After the Mathscript Probe Window appears, run your VI. Then explore all of the array information and plotting features this tool makes available.

3.3 DEBUGGING WITH ERROR LIST

Let's digress for a moment and explore one of LabVIEW's handy program debugging features. Intentionally put an error in your block diagram by popping up on the MathScript Node's y output and selecting **Choose Data Type>>All Types>> Scalar>>DBL**.

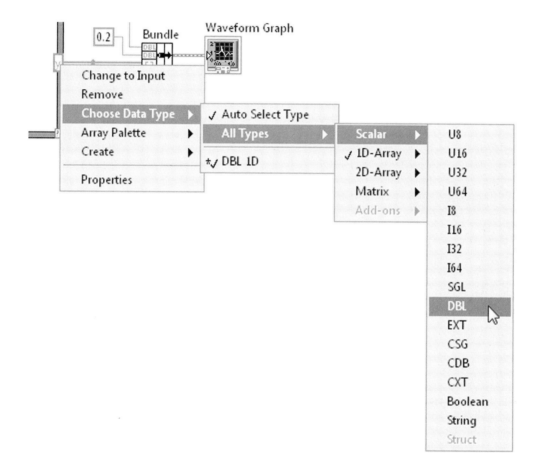

Now scrutinize the toolbar. There you will find that the **Run** button appears broken , which is LabVIEW's way of communicating that an error exists within your program. But this indicator isn't the only help available.

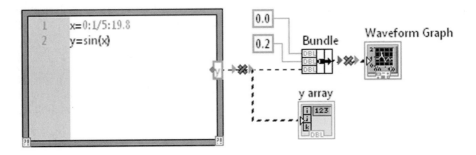

Click on the 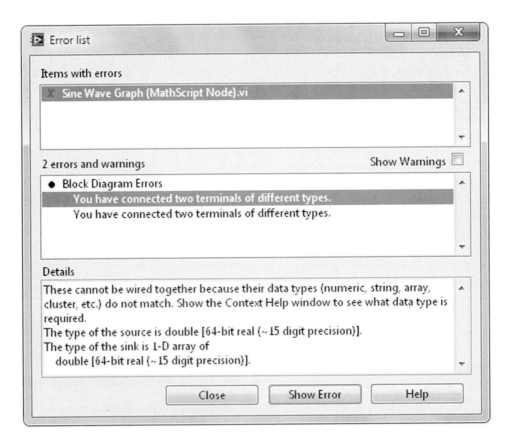 (a tip strip alerts you that this action will produce a list of program errors). A dialog window will appear, listing all of the errors in your program. In the present case, two errors exist, both related to a data-type mismatch, namely, a scalar wire connected to an input formatted to accept an array.

In the **Details** box, a detailed description is given of the error that is highlighted in the center box. To highlight a different error, simply click on that error in the center box. Clicking on the **Show Error** button will take you to the error's location (via a marquee) on the block diagram, where you can then make the required repair. With all of this available online help, quite commonly a LabVIEW program can be debugged without having to consult a user's manual.

Repair the error in your VI by popping up on the y output and selecting **Choose**

Data Type>>All Types>>1D-Array>>DBL 1D. You will note that the **Run** button ⇨ becomes whole again, signaling that the program is now executable.

3.4 WAVEFORM SIMULATOR USING A MATHSCRIPT NODE AND XY GRAPH

For our studies ahead, we will need a data simulator VI, that is, a program that outputs digitized data of the type produced by a data acquisition device (or digital oscilloscope) in a real experiment. In a typical laboratory experiment, a time-varying analog (continuous) signal $x(t)$ is input to an instrument, which is programmed to measure ("sample") this signal at N equally spaced times t_i. If Δt is the time interval between one sample and the next, then $t_i = i\,\Delta t$, where $i = 0, 1, 2, \ldots, N-1$. Since there will be one data sample taken per time interval Δt, the *sampling frequency* is defined to be

$$f_s = \frac{1}{\Delta t}$$

[1]

In this section, we will begin work on a data simulation VI based on a MathScript Node. Here, we will assume that a physical system under investigation is producing a pure sine-wave signal with a time-varying *displacement* given by $x(t) = A\sin(2\pi f t)$, where f and A are the sine wave's frequency and amplitude, respectively. Further, we will assume that this analog signal is input to a digitizing instrument, which samples the signal at N equally spaced times at a rate $f_s = 1/\Delta t$ and then outputs the result. Defining x_i to be the sample taken at time t_i, the resulting data set will be given by the following two relations:

$$t_i = i\,\Delta t$$

[2]

$$x_i = A\sin(2\pi f t_i)$$

[3]

where $i = 0, 1, 2, \ldots, N-1$ and $\Delta t = 1/f_s$. Note that the first and last data samples are taken at times $t_0 = 0$ and $t_{N-1} = (N-1)\Delta t$, respectively. Our goal then is to write a software program that creates the data set described by Equations [2] and [3] and thus simulates the (t, x) output of the above-mentioned digitizing equipment.

To display your VI's output, you will use an XY Graph. The XY Graph is a general-purpose graphing option that produces Cartesian-style plots. That is, each point on the plot is located by its (x, y) value. The entire set of (x, y) values is presented to this routine in the form of a cluster consisting of two 1D arrays of numerical values. The first and second arrays in the cluster correspond to the sequence of x- and y-values, respectively.

Open a new VI, and place an **XY Graph** (from **Controls>>Modern>>Graph**) on its front panel. Label the x- and y-axes as Time and Displacement, respectively. Autoscaling will be activated on each axis by default. Finally, place four **Numeric Control**s on the front panel labeled **Number of Samples, Sampling Frequency, Frequency,** and **Amplitude,** respectively. By default, the representation for these Numeric Controls is the double-precision floating-point data type **DBL**. Pop up on **Number of Samples** and change its data type to Long Integer by selecting **Representation>>I32**. Pop up on each of the other Numeric Controls and make sure that (its default) **Representation>>DBL** is selected. Finally, save this VI under the name **Waveform Simulator** in YourName\ **Chapter 3**.

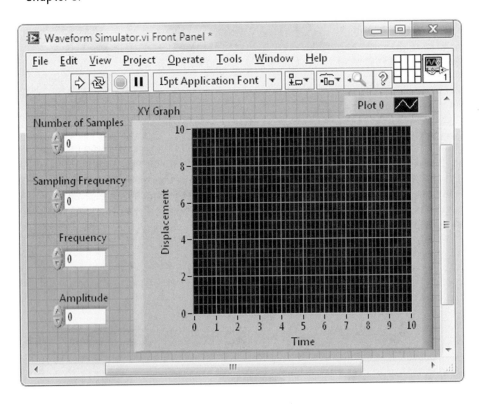

Switch to the block diagram and write MathScript-based code that evaluates Equations [1] through [3]. To produce the two arrays describing a time-varying sine wave's *x*- and *y*-coordinates, which we call *t* and *x*, respectively, the required code within the MathScript Node is as follows.

$$delta\_t = 1/f\_s$$
$$start = 0$$
$$step = delta\_t$$
$$stop = (N-1)*delta\_t$$
$$t = start : step : stop$$
$$x = A*\sin(2*pi*f*t)$$

After inputing this code, your block diagram should appear as shown next.

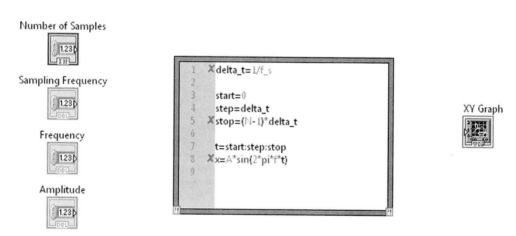

A ✖ appears at each line that involves an input variable that we have yet to assign. First, create an input for the variable *N*, which represents the value of **Number of Samples** within the MathScript Node, by popping up on the MathScript Node's left-hand border and selecting **Add Input**.

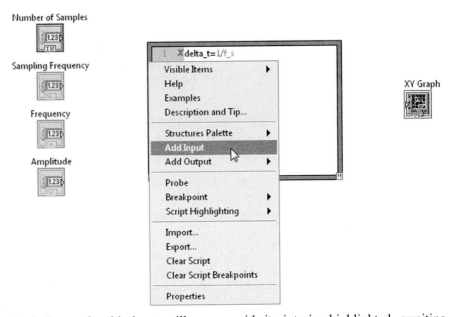

An indicator for this input will appear with its interior highlighted, awaiting your character-based name for this equation variable. If this highlighting disappears (because of an inadvertent click of the mouse), use the 🖑 to restore access to the input's interior. Label the input *N* and then wire it to the **Number of Samples** terminal. The input's indicator will turn blue, indicating that it has been formatted to receive the (**I32**) integer data from **Number of Samples**. Also, at the line involving *N*, the ✖ disappears.

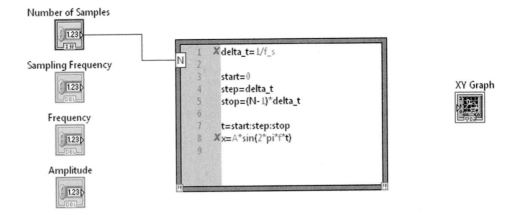

Similarly, wire the three other inputs that carry the values of **Sampling Frequency**, **Frequency**, and **Amplitude**, labeling them *f_s*, *f*, and *A*, respectively. Each of these inputs is an (orange) double-precision, floating-point numeric. Finally, create two outputs for the variables *t* and *x* at the right-hand border of the MathScript Node, as shown.

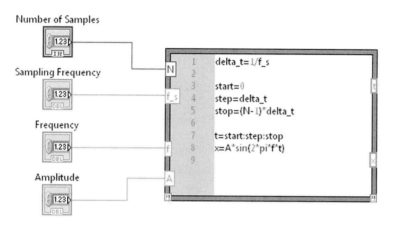

3.5 CREATING AN *XY* CLUSTER

To determine what type of input the XY Graph requires, activate its Help Window

by placing the mouse cursor over the XY Graph's icon terminal and pressing *<Ctrl+H>*. The Help Window will appear as shown next.

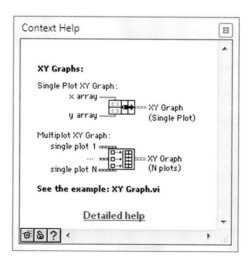

This Help Window indicates that to plot a single waveform, the XY Graph expects a cluster as input. This cluster is to contain the waveform's x and y arrays and is formed using the **Bundle** icon. Let's call this bundle of a waveform's x and y arrays its "*xy cluster.*"

For future reference, the Help Window also tells us that N waveforms can be plotted simultaneously on an XY Graph. In this situation, the data are passed to the XY Graph's terminal in the form of an "*array of clusters.*" To form this object, one must first form the xy cluster for each of the N waveforms as above. Then, using the yet-to-be-studied **Build Array** icon, an N-element array is constructed with each element being a particular waveform's xy cluster.

Refer to your block diagram. Here you will find the XY Graph's icon terminal. We note that, similar to the Waveform Graph, this terminal has a cluster data-type indicator (in its bottom center), but unlike the Waveform Graph, its border is brown (rather than pink). The brown color denotes that the default for this data input has the following special format: an array of clusters, where each cluster is composed of only numeric scalars. In our use of the XY Graph, we will follow the recommendation of its Help Window, and input a cluster of numeric arrays, a format that is more general than the default.

Complete the block diagram wiring as shown below. Use the **Bundle** icon (found in **Functions>>Programming>>Cluster**) to form the sine wave's xy cluster. Note that when this cluster is wired to the XY Graph's icon terminal, the icon's border changes from (the specialized default cluster) brown to (the general cluster) pink. The **Time** and **Displacement** array indicators can be created by popping up on the appropriate wires and selecting **Create>>Indicator**.

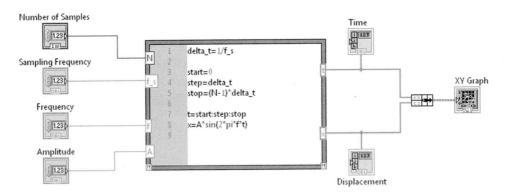

3.6   RUNNING THE VI

Return to the front panel, position the **Time** and **Displacement** indicators nicely, and then save your work. Set **Number of Samples** and **Sampling Frequency** equal to *100* and *1000*, and **Frequency** and **Amplitude** to *50* and *5*, respectively. These choices for the parameters simulate sampling a 50 Hz sine wave for 0.1 s (100 samples at a rate of 1000 samples per second) and so should result in outputting five cycles of the sine wave. Run your program, and if all goes well, you will see the five cycles. Marvel at the waveform you have created and reflect on how easy it would be to create other shapes by simply changing the equations within the block diagram's MathScript Node (which is exactly what you will do in a few moments).

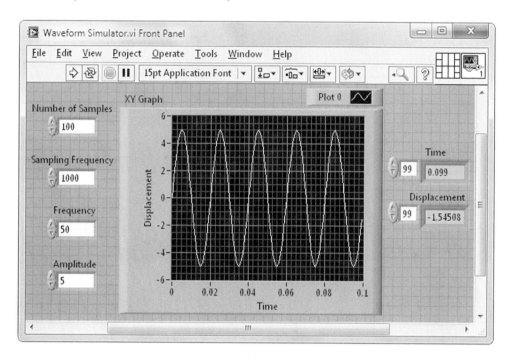

3.7 MATHSCRIPT INTERACTIVE WINDOW

An invaluable tool while developing MathScript-based code is the *MathScript Interactive Window*. Let's take a moment to learn how to use this tool. With a VI open or else in the **Getting Started** window, open the MathScript Interactive Window by selecting **Tools>>MathScript Window**... When the window appears, concentrate your attention on its left side. In the lower left you will find the *Command Window* box where MathScript commands can be entered using the keyboard. Once a command is loaded into the Command Window, it can be executed by pressing *<Enter>* on your keyboard, and the result of this executed command is displayed directly above in the *Output Window*.

Let's test drive the MathScript Interactive Window by executing commands similar to the ones just used in constructing **Waveform Simulator**. In the Command

Window, type *t = 0:0.01:0.99*, and then press *<Enter>*. In the Output Window, you will first see a copy of the command that you typed, followed by the result of its execution, which in this case is an array of 100 equally spaced *t*-values ranging from *t*=0 to *t*=0.99.

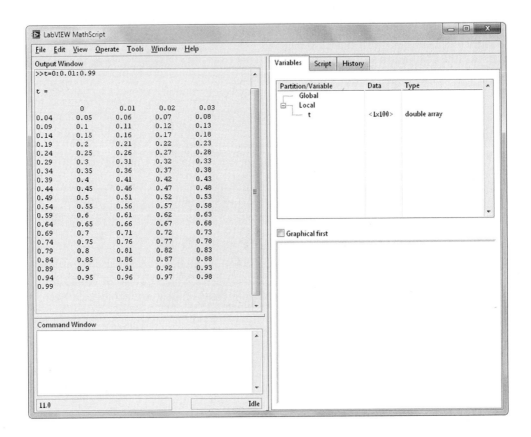

Next, in the Command Window, enter (i.e., type the command and then press *<Enter>*) *x=5 * sin (2 * pi *4 * t)*. In the Output Window you will find the sine wave with frequency 4 and amplitude 5 sampled at the 100 equally spaced *t*-values created by the previous command.

Finally, to observe this sine wave, enter *plot(t,x)* in the Command Window. A window with a plot of *x* vs. *t* will then appear as shown below, displaying four cycles of the given sine wave.

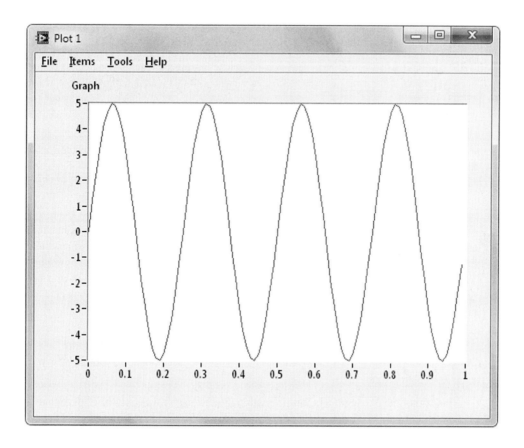

Next, consider a slightly more complicated task: Evaluate the expression $x = t \sin(t)$ at the three times $t = 1, 2, 3$. On the block diagram of **Waveform Simulator**, we multiplied a sine function by a scalar A; here, we are multiplying the sine function by a variable t. The expected three x-values, of course, are $1 \times \sin(1) = 0.841$, $2 \times \sin(2) = 1.819$, and $3 \times \sin(3) = 0.423$, where the argument of the sine function is assumed to be in radians. From the programming we just completed, it would seem that the correct commands to execute this task are

$$t = 1:1:3$$
$$x = t*\sin(t)$$

In the Command Window, enter $t = 1:1:3$, followed by $x = t*sin(t)$. In the Output Window, somewhat surprisingly, you will receive an error message related to the incompatibility of the matrices involved in the requested operation.

119

To understand the problem here, we must delve a bit deeper into the syntax of the MathScript language. In executing the commands you just input, MathScript produces the following two row vectors:

$$t = [1\ 2\ 3]$$
$$\sin(t) = [\sin(1)\ \sin(2)\ \sin(3)] = [0.841\ 0.909\ 0.141].$$

Then, in the syntax of MathScript, the command $t*\sin(t)$ requests that these two row vectors be multiplied together as matrices. That is, the command is interpreted as the following matrix multiplication operation:

$$[1\ 2\ 3] \times [0.841\ 0.909\ 0.141]$$

As you know, multiplication of two 1×3 matrices is an invalid operation, hence the error message in the Output Window. To demonstrate a valid matrix operation (which is not the task we are trying to accomplish), try entering the command $x = t*transpose(sin(t))$. The operation executed will then be the following valid matrix multiplication of a 1×3 row vector times a 3×1 column vector:

$$x = \begin{bmatrix} 1 & 2 & 3 \end{bmatrix} \times \begin{bmatrix} 0.841 \\ 0.909 \\ 0.141 \end{bmatrix} = (1)(0.841) + (2)(0.909) + (3)(0.141) = 3.08$$

Instead of the above matrix multiplication operation, we want MathScript to take the three-element row vectors t and $\sin(t)$ and form a three-element row vector, whose ith element is the product of the ith elements of original row vectors. This operation is called *element-wise multiplication* and its MathScript operator symbol is a period followed by an asterisk (.*). In the Command Window, enter $t = 1:1:3$, followed by $x = t.*sin(t)$. Remember to include the all-important period before the asterisk. You will find that the commands accomplish our originally stated task of evaluating $x = t\sin(t)$. Similarly, element-wise division and power operations can be executed by the operators ./ and .^, respectively.

The MathScript Interactive Window is also a handy source of help regarding use of MathScript functions. These (over 800) functions are cataloged into over 40 "classes" (e.g., basic mathematical functions, statistics, digital signal processing). To see a list of all of the classes, enter *help classes* in the Command Window. To then see a list of all of the functions in, say, the *dsp (digital signal processing)* class, enter *help dsp* in the Command Window. There you will find the available dsp functions, which include many that perform complex filtering, waveform generation, and spectral analysis operations.

To get some experience with a MathScript function, let's try using Periodic Signal Generator, which is called *gensignal* for short. Enter *help gensignal* in the Command Window. The help information that appears states that the gensignal command generates two arrays called x and t, where t is an array of time values starting at $t=0$ and stopping at $t = duration$ with the time step $\Delta t = interval$ and x is an array of associated periodic signal values that has the functional form and period given by the parameters *type* and *period*, respectively. The possible values for *type* are *'sin'*, *'square'*, and *'pulse'*. The syntax of the command is

$$[x, t] = gensignal(type, period, duration, interval)$$

If you only need the *x* array of associated periodic signal values, the command becomes

$$x = gensignal(type, period, duration, interval)$$

OK, let's try using *gensignal* to generate the same signal that we produced a few moments ago, which was four cycles of a sine wave with frequency and amplitude of 4 and 5, respectively. For a sine-wave frequency of 4, the *period* is $1/4 = 0.25$. For the time variable, we want the *duration* and *interval* to be 0.99 and 0.01, respectively. So, in the Command Window, enter *[x, t] = gensignal('sin', 0.25, 0.99, 0.01)*, and then you will find that both the *x* and the *t* arrays will be listed in the Output Window. To see a plot of these variables, enter *plot(t,x)* in the Command Window. In the plot window that appears, you should see the expected signal, except the amplitude is 1, rather than the desired 5. To remedy this problem, enter $x = 5*x$ in the Command Window, followed by *plot(t,x)*.

As a demonstration of the power of the *gensignal* command, produce a square wave by simply changing the command entered in the Command Window to *[x, t] = gensignal('square', 0.25, 0.99, 0.01)*. Plot the result. Next, plot a periodic pulse using the command *[x, t] = gensignal('pulse', 0.25, 0.99, 0.01)*. Finally, produce a single *x*-array of square-wave value with amplitude 5 using the command *x = 5*gensignal('square', 0.25, 0.99, 0.01)*.

Another useful MathScript function is *linramp(a,b,N)*, which generates *N* equally spaced samples between the lower value *a* and upper value *b*. Show that the command *t = linramp(0,0.99, 100)* generates the same array of values as the command *t = 0:0.01:0.99*. Also, show that a 100-element array, whose elements all equal 5, can be generated by *t = linramp(5,5, 100)*. Such an array could describe, for example, the DC voltage output of a battery at 100 equally spaced times. In the program you will soon write, we will call such a constant-valued array a *DC Level* waveform.

3.8 ADDING SHAPE OPTIONS TO WAVEFORM SIMULATOR

We are now positioned to turn **Waveform Simulator** into a versatile data simulator VI that can output a variety of digitized waveforms. In particular, we will program this VI to create five possible waveform shapes—*Sine, Cosine, DC Level, Square,* and *User-Defined*. So that a user can select the desired waveform, let's add a fifth control called **Shape** to the front panel of **Waveform Simulator**.

3.9 THE ENUMERATED TYPE CONTROL

For **Shape**, we will use an *Enumerated Type Control* (nicknamed an *Enum*). Go to **Controls>>Modern>>Ring & Enum**, select an **Enum**, and then place it on the front panel and label it **Shape**. Use the ⬦ to perform any necessary resizing and/or positioning.

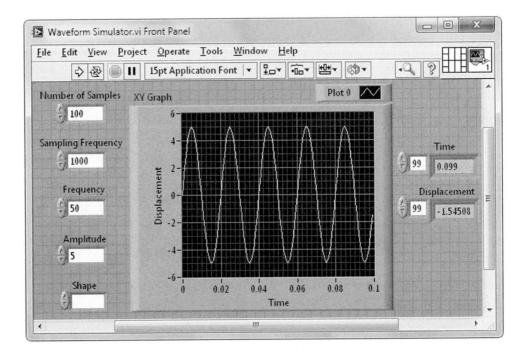

In **Controls>>Modern>>Ring & Emun**, you will find several types of *"ring"* style controls, each of which is useful when a programmer needs to present the user with a catalog of options. The **Enum** is a special type of ring control that links a sequence of text messages with a series of associated integers. By default, the data type of these integers is unsigned 16-bit (**U16**). When the user selects a particular text message on the front panel control, its associated integer is passed to a terminal on the block diagram. In our present circumstance, the **Enum** control will provide a handy method for selecting a desired waveform shape from among the available options.

The Enum is programmed with its available options as follows. Pop up on the **Enum** and select **Edit Items...** In the dialog window that appears, type *Sine <Enter> Cosine <Enter> DC Level <Enter> Square <Enter> User-Defined*, and then click the **OK** button. If you press *<Enter>* after typing *User-Defined*, you will accidentally create a sixth option, which be erased by clicking on the **Delete** button.

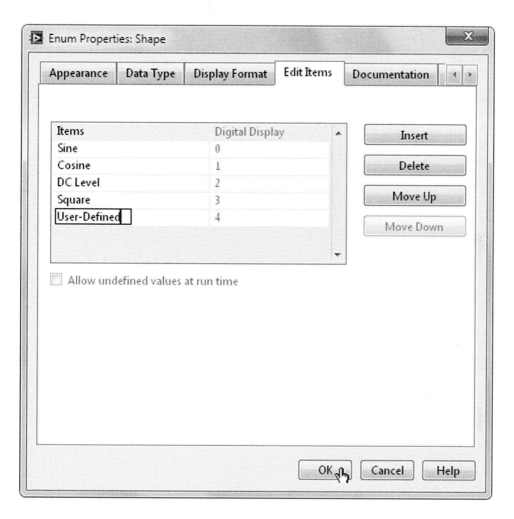

On the front panel, use the 🖐 to view the text message (Sine, Cosine, DC Level, Square, User-Defined) for each of the Enum's five programmed options. You can resize

the Enum using the ⬉. To make visible the integer associated with each of these text messages, pop up on the **Enum** and select **Visible Items>>Digital Display**. After viewing the Enum's Digital Display, hide it again (by selecting **Visible Items>>Digital Display**) and then save your work.

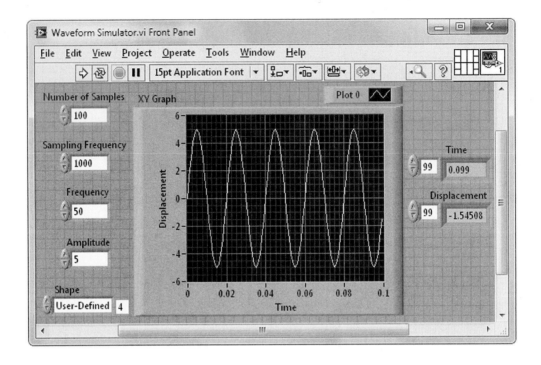

3.10 FINISHING THE BLOCK DIAGRAM

Switch to the block diagram and there you will find the **Shape** icon terminal, which is blue, denoting the **(U16)** integer data type emanating from this Enum control. Pop up on the left-hand border of the MathScript Node, select **Add Input** and label it *s*, and then complete the code as shown below.

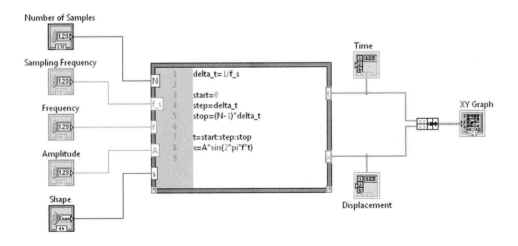

Finally, program the MathScript Node to produce the desired waveform alternatives. To accommodate the five possible options, we will implement MathScript's *if–else statement* syntax. For this type of statement, a logical condition is tested (in our situation, the equivalence of *s* to, say, zero via the command s==0; note that two equal signs denotes a test of logical equivalence whose outcome is either a Boolean TRUE or FALSE), and if the condition is TRUE, a given command (or list of commands) is executed. The code we require is as given below. For the User-Defined option, we have chosen the sum of two particular sinusoids as its default. A user can then modify this choice to produce any desired waveform shape.

$$delta\_t = 1/f\_s$$
$$start = 0$$
$$step = delta\_t$$
$$stop = (N-1)*delta\_t$$
$$t = start : step : stop$$
$$if\ s == 0$$
$$\qquad x = A*\sin(2*pi*f*t)$$
$$else\ if\ s == 1$$
$$\qquad x = A*\cos(2*pi*f*t)$$
$$else\ if\ s == 2$$
$$\qquad x = linramp(A, A, N)$$
$$else\ if\ s == 3$$
$$\qquad x = A*gensignal('square', 1/f, stop, step)$$
$$else\ if\ s == 4$$
$$\qquad x = 4.0*\sin(2*pi*100*t) + 6.0*\cos(2*pi*200*t)$$
$$end$$

Program the MathScript Node with the above code and then click outside the Node to secure it. Note that the *t* and *x* outputs are each assumed to supply 1D DBL arrays to the block-diagram objects to which they are wired.

In my version of LabVIEW, this MathScript Node coding resulted in a bad wire at the *x* output. Popping up on this output, we see that the source of this problem is that LabVIEW's automatic assignment for the variable *x* is **DBL 2D**, rather than the expected **DBL 1D**.

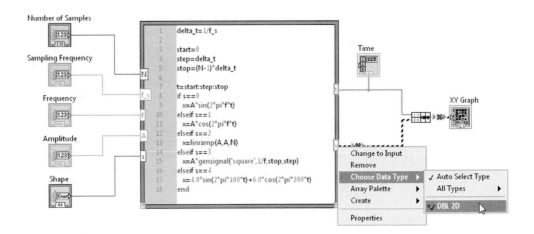

At this point, because you know that the data type for x should rightfully be a one-dimensional double-precision floating-point array, you may choose to override LabVIEW's assignment by popping up on the x output and selecting **Choose Data Type>>All Types>>1D-Array>>DBL 1D**. A red coercion dot will result, indicating that you have chosen a different data type than LabVIEW would auto-select. This coercion dot is there to strike some fear in your heart, reminding you that LabVIEW's choice might be smarter than yours.

To discover what has motivated LabVIEW's choice this time, activate Context Help via *<Ctrl+H>*, and then place the mouse cursor over the x variable in the several locations it appears within the MathScript Node's *if–else* statement. For example, when placed over the line where $x = A*sin(2*pi*f*t)$, we are informed that the data type of x is **1-D array** and **double**, the designation for a row vector of double-precision floating-point numbers.

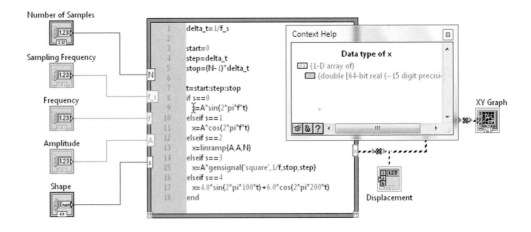

All other instances of x have this same data type, except at the line $x = A*$ *gensignal('square', 1/f, stop, step)*. There, as shown next, we find that the x created by gensignal is a column vector of double-precision floating-point numbers, rather than a row vector. It is this data-type mismatch between how x is created at various lines of the *if–else* statement that has caused LabVIEW to format x as a two-dimensional array.

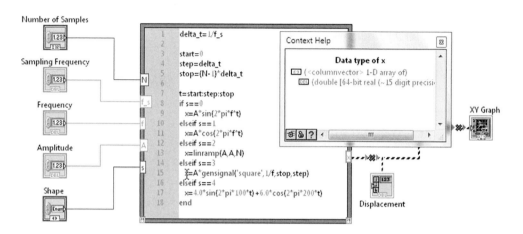

Now that we understand the source of our problem, we can correct it by transposing the column-vector array generated by gensignal, so that it is transformed into a row-vector array like all of the other x-related definitions. That is, simply replace $x = A*gensignal('square', 1/f, stop, step)$ with $x = transpose(A*gensignal('square', 1/f, stop, step))$, and then click somewhere outside the MathScript Node to secure this change. As planned, we find that LabVIEW now auto-assigns x to be **DBL 1D** and the bad wire is replaced by a good one.

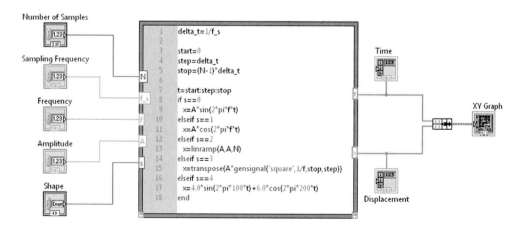

3.11 RUNNING THE VI

Save your work. Return to the front panel; set **Number of Samples** and **Sampling Frequency** to be *100* and *1000*, and **Frequency, Amplitude,** and **Shape** as *100*, *5*, and *Sine*, respectively; and then run your program. Assuming that the **Time**-axis is calibrated in seconds, with this choice of parameters, neighboring data samples are separated by a time interval of 0.001 s, and we are simulating an experiment that acquires 100 data samples over the time span from $t = 0.000$ to $t = 0.099$ s. Thus, for a 100 Hz sine-wave signal whose period is 0.01 s, we expect to observe ten of its cycles on the XY Graph. Each cycle is described by a succession of ten samples, as can be seen by popping up on the **Plot Legend** and selecting **Point Style>>Solid Dot**.

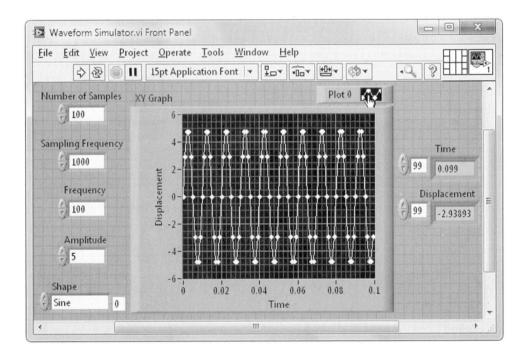

Next, without changing **Frequency, Amplitude,** and **Shape,** set **Number of Samples** and **Sampling Frequency** to *100* and *10,000*, respectively, and then run your program. With this choice of parameters, neighboring data samples are separated by a time interval of 0.0001 s, so now the 100 samples are acquired over the time span from $t = 0.0000$ to $t = 0.0099$ s. Thus, for the 100 Hz sine-wave signal with period of 0.01 s, we expect to observe only one of its cycles on the XY Graph, where this single cycle is well resolved because it is described by a succession of 100 samples.

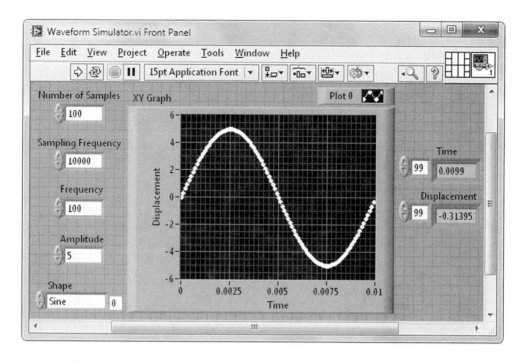

Finally, run your VI with **Shape** set to its other available options and enjoy the results.

3.12 CONTROL AND INDICATOR CLUSTERS

Let's take this opportunity to add to your bag of LabVIEW programming tricks. We will explore a "clustering" technique that improves the user-friendliness of a VI, both for a user operating its front panel as well as a programmer (including yourself) trying to decipher its block diagram.

We first note that the Numeric Controls on our VI's front panel can be organized into two categories: *Digitizing Parameters* (**Number of Samples**, **Sampling Frequency**) and *Waveform Parameters* (**Frequency**, **Amplitude**, **Shape**). Likewise, both **Time** and **Displacement** Array Indicators are related to the category of *Waveform Output*. This organizational structure within your program can be made explicit by grouping these controls and indicators on the front panel and block diagram. This feat will be accomplished by creating a cluster associated with each of the three categories through the following steps. In **Controls>>Modern>>Array, Matrix & Cluster**, obtain a **Cluster** shell and place it in an empty region of the front panel near the Numeric Controls. Label this **Cluster** shell **Digitizing Parameters**. Similarly, place another **Cluster** shell near the Numeric Controls and label it **Waveform Parameters**. Finally, place a third **Cluster** shell near the Array Indicators and label it **Waveform Output**.

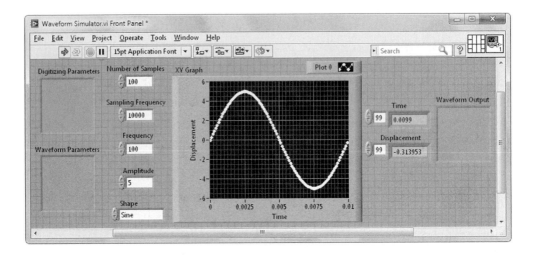

Now, using the 🔲, drag the **Number of Samples** and **Sampling Frequency** controls (in that order) into the interior of the **Digitizing Parameter** cluster shell. If needed, the

Cluster shell can be resized at its edges using the 🔲 or (more simply) by popping up on its border and selecting **AutoSizing>>Size to Fit** in the pop-up menu (if you pop up within the interior of the Cluster shell, a Controls Palette will appear, rather than the desired pop-up menu).

Likewise, drag the **Frequency** control, then the **Amplitude** control, and then the **Shape** control into the interior of the **Waveform Parameters** Cluster shell and the **Time** and **Displacement** indicators (in that order) into the **Waveform Output** Cluster shell.

Finally, using the 🔲, arrange the front-panel objects in some pleasing pattern.

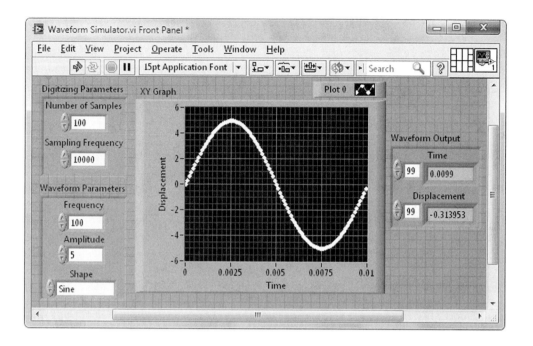

The controls or indicators, which we will generically call *elements*, within a cluster are ordered. This ordering is not based on the position of the elements within the Cluster shell, but initially is determined by the order in which the elements were placed in the Cluster shell during programming. The first element placed in the Cluster shell is indexed zero, the second element is indexed one, and so on. Thus, for example, if you followed the directions above, the **Number of Samples** and **Sampling Frequency** controls within the **Digitizing Parameters** control cluster are indexed 0 and 1, respectively. To verify this ordering, pop up on the border of the **Digitizing Parameters** control cluster and select **Reorder Controls In Cluster...**

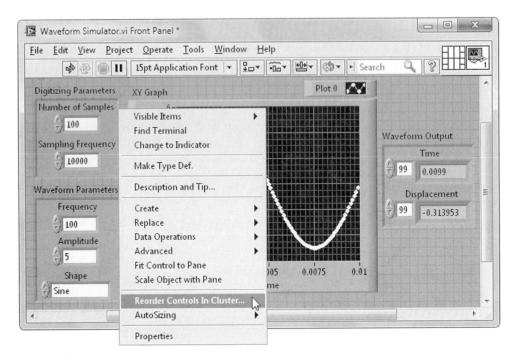

The front panel will change its appearance as shown in the following image.

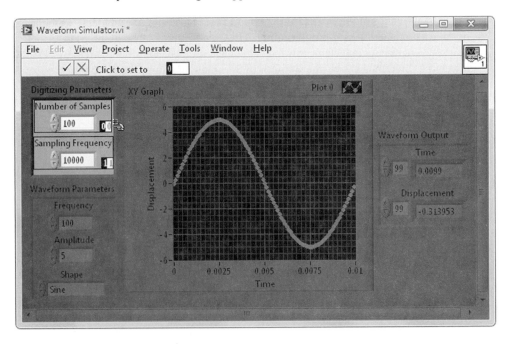

At the bottom right corner of each cluster element, its current *cluster index* is given in the white box. The mouse cursor has morphed into the *cluster order cursor* ⌗⇲, which can be used to change the indexing of the cluster elements by clicking on their associated black boxes. Once the new index assignments have been made, they can be secured by clicking on the **Confirm** button ✓ or reverted to the original settings with the **Cancel** button ✗. You might wish to experiment with this tool for a few moments. Then, after assigning **Number of Samples** and **Sampling Frequency** as the index 0 and 1 elements, respectively, click on the **Confirm** button.

Switch to the block diagram and select **Edit>>Remove Broken Wires**. From **Functions>>Programming>>Cluster, Class, & Variant**, select **Unbundle By Name** ➡☐ and place it near the **Digitizing Parameters** icon terminal. Wire this icon terminal to the left-hand input terminal of the ➡☐ and when the cluster–wire connection is made, the **Unbundle By Name** icon will morph so that its output (on its right side) contains the name of, and gives access to, the index-zero element in the cluster ➡ Number of Samples . Using the ↖, resize (at its bottom center) **Unbundle By Name** so that it has two outputs, thus giving access to both elements ➡ Number of Samples / Sampling Frequency .

The particular element appearing in a given output terminal can be changed either by clicking on it with the 🖑 or by popping up on it and choosing **Select Item**. The output terminal ordering you select for **Unbundle By Name** is not required to have the same ordering as the indexing of the cluster wired to its input: For example, the top output terminal does not have to correspond to the input cluster's index-zero element; it can be programmed to be any one of the input cluster's elements. Complete the wiring as shown below.

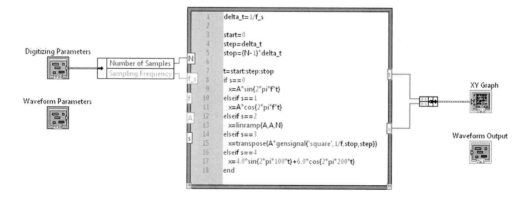

In a similar fashion, use an **Unbundle By Name** to wire the elements within the **Waveform Parameters** cluster to the *f*, *A*, and *s* inputs of the MathScript Node. Finally, since the XY Graph requires a cluster to be formed containing the *t* and *x* arrays, wire the output of the already present **Bundle** icon to the **Waveform Output** cluster's icon terminal.

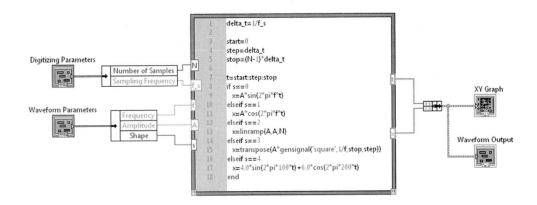

As you can see, clustering enhances the readability of a block diagram, especially because of the labeling that appears within the **Unbundle By Name** icons.

Return to the front panel. Run your VI to check that it functions as expected, and then save your work.

Finally, save the customized front-panel objects that you have created so that you can reuse them in future programs. To save the **Digitizing Parameters** control cluster, pop up on the border of this cluster and select **Advanced>>Customize...** in the pop-up menu.

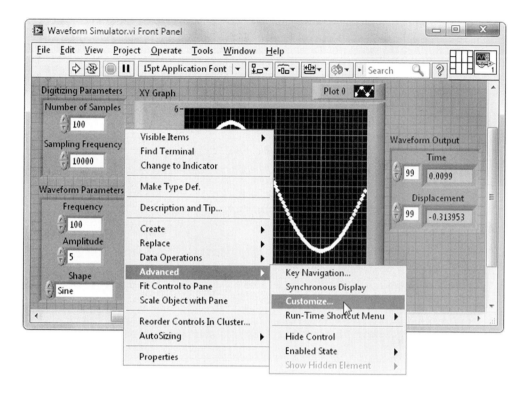

A *Control Editor* window, which looks similar to the front panel of a VI, will open with your custom-made **Digitizing Parameters** control cluster on it.

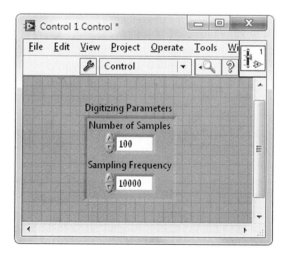

Make sure that the **Control** option (as opposed to **Type Def.** and **Strict Type Def.**) is selected in the toolbar's **Control Type** pull-down menu.

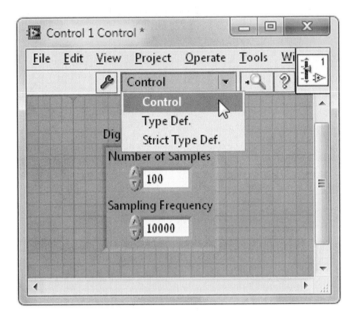

Then, use **File>>Save** to first create a folder named Controls within the YourName folder, and then save your control cluster under the name Digitizing Parameters in YourName\Controls. The extension **.ctl** will be automatically appended to the control's name. In a similar manner, save your Waveform Parameters control cluster and Waveform Output indicator cluster under the names Waveform Parameters and Waveform Output in YourName\Controls.

3.13 CREATING AN ICON USING THE ICON EDITOR

A well-written LabVIEW program is hierarchical in nature. It consists of a top-level VI whose front panel accepts inputs and displays outputs, while its block diagram is constructed from lower-level *subVIs*. These subVIs, which are analogous to a subroutine in a text-based language such as C, may call even lower-level subVIs, which in turn may call still lower-level subVIs. Just as there is no limit to the level of subroutine layering in a C program, there is no limit to the layers of subVIs used in a LabVIEW program. This modular approach to programming makes programs easy to read and debug.

Your program's subVIs may either be taken from LabVIEW's extensive libraries of built-in icons (found in the Functions Palette) or be custom written by you. It is this latter point that we now wish to focus our attention upon. The importance of what you are about to learn is this: Any VI that you write can then be used as a subVI in the block diagram of a higher-level VI.

To use a program as a subVI, it must have an icon to represent it in the block diagram of the higher-level ("calling") VI. There are two steps in creating this icon: designing its appearance and assigning its terminals. Let's learn these skills by creating an icon for our **Waveform Simulator** program.

3.14 ICON DESIGN

First, here is the procedure for designing the icon's appearance. Position the mouse cursor (it doesn't matter which tool it is manifesting at the moment) over the *icon pane* in the upper right-hand corner of the front panel.

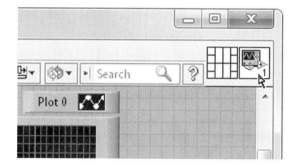

Pop up on the icon pane and select **Edit Icon...** from the pop-up menu.

The *Icon Editor* window, shown in the following illustration, will appear. Note the **Icon Text** tab is selected.

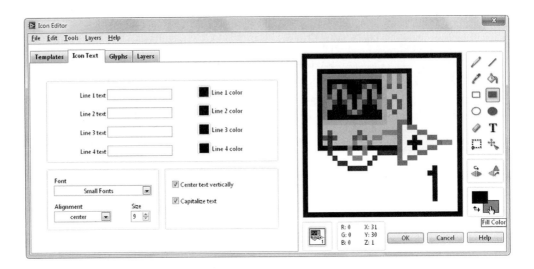

Within this window, you will find a default picture of the icon and a palette of tools that may be used to redesign its appearance. Some of the useful tools in the palette, many of which may be familiar to you from computer-based drawing programs, have the following functions:

Pencil	Draws individual pixels in the **Line Color**.
Line	Draws straight lines in the **Line Color**.
Dropper	Sets the **Line Color** to the color of a pixel you left-click or the **Fill Color** to the color of a pixel you right-click.
Fill	Fills an outlined area with the **Line Color**.
Rectangle	Draws a rectangular border in the **Line Color**. Double-click to add a one-pixel border to the entire icon in the **Line Color**.
Filled Rectangle	Draws a rectangle with a border in the **Line Color** and filled in the **Fill Color**. Double-click to add a one-pixel border to the entire icon in the **Line Color** and to fill the icon in the **Fill Color**.
Eraser	Draws individual pixels as transparent.
Select	Selects an area of the icon to cut, copy, or move. Double-click to select the entire icon.
Text	Enters text into the icon design. Double-click to access font selection dialog window.

Move

Moves all pixels you have selected.

Color

Displays the current **Line Color** and **Fill Color**. Click on each to get a palette from which to choose new colors.

You may wish to explore the use of these tools and create a sophisticated design for your icon (within the constraint of its 32×32 pixel size). The following step-by-step description will result in an icon with only rudimentary features.

First, click on the **Color** tool and set the **Line Color** and **Fill Color** to black and white, respectively.

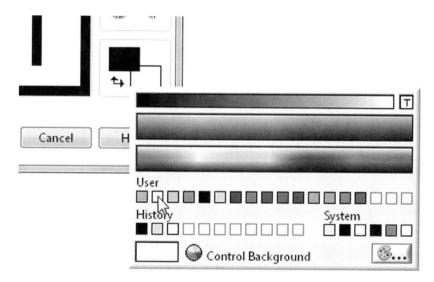

Then, place the mouse cursor over the **Filled Rectangle** tool and double-click.

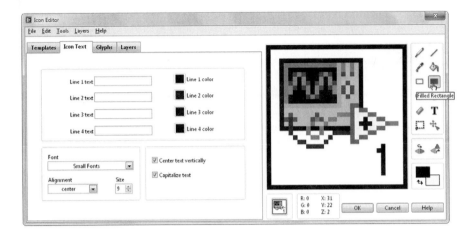

This action will frame the icon in the **Line Color** (black) and fill it with the **Fill Color** (white). You now have a framed blank canvas on which to create your icon design.

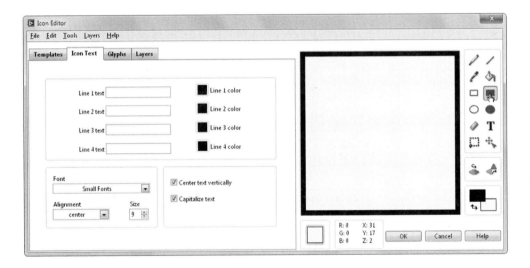

We will now enter the vertically centered text *Wave Sim* in the icon's interior as follows. In the **Line 1 text** box enter *Wave*, and in the **Line 2 text** box enter *Sim*. Keep the **Center text vertically** box checked, but uncheck **Capitalize text**. Set the font **Size** to *11*.

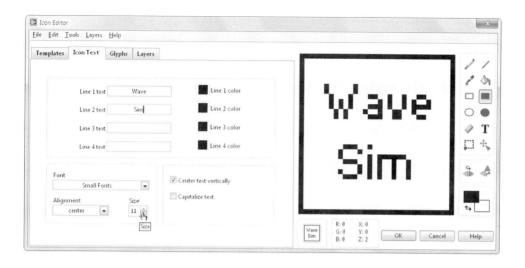

Click on the **OK** button to save the icon design. The Icon Editor window will close, returning you to **Waveform Simulator**'s front panel, with your new icon design now in the icon pane.

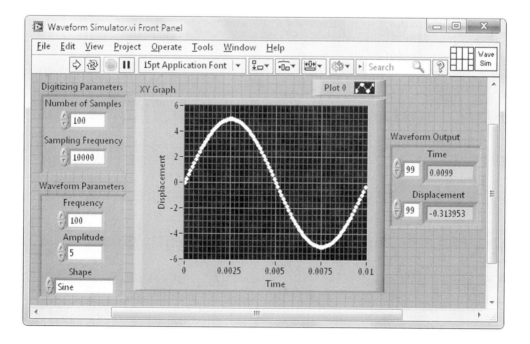

3.15 CONNECTOR ASSIGNMENT

Second, here's how to assign the icon's connector pane terminals. In the latest versions of LabVIEW, the connector pane, which consists of a pattern of rectangular terminals, is permanently visible just to the left of the icon pane; in earlier LabVIEW versions, the connector pane is accessed by popping up on the icon pane and selecting **Show Connector**. By default, LabVIEW chooses the $4\times2\times2\times4$ pattern of terminals shown in the connector pane below. We now wish to associate each of our program's inputs and outputs to a particular terminal in this pattern.

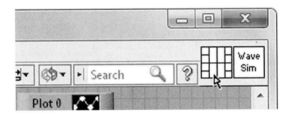

When we use **Waveform Simulator** as a subVI in the future, two quantities will be passed to it—the **Digitizing Parameters** and **Waveform Parameters** control clusters—and one quantity will be taken from it—the **Waveform Output** indicator cluster. So, with the convention that inputs are on the left and outputs are on the right, we will assign two of the leftmost terminals on the $4\times2\times2\times4$ pattern as the **Digitizing Parameters** and **Waveform Parameters** input terminals and one of the rightmost terminals as the **Waveform Output** output terminal.

Alternately, you may wish to override LabVIEW's default $4\times2\times2\times4$ connector pane pattern and choose a pattern more appropriate to the present VI, say, a connector with only two inputs and one output. To change its pattern, pop up on the connector pane and choose the desired arrangement from the **Patterns** palette.

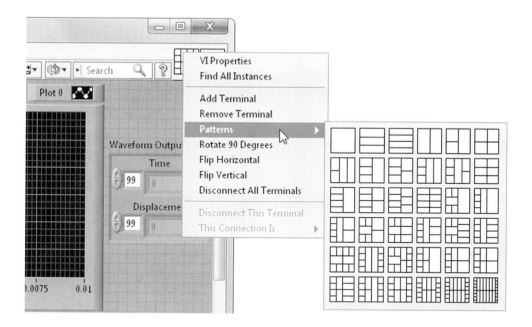

If the exact input–output pattern you desire isn't directly on the palette, you can create it by first selecting a related pattern and then applying some operation from the menu such as **Flip Horizontal**. Or, you can create your own pattern of inputs and outputs by selecting **Add Terminal** and/or **Remove Terminal** in the pop-up menu (you might take a few moments to explore the operation of these two commands).

In this book, I will always use LabVIEW's default $4\times2\times2\times4$ connector pane pattern. Many LabVIEW experts advocate this approach because it promotes uniformity between subVI icons and, as in our present case, leaves some unused terminals that may be handy if the VI is subsequently expanded. Feel free to deviate from this policy if some pattern from the **Patterns** palette better pleases you.

Now assign the **Digitizing Parameters** control cluster to one of the pattern's left-hand terminals through the following steps. Place the mouse cursor (it doesn't matter which tool it is manifesting) over one of the left-hand terminals and then click.

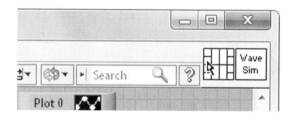

The cursor will morph into the Wiring Tool and the terminal will turn dark, as shown below.

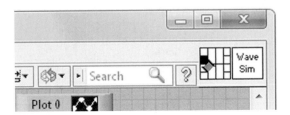

Click on the **Digitizing Parameters** control cluster. A moving dashed-line marquee will then frame the control and the selected terminal will be colored, indicating the data type of the control to which it has been assigned.

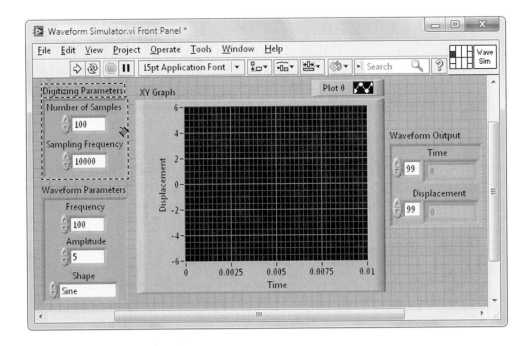

In a similar manner, assign another left-hand terminal to the **Waveform Parameters** control cluster and one of the right-hand terminals to the **Waveform Output** indicator cluster, respectively. The connector pane will then appear as shown next.

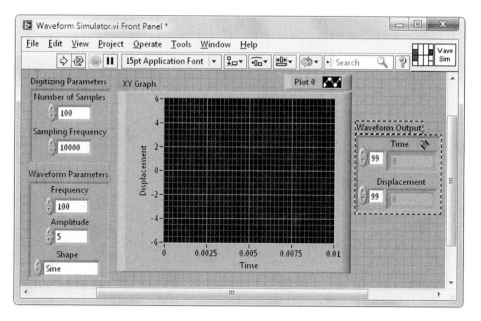

End the assignment procedure by clicking on a blank region of the front panel and then saving your work. If you have an earlier version of LabVIEW, you can return to the icon pane by popping up on the connector pane and selecting **Show Icon**.

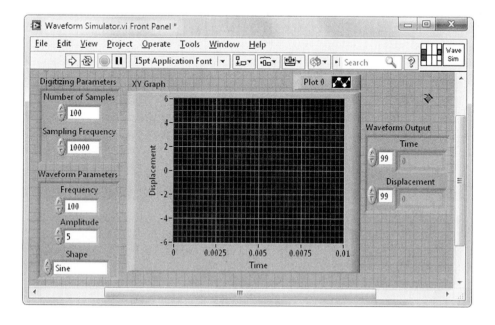

To demonstrate the success of your icon creation, place the mouse cursor over the icon pane and activate the Help Window via <*Ctrl+H*>.

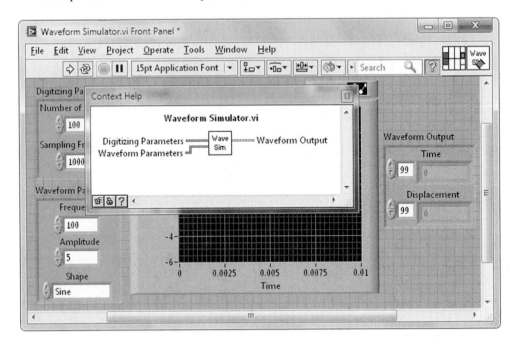

Congratulations—you've now completed your first custom-made VI that, if desired, can be used as a subVI in another program. Using **File>>Save**, store this final version of **Waveform Simulator** in **YourName\Chapter 3**. In later chapters (e.g., see Section 4.13), we'll find out how to place a custom-made VI as a subVI on the block diagram of a higher-level program.

DO IT YOURSELF

The displacement x of an amplitude-modulated (AM) wave obeys the following relation

$$x = A\left[1 + \sin\left(2\pi f_{\mathrm{mod}}t\right)\right]\sin\left(2\pi f_{\mathrm{sig}}t\right) \qquad [4]$$

where A is the amplitude, and f_{mod} and f_{sig} are the modulation and signal frequencies, respectively. Write a MathScript-based program called **AM Wave** that produces and plots an N-element array of displacement values for this AM wave. Remember to use MathScript element-wise operators when appropriate. The front panel and connector assignment for **AM Wave** should appear as shown in the following illustration.

Run your program with $A = 1$ and some sensible choice of values for f_{mod} and f_{sig}. For a typical AM wave, $f_{sig} \gg f_{mod}$, so you might try $f_{sig} = 10\,f_{mod}$.

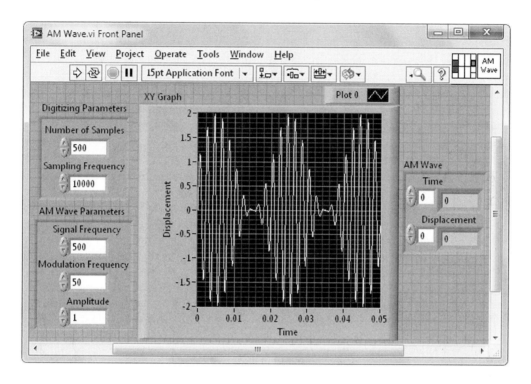

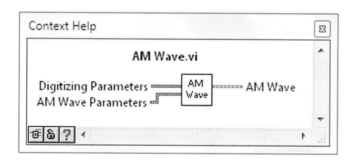

PROBLEMS

1. A unity-amplitude 1 Hz square wave y (t) can be approximated by the following sum of sine waves:

$$y(t) \approx \frac{4}{\pi} \left\{ \sin[2\pi t] + \frac{1}{3}\sin[6\pi t] + \frac{1}{5}\sin[10\pi t] + \frac{1}{7}\sin[14\pi t] \right\}$$ [5]

Write a VI named **Square Wave** that uses a MathScript Node to evaluate this expression for y in the range $0 \leq t \leq 3$s, and then plot y vs. t on an XY Graph.

2. In polar coordinates (r, θ), a particular spiral is defined by the relation r = θ, where $0 \leq \theta \leq 6\pi$. Write a VI called **Spiral**, which uses a MathScript Node to describe this spiral in terms of Cartesian coordinates (x, y), and then displays the spiral on an XY Graph. You may find the MathScript function *polar_to_cart* useful.

3. Open a MathScript Interactive Window, and then type *help roots* in the Command Window. After learning the syntax of the MathScript command *roots*, use it in the MathScript Interactive Window to find the two roots of the second-order polynomial $x^2 + x - 1 = 0$. To check that you used *roots* correctly, you can compare your result with that found using the quadratic formula.

4. Write a MathScript-based VI called **Noisy Sine**, which generates three cycles of a 100 Hz sine wave with added random "Gaussian" noise and then plots this waveform on an XY Graph. Calculate N samples of the sine wave at equally spaced times from $t = 0$ to $t = 0.03$ s. Then add a unique "Gaussian" random number to each of these sine-wave samples to create the noisy waveform. A collection of Gaussian random numbers is distributed about zero with probabilities given by a Gaussian ("bell") curve with a standard deviation of σ. This distribution accurately models the noise in many experimental situations. In your VI, use the MathScript command is *randnormal(1,N)*, which produces a single row vector of N Gaussian random numbers with $\sigma=1$.

5. It is easy to show analytically that the integral $\int_0^1 5x^4 \, dx$ equals exactly 1. Open a MathScript Interactive Window and then type *help quadn_trap* in the Command Window. After learning the syntax of the MathScript command *quadn_trap*, write a VI called **Numerical Integral** that uses *quadn_trap* within a MathScript Node to calculate the integral $\int_0^1 5x^4 \, dx$ numerically and then displays the result in a front-panel indicator with **Digits of precision** equal to 5. For your numerical result to be correct to five decimal places (i.e., the first five decimal places are all zero), how many x-values must be supplied to *quadn_trap* within the integration range $x = 0$ to $x = 1$?

6. Quite commonly, a projectile lands at a level that is different than that from which it is launched. As shown below, consider a projectile launched with speed v (m/s)

at angle θ above the horizontal at a height H above the ground and define range R to be the horizontal distance it travels during its flight (Figure 3.1).

For this situation, it can be shown that

$$R = \frac{v^2}{2g}\left[\sin(2\theta)+\sqrt{\sin^2(2\theta)+\frac{8gH}{v^2}\cos^2\theta}\right] \qquad [6]$$

where θ is in the range from 0° to 90°.

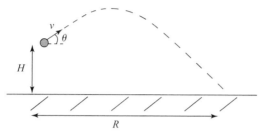

FIG. 3.1

(a) Write a MathScript-based program called **Projectile** that, given v and H, calculates R for a sequence of 91 equally spaced angles in the range from $\theta=0°$ to $\theta=90°$ and then displays R vs. θ on an XY Graph as well as in an indicator cluster. The front panel of **Projectile** is shown next.

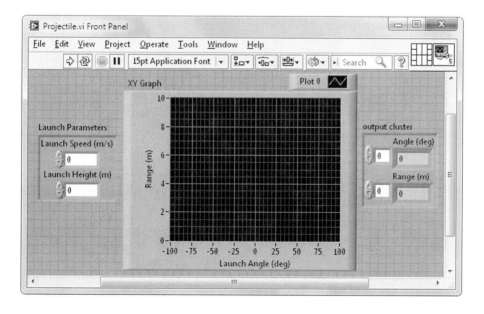

(b) At a track-and-field meet, a shot putter launches a shot (heavy metal ball) with speed 13.5 m/s. Run **Projectile** with this launch speed, assuming the shot lands at the same level from which it was launched. Use the resulting XY Graph and output cluster to demonstrate the well-known fact that the maximum range results if the launch angle equals 45°.

(c) In reality, a shot putter launches the shot from a height close to his or her head. With $v = 13.5$ m/s and $H = 2.1$m, run **Projectile** and then determine the launch angle that produces the maximum range.

(d) Finally, if the shot putter launched the shot with $v = 13.5$ m/s from a cliff of height $H = 100$ m, run **Projectile** and then determine the launch angle that produces the maximum range.

7. Write a VI called **Waveform Simulator (Express)**, which implements one of LabVIEW's easy-to-use Express VIs to generate a waveform. Place a **Waveform Graph** on the front panel, and then switch to the block diagram and place **Simulate Signal** there, which is found in **Functions>>Express>>Input**. Immediately after placing this Express VI on the block diagram, its dialog window will open. In this window, choose **Sample per second (Hz)** and **Number of samples** to be *1000* and *100*, respectively, and then click the **OK** button near the bottom of the window. When you are returned to the block diagram, complete the code shown below. You will find that a new type of wire called *dynamic data type* emanates from the Express VI's **Sine** output and that the **Waveform Graph** automatically adapts to this format.

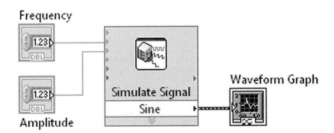

(a) Return to the front panel and run **Waveform Simulator (Express)** with **Frequency** and **Amplitude** equal to *50* and *5*, respectively. Note that the Waveform Graph's *x*-axis is automatically scaled correctly, a benefit of the dynamic data type wire.

(b) On the block diagram, pop up on the **Simulate Signal** icon and select **Properties**. The Express VI's dialog window will reopen. Select **Signal type>>Sawtooth**, and then click the **OK** button. Rerun your VI to view the new waveform shape.

8. In this problem, you will use the MathScript command $[p, s] = polyfit(x, y, n)$ to fit a given set of data to a second-order polynomial function of the form $y = a_0 + a_1 x + a_2 x^2$, where a_0, a_1, and a_2 are constants. A description of this MathScript command can be found by typing *help polyfit* in the MathScript Interactive Window. Assume that a quantity y is measured as a function of another quantity x, resulting in the following data set.

x	1.0	2.0	3.0	4.0	5.0	6.0
y	11.0	15.1	19.8	25.1	31.0	37.5

Construct a VI called **Polynomial Fit (MathScript)** as follows: Enter the given y vs. x data as row vectors in a MathScript Node. The y row vector is created with the command $y = [11.0\ 15.1\ 19.8\ 25.1\ 31.0\ 37.5]$. Then use the command $[p, s] = polyfit(x, y, n)$ to fit these given data to a second-order polynomial. The values of a_2, a_1, and a_0 will be given as the first, second, and third elements of the row vector p, which can be found by the commands $p(1) = a_2$, $p(2) = a_1$, and $p(3) = a_0$ (the integer in parentheses is the index of the desired element of p; unlike LabVIEW arrays, MathScript-array indexing begins with 1, rather than 0). Display the resulting values of a_2, a_1, and a_0 on the front panel.

CHAPTER 4

Data Acquisition Using DAQ Assistant

> In this chapter, you will learn how to control a National Instruments data acquisition (DAQ) device connected to your computer. Here, it is assumed that you have such a DAQ device available, along with a function generator, voltmeter, oscilloscope, and required cabling.

Outfitted with the skills developed in the previous three chapters, you are now equipped to explore one of LabVIEW's most powerful capabilities, computer-controlled data acquisition and generation. In this chapter, you will learn how LabVIEW allows a scientific investigator to reach out through a PC to monitor, as well as create, events in the outside world. In particular, you will carry out this investigation by writing programs based on *DAQ Assistant*, an easy-to-use *Express VI* that controls the operation of a multifunction data acquisition (DAQ) device.

4.1 DATA ACQUISITION VIs

LabVIEW's built-in data acquisition functions are designed to operate all of the features on any DAQ device manufactured by National Instruments (NI), the company that also produces LabVIEW. These features include analog-to-digital conversion, digital-to-analog conversion, event timing, pulse counting, and digital input/output operations. Once a NI DAQ device is connected to your computer (e.g., plugged into a PCI expansion slot or attached via a USB cable), LabVIEW offers you both a low-level and a high-level approach to control its operation.

In the low-level approach, one develops programs based on the *DAQmx VIs*, which are found in **Functions>>Measurement I/O>>NI-DAQmx**. Each of the DAQmx VIs executes a key step (such as setting the sampling rate or storing acquired data in a memory buffer) necessary in performing a complete data acquisition or data generation process. By properly configuring these software building blocks on a block diagram, a

LabVIEW programmer can fully control every operation involved in the data acquisition process and so write programs that provide the ultimate in flexibility and power to the user. This control, of course, comes at the price of added complexity to the block-diagram programming.

At the other extreme, LabVIEW offers the programmer a suite of high-level functions that perform common measurement tasks such as communicating with an instrument, storing acquired data in a file, and performing fast Fourier transform analysis. These functions, called *Express VIs*, are found in **Functions>>Express**. Express VIs are straightforward to use and perform many sophisticated tasks automatically, but have limited capabilities. If your programming requirements are simple, you may often find that an Express VI will do your job quite nicely.

For the high-level method of controlling the data acquisition or generation process, LabVIEW offers an Express VI called **DAQ Assistant**. In this chapter, you will explore the fundamentals of controlling a DAQ device using DAQ Assistant to perform three basic DAQ device operations—analog-to-digital conversion (acquiring a digitized representation of an incoming analog voltage), digital-to-analog conversion (generating an outgoing analog signal from a sequence of digitized numbers), and digital port control (setting and reading the port's HIGH/LOW state). In LabVIEWspeak, these three operations are called *Analog Input (AI)*, *Analog Output (AO)*, and *Digital Input/Output (DIO)*, respectively. Later, in Chapter 11, we will revisit this topic and learn to control a DAQ device using the more advanced DAQmx VIs.

4.2 DATA ACQUISITION HARDWARE

National Instruments, based in Austin, Texas, offers a wide variety of computer-based data acquisition devices. The operations these devices are designed to perform include Analog Input, Analog Output, Counter Input (event timing, pulse counting), Counter Output (pulse generation), and Digital Input/Output. Some of the available DAQ devices are quite specialized (for instance, only designed to process digital signals), while others are multi-purpose, performing some or all of the above-mentioned operations. There are high-speed devices, which digitize incoming signals at rates over 10^9 samples per second (S/s), and more garden-variety systems with maximum sampling frequencies on the order of 10^6 S/s. Digital resolution varies from 8 to 24 bits, and each device connects to a PC through one of about ten possible ways (including the PCI and USB interfaces).

In this chapter, I will use four popular multifunction DAQ devices—PCI-6251, USB-6009, myDAQ, and ELVIS II—to illustrate the generic features of the particular NI data acquisition product that is present in your system. A brief description of each of these representative devices follows.

4.2.1 PCI-6251

This data acquisition board is one of NI's professional-grade multifunction *M Series* DAQ devices. It resides within a desktop computer, plugged into a PCI expansion slot. The 6251 has 16 analog input channels that can perform 16-bit analog-to-digital conversion operations at rates up to 1.25×10^6 S/s. Also, there are two 16-bit analog output channels, which can update output voltages at a maximum rate of 2.8×10^6 S/s. Additionally, this DAQ device has 24 digital I/O ports and two 32-bit counters/timers.

4.2.2 USB-6009

This low-cost multifunction DAQ device provides basic data acquisition functionality for applications such as simple data logging, portable measurements, and academic lab experiments. It is a small, stand-alone unit that connects to your computer via a USB port. The 6009 has eight analog input channels that can perform 14-bit analog-to-digital conversion operations at rates up to 4.8×10^4 S/s. Also, there are two 12-bit analog output channels, which can update output voltages at a maximum rate of 150 S/s. Additionally, the device has 12 digital I/O ports and a 32-bit counter (with no time-measurement capabilities).

4.2.3 myDAQ

Priced to be affordable for students, this USB-interfaced unit is designed for hands-on experimentation anywhere you can take a laptop. The myDAQ has only two analog input channels, but they can perform 16-bit analog-to-digital conversion operations at rates up to 2×10^5 S/s. Impressively, it also has two 16-bit analog output channels that can update output voltages at a maximum rate of 2×10^5 S/s, which is over 1300 times faster than the AO capabilities of the USB-6009. Additionally, the myDAQ has eight digital I/O ports and a 32-bit counter and can be configured as a digital multimeter (DMM).

4.2.4 ELVIS II

Designed specifically for instructional lab use, the ELVIS II has a built-in DAQ device with specifications comparable to a PCI-6251 (sixteen 16-bit analog input channels that operate at rates up to 1.25×10^6 S/s, and so on). ELVIS II communicates with a computer via a USB connection and includes a powered prototyping breadboard for use in the study of electronic circuits.

The exercises in this chapter are designed so that they can be completed using a DAQ device with specifications similar to that of any of the four above-mentioned units. At the time of this writing, however, myDAQ does not have hardware digital triggering capabilities, which is a requirement for the **Digital Oscilloscope (Express)** VI we will write in Section 4.8 (myDAQ owners, see Problem 8 at the end of this chapter for a software work-around to this limitation). In addition, the USB-6009 is not equipped to perform hardware-timed waveform generation and so cannot be used for the work in Sections 4.12–4.14; a DAQ device similar to the PCI-6251, myDAQ, or ELVIS II is needed there.

A conduit carrying signals to and from the outside world must somehow be connected to a DAQ device. For example, among its various capabilities, a PCI-6251 has 16 analog input channels, each standing ready to digitize an incoming analog signal. So how does one get the meaning-laden analog signals from an experiment to the DAQ unit's channel inputs? Typically, an experimenter connects wires from his or her experiment to appropriate terminals on an *I/O Terminal Block* (various styles of this item may be purchased from National Instruments). An I/O Terminal Block has screw (or BNC) terminals to which the experimental wires can be attached. A cable is then connected from the I/O Terminal Block to a connector at the end of the DAQ device. This connector has (depending on the particular device) 50, 68, or 100 pins, where each pin is associated with a particular function of the DAQ device. The PCI-6251 employs the above-described connection method. Alternately, some DAQ devices are configured with an on-board terminal block or prototyping breadboard to which wires from an experiment are directly attached. The USB-6009, myDAQ, and ELVIS II are outfitted in this scheme. Regardless of connection method, the function of each DAQ-device pin can be determined from the device's pinout diagram. The pinout diagrams for the PCI-6251, USB-6009, and myDAQ are reproduced in Figure 4.1.

4.3 ANALOG INPUT MODES

With regard to analog input (AI) operations, the pinout diagrams in Figure 4.1 can be deciphered by noting that the PCI-6251 has 16 AI channels, while the USB-6009 has 8 (let's set aside the myDAQ and ELVIS II for a moment). Through software settings, these inputs can be configured to operate in two distinct analog input modes: *single-ended* and *differential*.

In the single-ended mode, each available analog input pin is an AI channel, with all of these channels referenced to the same common ground. You may choose this common ground to be the building ground, termed the *referenced singled-ended (RSE)* mode, or supply your own ground level at the AI SENSE pin, called the *nonreferenced singled-ended (NRSE)* mode. As an example, when operated in the RSE input mode, the pin assignments for the AI channels and common ground (called AI GND) on the DAQ devices cited above are given in Table 4.1.

For the NRSE mode on the PCI-6251, the pin assignment of each channel is the same as given in Table 4.1; however, a ground level must be applied to pin 62 (AI SENSE). The NRSE mode is not available on the USB-6009, hence the absence of an AI SENSE pin on its pinout diagram.

In differential input mode, available analog input pins are paired to form eight (PCI-6251) or four (USB-6009) independent AI channels, each of which is sensitive only to the voltage difference between paired pins. On the PCI-6251, differential channel 0 is AI 0 paired with AI 8, differential channel 1 is AI 1 paired with AI 9, and so on. In this "differential amplifier" configuration, noise pickup by the AI channels is suppressed. Thus, if it is not a problem in your situation to halve the number of available measurement channels, the differential input mode is the preferred *modus operandi*. Again, as an example, the pin assignments for the differential input mode AI channels for our representative boards are given in Table 4.2.

For the myDAQ and ELVIS II, the situation is simplified because the input channels are clearly marked with their names, rather than number coded. For example, on both

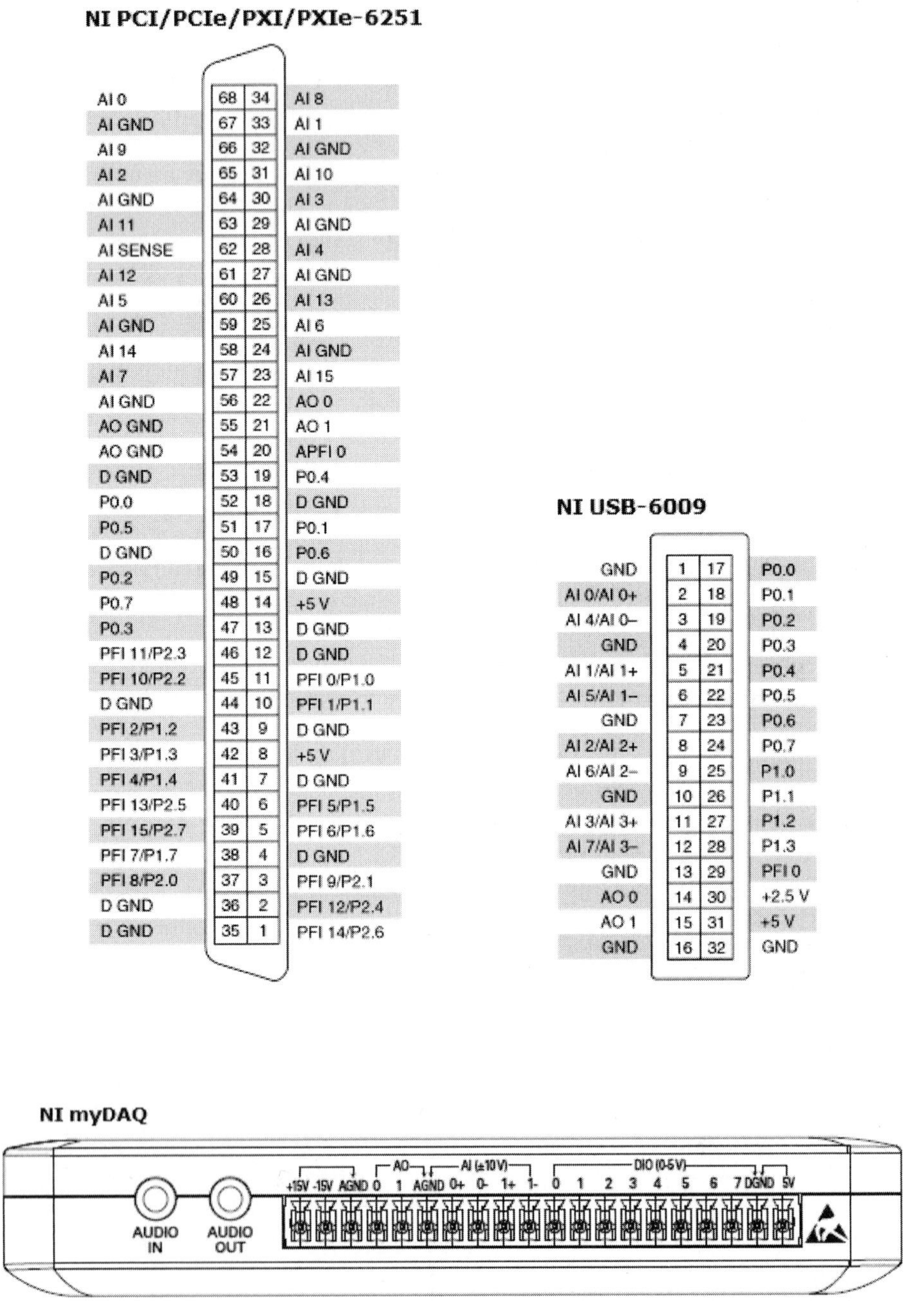

NI PCI/PCIe/PXI/PXIe-6251

AI 0	68	34	AI 8
AI GND	67	33	AI 1
AI 9	66	32	AI GND
AI 2	65	31	AI 10
AI GND	64	30	AI 3
AI 11	63	29	AI GND
AI SENSE	62	28	AI 4
AI 12	61	27	AI GND
AI 5	60	26	AI 13
AI GND	59	25	AI 6
AI 14	58	24	AI GND
AI 7	57	23	AI 15
AI GND	56	22	AO 0
AO GND	55	21	AO 1
AO GND	54	20	APFI 0
D GND	53	19	P0.4
P0.0	52	18	D GND
P0.5	51	17	P0.1
D GND	50	16	P0.6
P0.2	49	15	D GND
P0.7	48	14	+5 V
P0.3	47	13	D GND
PFI 11/P2.3	46	12	D GND
PFI 10/P2.2	45	11	PFI 0/P1.0
D GND	44	10	PFI 1/P1.1
PFI 2/P1.2	43	9	D GND
PFI 3/P1.3	42	8	+5 V
PFI 4/P1.4	41	7	D GND
PFI 13/P2.5	40	6	PFI 5/P1.5
PFI 15/P2.7	39	5	PFI 6/P1.6
PFI 7/P1.7	38	4	D GND
PFI 8/P2.0	37	3	PFI 9/P2.1
D GND	36	2	PFI 12/P2.4
D GND	35	1	PFI 14/P2.6

NI USB-6009

GND	1	17	P0.0
AI 0/AI 0+	2	18	P0.1
AI 4/AI 0−	3	19	P0.2
GND	4	20	P0.3
AI 1/AI 1+	5	21	P0.4
AI 5/AI 1−	6	22	P0.5
GND	7	23	P0.6
AI 2/AI 2+	8	24	P0.7
AI 6/AI 2−	9	25	P1.0
GND	10	26	P1.1
AI 3/AI 3+	11	27	P1.2
AI 7/AI 3−	12	28	P1.3
GND	13	29	PFI 0
AO 0	14	30	+2.5 V
AO 1	15	31	+5 V
GND	16	32	GND

NI myDAQ

+15V -15V AGND   0   1   AGND   0+   0−   1+   1−   0   1   2   3   4   5   6   7 DGND 5V

AUDIO IN   AUDIO OUT

AO   AI (±10 V)   DIO (0-5 V)

FIG. 4.1 Pinout diagrams for DAQ devices.

TABLE 4.1 Channel Assignments for Referenced Single-Ended (RSE) Input Mode

Channel	0	1	2	3	4	5	6	7	8	9	10	11	12	13	14	15
PCI-6251 pin (AI GND is pin 67, 32, 64, 29, 27, 59, 24, 56)	68	33	65	30	28	60	25	57	34	66	31	63	61	26	58	23
USB-6009 pin (AI GND is pin 1, 4, 7, 10)	2	5	8	11	3	6	9	12								

units the positive and negative inputs for analog input channel 0 are labeled AI 0^+ and AI 0^-, respectively. On the myDAQ, each of the two analog input channels can be operated in differential or RSE mode (AI GND is labeled AGND), while the ELVIS II's AI channels allow differential, RSE, or NRSE modes (AIGND and AI SENSE are labeled as such).

4.4 RANGE AND RESOLUTION

When measuring analog input signals with a DAQ device, there are several issues to keep in mind. First, a DAQ device can only measure voltages that fall within an allowed *range*. An accepted range extends from a minimum voltage V_{min} to a maximum voltage V_{max} and a NI DAQ device typically offers the choice of several ranges (such as 0 V to +10 V, −10 V to +10 V, and −5 V to +5 V), with the choice being made by a software setting. Check the spec sheet for your particular DAQ device to find the available ranges.

TABLE 4.2 Channel Assignments for Differential Input Mode

Interface Board	Channel Number	Positive Input Pin	Negative Input Pin
	0	68	34
	1	33	66
	2	65	31
PCI-6251	3	30	63
	4	28	61
	5	60	26
	6	25	58
	7	57	23
	0	2	3
USB-6009	1	5	6
	2	8	9
	3	11	12

Next, the DAQ device's analog-to-digital converter has a built-in *resolution* of *n*-bits, which means that this digitizer represents the analog voltage level being sampled with a binary number of *n* digits. This binary resolution places a limit on the resulting *voltage resolution* (i.e., the smallest detectable voltage difference ΔV). An *n*-bit digitizer divides the measurable voltage span $V_{span} \equiv V_{min} - V_{min}$ ($V_{span} = 20$ V when the measurable range is from $V_{min} = -10$ V to $V_{max} = +10$ V) into 2^n divisions, and so the resulting voltage resolution of the analog input signal being sampled is given by

$$\Delta V = \frac{V_{span}}{2^n} \tag{1}$$

For the typical value of $V_{span} = 20$ V, the voltage resolution provided by a 14-bit (USB-6009) and 16-bit (PCI-6251) DAQ device is 1 mV and 0.3 mV, respectively.

4.5   SAMPLING FREQUENCY AND THE ALIASING EFFECT

Another important issue peculiar to digitized data is related to the sampling frequency f_s, where f_s is the rate at which analog-to-digital conversions take place. The sampling frequency places an upper limit on the range of frequencies allowed within the original analog signal, if the digitizing process is to result in a faithful representation of the input. This upper threshold is called the *Nyquist frequency* $f_{nyquist}$, which we will show equals one-half of the sampling frequency, that is, $f_{nyquist} = f_s/2$. The physical significance of the Nyquist frequency is this: For a given sampling rate f_s, an input sine wave of frequency $f_{nyquist}$ will be sampled in the minimal manner necessary to represent the sinusoid's peaks and valleys—just twice per cycle. If the input frequency exceeds this threshold, the sampling rate becomes insufficient and the analog-to-digital conversion process becomes inaccurate in a way described next.

Fairly often, the bandwidth limitation placed on a "to-be-digitized" input analog signal does not cause you—the experimentalist—much of a problem. For example, if dealing with audio-produced electrical signals, you might know on physical grounds that the frequency content of the analog input is bracketed by 20 and 20,000 Hz. Then, with a sampling rate of at least 40 kHz, this input can be properly acquired. Alternately, an analog signal may have passed through an amplifier that behaves as a low-pass filter because of its finite bandwidth response. In this case, the sampling rate must be twice the maximum frequency passed by the amplifier. If no natural frequency bracketing exists in your experiment, then you must impose a high-frequency cutoff by placing a low-pass filter in your data-gathering circuitry. Given the available sampling rate f_s, the filter's components then are chosen so that frequencies higher than $f_s/2$ are not passed.

What happens if a frequency higher than the Nyquist limit accidentally is input to your digitizer? Something much worse than you might expect. In a process called

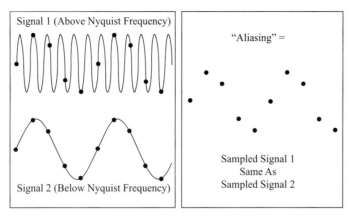

FIG. 4.2 Aliasing effect when signal with frequency greater than the Nyquist frequency is digitally sampled.

aliasing, that too-high frequency appears falsely as a lower-than-Nyquist frequency when processed by the digitizer. This phenomenon, which is peculiar to discrete sampling, is illustrated in Figure 4.2.

Quantitatively, assume a sine wave with (high) frequency f is digitally sampled at N discrete times $t_i = i\Delta t$, where $i = 0, 1, 2, \ldots, N-1$ and Δt is the time increment between neighboring data samples. Then, the sampling frequency and Nyquist frequency are $f_s = 1/\Delta t$ and $f_{nyquist} = f_s/2$, respectively. If $f > f_{nyquist}$, it can be shown that at every digitizing time t_i, the displacement of the sine wave $\sin(2\pi f t_i)$ is equal to plus or minus the displacement of a sine wave with the "alias frequency" f_{alias}, where $0 \le f_{alias} \le f_{nyquist}$. By applying the condition $\sin(2\pi f t_i) = \pm \sin(2\pi f_{alias} t_i)$ at every t_i (see Problem 1), one finds that

$$f_{alias} = |f - n f_s| \qquad \text{(Aliasing Condition)} \qquad [2]$$

where $n = 1, 2, 3 \ldots$

Demonstrate the aliasing effect by opening **Waveform Simulator** (located in **YourName\Chapter 3**) and programming it to produce a 100 Hz sine wave of amplitude 5. That is, in the **Waveform Parameters** control cluster, set **Frequency, Amplitude,** and **Type** equal to *100, 5,* and *Sine,* respectively. Then, in the **Digitizing Parameters** control cluster, taking **Number of Samples** and **Sampling Frequency** to be *20* and *2000,* respectively, run this VI to demonstrate what a digitized sinusoid with frequency less than the Nyquist frequency (in this case, $f_{nyquist} = 1000$ Hz) looks like. Note that one cycle occurs over the course of 0.01 second, so the frequency of the wave is 1/0.01 s = 100 Hz, as expected. Now, with the **Digitizing Parameters** unchanged, program **Waveform Simulator** to produce a 2100-Hz sine wave, and then rerun the VI. You will find that,

159

when digitized, this "higher-than-Nyquist" frequency wrongly appears as a 100 Hz sine wave with period of 0.01 second, consistent with the prediction of Equation [2] with $n = 1$, that is, $f_{alias} = \left|2100 \text{ Hz} - (1)(2000 \text{ Hz})\right| = 100 \text{ Hz}$. Predict what will happen if you program **Waveform Simulator** with some other "higher-than-Nyquist" frequencies such as 1800 Hz and 4100 Hz, and then see if your expectations are realized via the use of **Waveform Simulator**.

4.6   MEASUREMENT & AUTOMATION EXPLORER (MAX)

To carry out the exercises in this chapter, you must have a National Instruments data acquisition device connected to your computer, and its driver software (called *NI-DAQmx*) must be correctly installed. To verify that these conditions are met, we will use the handy utility *Measurement & Automation Explorer*, which is nicknamed *MAX*.

To open MAX, either select **Tools>>Measurement & Automation Explorer...** (if you have an open VI or **Getting Started** window) or double-click on MAX's desktop icon (if available). After MAX opens, double-click on **Devices and Interfaces** in the **Configuration** box. This action will command MAX to determine all of the data acquisition devices present within your computing system.

MAX will return the findings of its device survey in hierarchical tree fashion as shown next. For the system used here, we see that a *NI PCI-6251* device was found to be connected to the computer (for older versions of MAX, you may have to double-click on a folder called **NI-DAQmx Devices** to view the DAQ devices that have been found). Note that the driver software has (automatically) given this device the shorthand name *Dev1*. The DAQ device on your system may have a different shorthand name. When you go through this procedure and for some reason a device that you believe is connected to your computer does not appear on the **Devices and Interfaces** list, try selecting

View>>Refresh (or press *<F5>* on your keyboard) to command MAX to perform its device survey again. If the device in question doesn't appear then, there is something wrong with the device's connection that must be repaired.

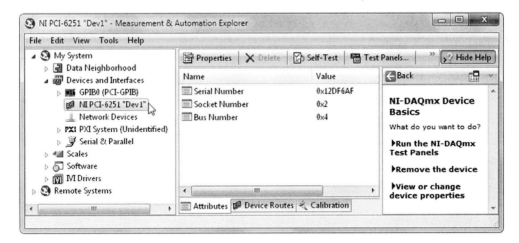

To verify that your DAQ device is properly functioning, pop up on its name and select **Self-Test** from the pop-up menu. Alternately, you can click on the **Self-Test** button in the toolbar near the top of the window.

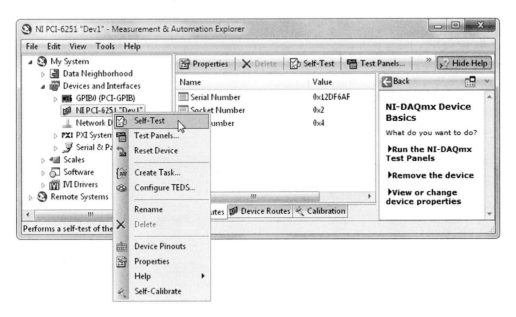

A brief test of your device's functionalities will be performed, and ideally you will receive a dialog-box message stating that all is well.

In a moment, we will program your DAQ device to digitize an analog voltage signal applied at its input. Before we execute this analog input operation, we must secure the connection from the source of analog voltage to the correct pins on your DAQ device. Let's perform this AI operation in differential mode, using differential channel 0, called *ai0* for short. For a DAQ device with a total of 8 AI channels, the differential positive (ai0+) and negative (ai0⁻) channels are pins AI 0 and AI 4, respectively. For a DAQ device with 16 AI channels, ai0+ and ai0⁻ are pins AI 0 and AI 8, respectively. To determine the location of these pins, pop up on the name of your DAQ device and select **Device Pinouts** from the pop-up menu.

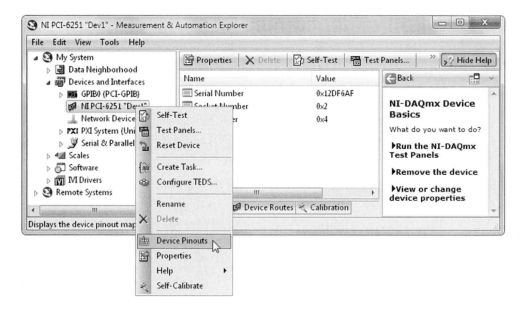

A window will appear with the pinout diagram for your device.

Connect a voltage difference of (approximately) +5 V to the differential analog input channel 0 of your DAQ device. For example, on a PCI-6251 in the differential mode, connect +5 V to pin 68 and GND to pin 34, while on a USB-6009, attach +5 V to pin 2 and GND to pin 3. On a myDAQ and NI ELVIS II, connect +5 V and GND to the pins labeled AI 0⁺ and AI 0⁻ , respectively.

Next, pop up on the name of your DAQ device and select **Test Panels...** (or press the **Test Panels...** button in the toolbar). A diagnostic window will appear, which will allow you to test the functioning of your setup. With the mouse cursor, select **Analog Input** tab, and then enter the settings as shown below. In the **Channel Name** box, the text *Dev1/ai0* selects (differential) AI channel 0 on the DAQ device with shorthand name *Dev1* (on your Test Panel, use the shorthand name of your particular DAQ device, which may be different than *Dev1*).

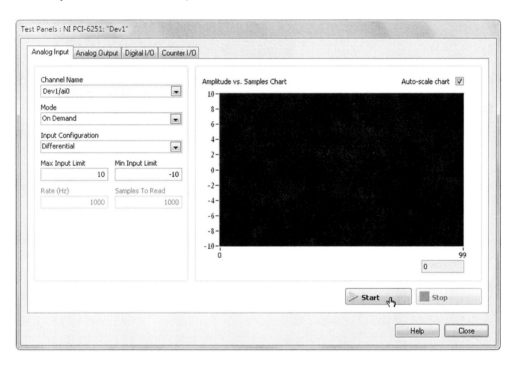

Then, press the **Start** button. If the connections from the voltage source to your DAQ device are properly made, the digitized value of (approximately) +5 V obtained by the DAQ device will appear in the indicator below the chart. Until you press the **Stop** button, this measurement will be repeated several times per second and the results plotted on the chart.

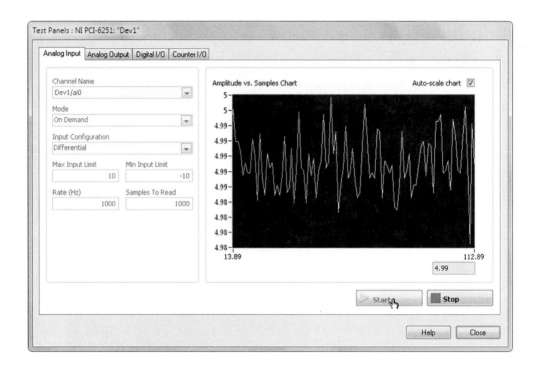

Small fluctuations in your measured voltage are expected because of experimental noise, as shown above. However, if your measured voltage undergoes very large variations as time goes on, you may be having problems resulting from the use of a "floating" voltage source. Examples of floating voltage sources are batteries, transformers, and non-referenced power supplies. To remedy this predicament, reference your "floating" voltage source to the DAQ device's ground by connecting an extra wire from the source's negative terminal to your DAQ device's AI GND pin.

After pressing the **Stop** button, you may want to take a moment to explore the diagnostic features available under the other tabs (e.g., Analog Output) before closing MAX.

4.7 SIMPLE ANALOG INPUT OPERATION ON A DC VOLTAGE

Let's turn your computer into a DC voltmeter by building a VI that carries out the same Analog Input operation that we just ran using MAX. For this exercise, keep the 5 V voltage difference connected to differential channel ai0 of your DAQ device.

Construct the front panel shown below in which an acquired voltage is displayed in the **Numeric Indicator** labeled **Voltage**. Through Equation [1], you can determine the voltage resolution ΔV of the analog-to-digital converter in your DAQ device and use this information to select the proper number of **Digits of precision** on the **Voltage** indicator (using the pop-up menu's **Display Format...** option). For example, if $\Delta V = 0.3\,\text{mV}$, **Digits of precision** should be *4*. Using **File>>Save**, first create a new folder named **Chapter 4** within the **YourName** folder, and then save this VI under the name **DC Voltmeter (Express)** in YourName\Chapter 4.

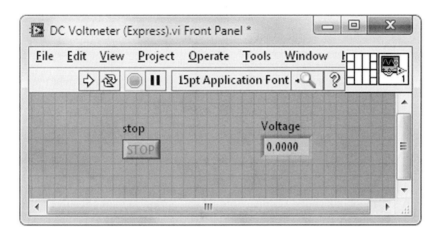

Switch to the block diagram. There, place **DAQ Assistant**, found in **Functions>> Programming>>Express>>Input**.

A moment after you place DAQ Assistant on the block diagram, a dialog window called **Create New Express Task...** will open automatically. DAQ Assistant can be configured to perform any of the data acquisition operations within the capabilities of your DAQ device. In this dialog window, you will select the particular operation that you would like DAQ Assistant to perform for this VI. Since we want to digitize ("acquire") an incoming analog voltage, click on **Acquire Signals**.

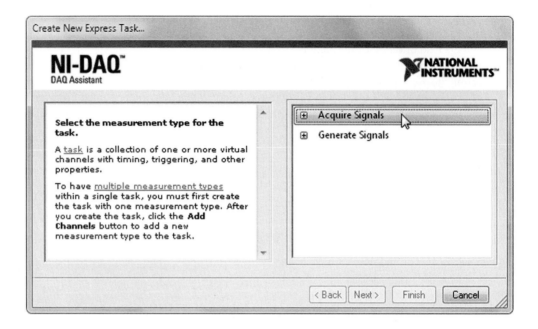

A list of options under **Acquire Signals** will appear in hierarchal tree fashion. Click on **Analog Input**.

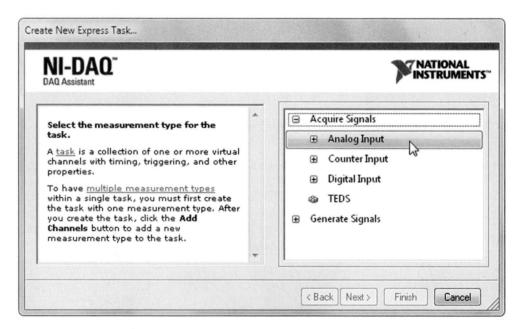

Then, among the measurement types available under **Analog Input**, select **Voltage**.

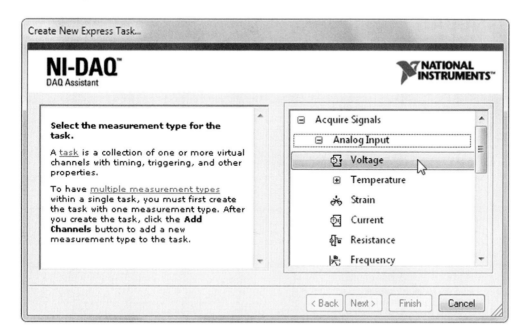

Finally, the available AI channels on your DAQ device (called *physical channels*) will be listed. From this list, select channel **ai0**. Then click on the **Finish** button. From now on, our shorthand for such a sequence of selections will be **Acquire Signals>>Analog Input>>Voltage>>ai0**.

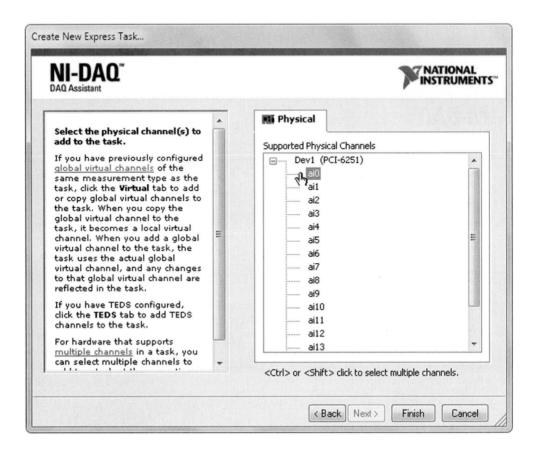

Above, we selected a particular data acquisition operation (analog input voltage measurement) to be performed on a particular physical channel (ai0). This combination of selections becomes part of the definition of a newly created *task*. The **DAQ Assistant** dialog window will now appear automatically. In this window we will configure data-taking details that will complete the task's definition.

The **DAQ Assistant** dialog window opens with the **Express Task** tab selected and the (automatically given default) name of our task—**Voltage**—is highlighted. Under the **Settings** tab, enter the values shown in the following diagram. Under **Timing Settings**, select **Acquisition Mode>> 1 Sample (On Demand)**. In this mode, DAQ Assistant will acquire (only) one voltage reading every time it is executed.

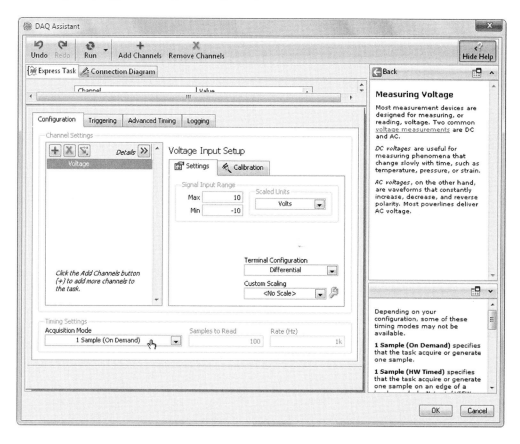

To familiarize yourself with some of the available information within this dialog window, click on the **Connection Diagram** tab. There (after possibly having to input the I/O Terminal Block used by your system) you will find a helpful list and diagram outlining the proper DAQ device pins to use when connecting to differential channel ai0.

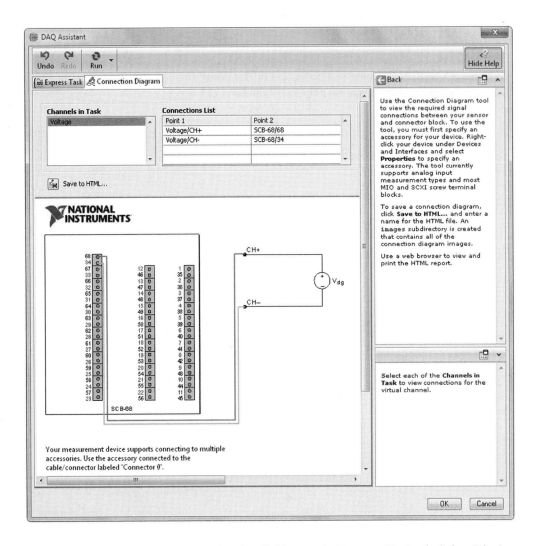

Return to the original window by clicking on the **Express Task** tab. Select **Display Type>>Chart**, and then click on the **Run** button near the upper left corner of the window. DAQ Assistant will execute repeatedly, acquiring one voltage sample per execution, and the sequence of acquired voltage readings is plotted on the chart. If this plot consists of a sequence of values, each (approximately) equal to +5 V, then you have successfully configured DAQ Assistant for our intended task.

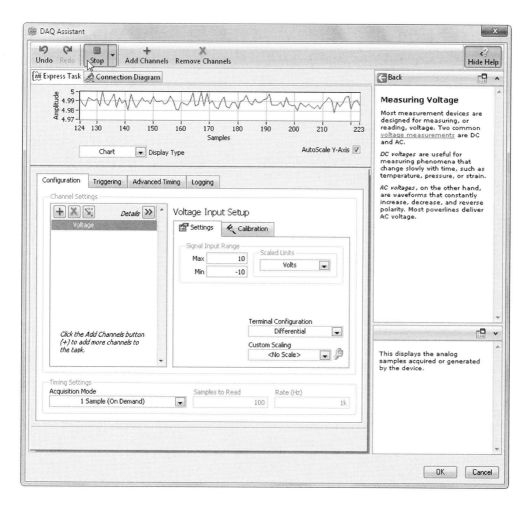

Click **OK** near the bottom right corner to finalize this configuration process. When the dialog window disappears, **DAQ Assistant** will appear on your block diagram as an *expandable node*, with inputs and outputs appropriate for the selections that you made within the DAQ Assistant dialog window. In an expandable node, some of its terminals are "unexpanded" unnamed arrows (inward direction for inputs, outward directed for outputs), while other terminals are "expanded" in their own named band near the bottom of the icon. In the next diagram, only the **data** output terminal is expanded. This terminal reports the acquired voltage reading each time DAQ Assistant executes.

stop

DAQ Assistant
data

Voltage

To expand all of DAQ Assistant's terminals, place the ⬚ at the bottom center of the expandable node. The tool will morph into a resizing handle, which you can click and drag downward until the node has expanded to its full extent.

stop

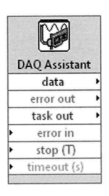
DAQ Assistant
data

Voltage

When you release the mouse button, all of the terminals will appear expanded.

stop

DAQ Assistant
data
error out
task out
error in
stop (T)
timeout (s)

Voltage

Code the block diagram as shown next. Here, the While Loop iterates once every 0.1 s until the **Stop Button** is pressed. With each iteration, DAQ Assistant acquires a single voltage reading from channel ai0, and this value is displayed on the front panel **Voltage** indicator.

172

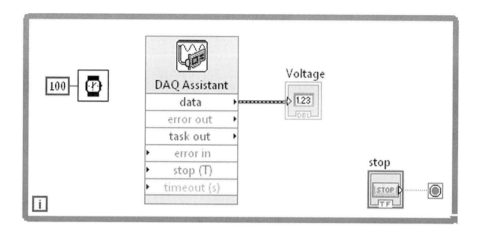

There are a few subtle features of this block diagram, which we will pause for a moment to discuss. First, the **Wait (ms)** icon is included within the While Loop so that, when executing, this program does not monopolize your computer's resources. Because DAQ Assistant alone executes very quickly, without the delay produced by Wait (ms), the loop would iterate at such a high rate that your processor would have to devote most of its efforts to running this program, leaving little time for the other operations for which it is responsible.

Second, look closely and you'll find that the wire connecting the **data** output of DAQ Assistant and Voltage's icon terminal is dark blue and banded, a form that we have not encountered up to this point. The format of this wire is called *dynamic data type*, a data type used by Express VIs. In addition to the data associated with a signal (in our present case, a single digitized voltage value), a dynamic data type wire includes attributes of the signal such as its name and timestamp (i.e., the date and time that the data were acquired). Note, that because of the polymorphic nature of a **Numeric Indicator**, we are able to wire the dynamic data type wire directly to Voltage's icon terminal. A red coercion dot appears, however, indicating that only the data value is being passed to the icon terminal (and all additional information about this signal's attributes is being ignored). Later in this chapter, we will see that these attributes can be quite useful when plotting an acquired signal on a Waveform Graph.

Finally, all LabVIEW data acquisition icons (including DAQ Assistant) have an *error cluster* input and output that you can elect to use for runtime error reporting. Include this useful feature in your VI by popping up on the **error out** terminal and selecting **Create>>Indicator**.

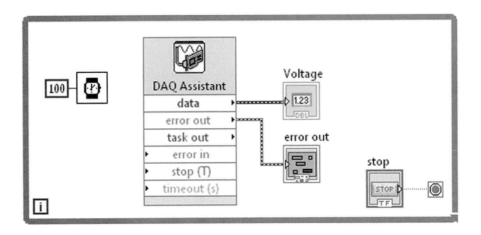

Return to the front panel, where you will now find an **error out** cluster included. The error cluster consists of three elements: **status** (Boolean TRUE if there is an error, FALSE if no error), **code** (integer identifying the error), and **source** (descriptive text identifying the function within which the error occurred). When an error does occur, its integer code can be deciphered by popping up on **code** and selecting **Explain Error**.

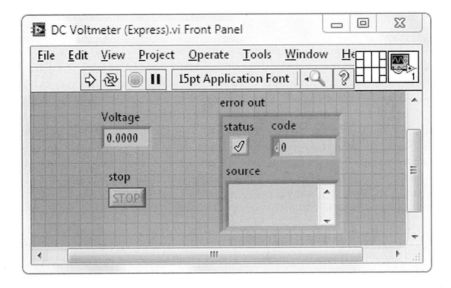

Use the ⬆ to position all the front-panel objects nicely. Then, with the 5 V voltage difference connected to differential channel ai0, run **DC Voltmeter (Express)**. Does it read the input voltage correctly? Save your work on this VI.

4.8 DIGITAL OSCILLOSCOPE

In the previous exercise, you configured DAQ Assistant so that with each execution it acquired a single voltage reading of the analog signal at a channel input. By putting this icon within a repetitive loop and controlling the time per loop iteration with Wait (ms), you wrote DC Voltmeter (Express), a program that, with the simple addition of some data storage and plotting capability, one might consider using to sample and display N data samples of a time-varying waveform. The last sentence outlines the schematic for one of the most useful laboratory monitoring systems—the *digital oscilloscope*. Perhaps you've already thought of some cool things to do with this idea and are ready to go back and start to make the needed modifications to DC Voltmeter (Express) ... but unfortunately Wait (ms) is an Achilles' heel in this plan. Wait (ms) measures time by accessing a clock within your computer, which has an accuracy of one millisecond. Thus, by using Wait (ms) to mark time in the data-taking process, the moment of each voltage digitization will be determined with an uncertainty on the order of 0.001 second. For very-low-frequency inputs (say, less than 10 Hz), this level of uncertainty in x-axis values might possibly be acceptable for the envisioned oscilloscope's *Voltage* vs. *Time* output, but it will be useless over the higher range of frequencies in which you'd want an oscilloscope to operate.

Thankfully, a much better clock, with precision timing on the order of microseconds (or, in the case of the USB-6009, 20 μs), exists on a National Instruments DAQ device. And, even better, LabVIEW provides access and control of this clock through DAQ Assistant. Let's explore this feature of DAQ Assistant as we write a useful digital oscilloscope program. The program we wish to write, called Digital Oscilloscope (Express), will acquire N equally spaced voltage samples of a time-varying analog input and then quickly plot this array of data values. By repeating this process over and over, we'll achieve a real-time display of the waveform input.

Open a blank VI and, using **File>>Save**, store it in YourName\Chapter 4 under the name Digital Oscilloscope (Express). On the front panel of Digital Oscilloscope (Express), place a **Stop Button** and a **Waveform Graph** with its x- and y-axes labeled Time and Voltage, respectively. If you'd like, hide the Stop Button's label by toggling it off via **Visible Items>>Label**. Also, in the Controls Palette, choose **Select a Control...**, and then navigate to the YourName\Controls folder in the **Look in:** box. There, select a **Digitizing Parameters.ctl** control cluster (by either double-clicking or highlighting and then pressing the **OK** button), and then place this control cluster on the front panel. (If you haven't created and saved **Digitizing Parameters.ctl** previously, refer to Section 3.12, "**Control and Indicator Clusters**," and create this control cluster now.)

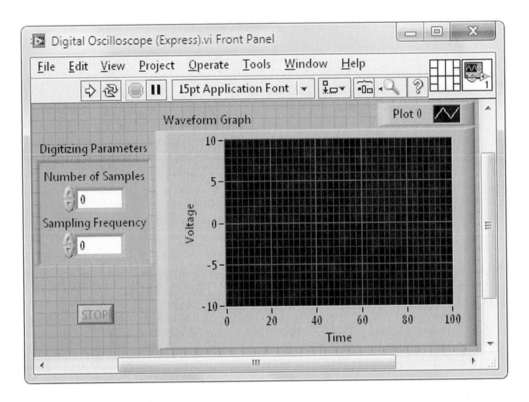

Switch to the block diagram. Place a **While Loop** there and then place a **DAQ Assistant** (from **Functions>>Express>>Input**) icon within the loop. When the **Create New Express Task...** dialog window opens automatically, select **Acquire Signals>>Analog Input>>Voltage>>ai0**. Then, when the **DAQ Assistant** dialog window appears, configure it as shown below. Note that we have set **Acquisition Mode** to **N Samples** and that in this mode the settings for **Samples to Read** and **Rate (Hz)** are activated. These settings were disabled in the **1 Sample (On Demand)** mode we used for DC Voltmeter [Express].

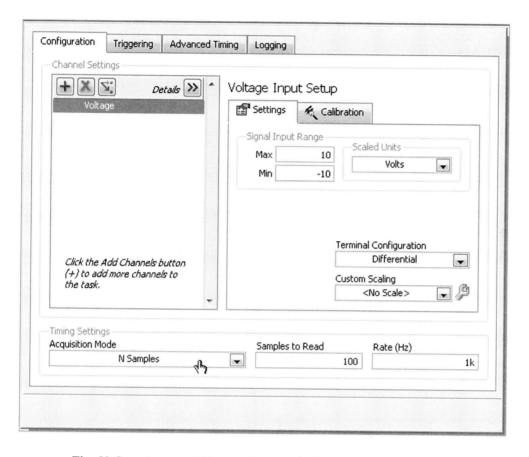

The **N Samples** acquisition mode is a *buffered hardware-timed analog input* operation. In this process, a time-dependent analog signal is digitized at N equally spaced times. The time interval between neighboring samples $\Delta t = 1/f_s$, where f_s is the sampling frequency, is very accurately controlled by a hardware clock on the DAQ device. As the data samples are acquired, they are first placed in a memory buffer on the DAQ device and then transferred from there into the computer's memory. This method assures that no data samples are lost when the computer multitasks (i.e., switches among its various chores) during the data-taking process. In the dialog window's **Timing Settings** section, **Samples to Read** and **Rate (Hz)** are used to select the values for N and f_s, respectively.

Once you have programmed the appropriate values in the DAQ Assistant dialog window, close it by pressing the **OK** button. If later you wish to reopen this dialog window, you can do so by either popping up on **DAQ Assistant** and selecting **Properties** in its pop-up menu or simply placing the ⬉ over **DAQ Assistant** and double-clicking.

After you are returned to the block diagram, observe the available input and output terminals for DAQ Assistant when configured in the **N Sample** mode by resizing the DAQ Assistant icon so that all of its terminals are expanded.

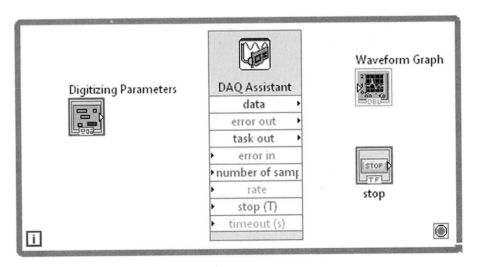

We could now proceed directly to wiring, but some convoluted wire paths would result from the present ordering of the DAQ Assistant terminals. Let's first change the terminal ordering to something more desirable. Pop up on the (top) **data** terminal, and choose **Select Input/Output>>number of samples**. DAQ Assistant will then appear as shown next.

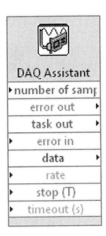

Through a similar procedure, order the terminals as shown below and then resize DAQ Assistant so that **error in**, **task out**, and **timeout (s)** are no longer expanded.

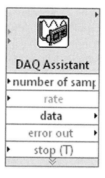

Finally, complete the block diagram as shown next. When **Unbundle By Name** (found in **Functions>>Programming>>Cluster, Class, & Variant**) is first placed on the block diagram, it will only have one output terminal. Use the ⟨ to make both the **Number of Samples** and the **Sampling Frequency** terminals visible, and then wire these outputs to DAQ Assistant's **number of samples** and **rate** inputs, respectively. Also, create the **error out** cluster by popping up on DAQ Assistant's **error out** terminal and selecting **Create>>Indicator**. Note that when you connect DAQ Assistant's **data** output to the Waveform Graph's terminal, the resulting wire is of the dynamic data type.

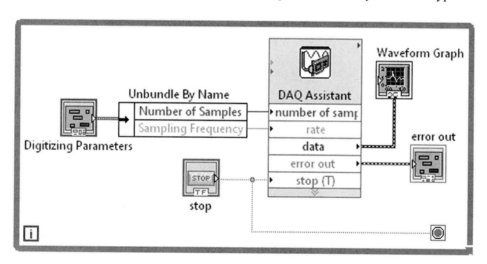

Besides being wired to the While Loop's conditional terminal ◉, the **Stop Button** is wired to DAQ Assistant's **stop (T)** input for the following reason. Within DAQ Assistant several operations are carried out that either start up or shut down the requested data acquisition task. Thus, these particular operations need only be called during the first or last of the multitude of While Loop iterations. However, by default, the **stop (T)** input of DAQ Assistant is TRUE, causing these start-up and shut-down (collectively termed "*overhead*") operations to be executed each time DAQ Assistant executes. Since your data-taking system is not able to digitize the incoming stream of real-time data during overhead operations, needlessly including them increases the program's *dead time*, the percentage of each iteration period during which the system is not sensitive to the incoming data. With increasing dead time, more and more cycles of the incoming repetitive data will stream past your system's input undetected. In certain situations, such as when collecting a large number of data cycles with the intent of adding them together so as to average out random noise, a programming inefficiency that causes you to miss a significant percentage of the incoming data cycles can easily extend the completion time of your experiment by minutes (or sometimes even hours). By wiring **stop (T)** to the **Stop Button**, this input will be FALSE during all but the last While Loop iteration. Then, DAQ Assistant will perform its start-up operations only during the first iteration and its shut-down operations solely during the last, thus enhancing the performance of this program.

Return to the front panel, neaten the arrangement of objects using the ⬛ and then save your work.

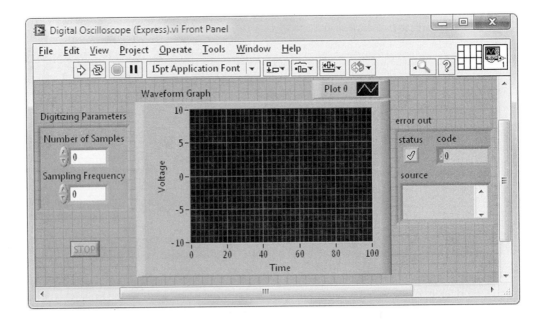

Let's test drive **Digital Oscilloscope (Express)**. Adjust the settings on a function generator so that it is outputting an analog sine wave with amplitude less than 10 V and a frequency of about 50 Hz. Using the appropriate method for your particular system, connect the positive and negative (ground) outputs from the function generator to the channel inputs ai0+ and ai0⁻ on your DAQ device (Note: most function generators are referenced to building ground and so are not "floating" voltage sources). Then, set **Number of Samples** and **Sampling Frequency** equal to *100* and *1000*, respectively, and run **Digital Oscilloscope (Express)**. If all goes well, you will see a plot of about five cycles of the 50 Hz sine wave.

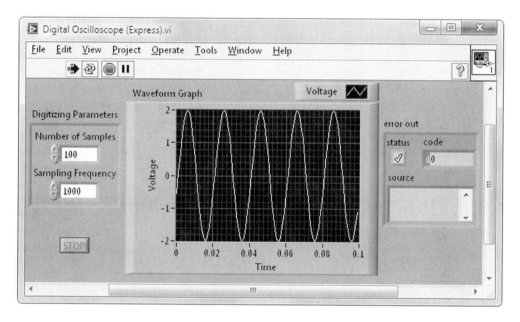

Note that, without any coding effort on our part, the **Plot Legend** is now labeled with the task name **Voltage** and the *x*-axis is properly scaled to reflect the time increment Δ*t* between successive sample acquisitions. These automatic actions are a benefit of the dynamic data type format used to pass the acquired data from DAQ Assistant to the Waveform Graph. Attribute information about these data is included within the dynamic data type wire and is used by the Waveform Graph to accomplish these automatic actions.

Be impressed by the relative ease with which you wrote this program, which acquires and plots a succession of digitized sine waves. However, I'm sure that one flaw in this computer-based instrument is apparent to you. Most likely, the sine wave trace you are observing does not appear stationary but instead is moving either to the right or to the left. And, by slightly changing the function generator's sine-wave frequency, the plotted sine wave can be made to move first one direction and then the other. Why is this so?

In your program, DAQ Assistant is configured so that with each execution, it outputs an array of *N* equally spaced data samples. Let's call this array of *N* samples a *trace*. As you know from coding the block diagram, the sine-wave image viewed on the Waveform Graph is not just a single trace, but is actually a sequence of many traces, where a new trace is generated and plotted with each iteration of the While Loop. Since nothing in your code forces each trace to begin at an equivalent point on the repetitive incoming waveform, the successive traces are displaced from each other along the *x*-axis, resulting in a motion picture–style moving image or, worse, a jumbled mess. The issue we are broaching here is called *triggering* and, for **Digital Oscilloscope (Express)** to be a useful program, we must build in this capability.

In a commercially available oscilloscope, triggering is accomplished in the following manner. The input signal is monitored by an analog "*level-crossing*" circuit. The purpose of this circuit is to determine each time the incoming signal passes through a specified voltage level and then immediately trigger the scope's data acquisition process. Using knobs on the scope's front panel, the oscilloscope user sets the circuit's threshold level and specifies whether acquisition should be initiated when the level is passed through starting from above (*negative*, or *falling, slope*) or starting from below (*positive*, or *rising, slope*).

Many National Instruments DAQ devices are capable of performing the analog level-crossing triggering procedure just described. For example, the PCI-6251 and ELVIS II have this capability; however, the lower-cost USB-6009 and myDAQ do not. So that the broadest range of readers can include triggering in their **Digital Oscilloscope (Express)** VI, we'll implement an alternate mode—*digital edge triggering*—in our program because this mode is available on (almost) all NI DAQ devices. At the time of this writing, the exception to this rule is the myDAQ; myDAQ owners, please see Problem 8 at the end of this chapter for a software triggering alternative to the hardware digital triggering described here.

A digital signal can be in one of two possible states, termed HIGH and LOW. For the digital ports on NI DAQ devices, these states conform to the *transistor–transistor logic (TTL) standard* with HIGH (close to) 5 V and LOW (close to) 0 V. As a digital signal alternates between its two states, the transition from LOW to HIGH is called the *rising edge*, while the HIGH to LOW transition is the *falling edge*. To execute a data acquisition operation under digital edge triggering, one connects a digital signal to the appropriate pins on a DAQ device. Then, by choosing the proper software settings, the desired data acquisition operation can be begun ("triggered") whenever the digital signal has a rising or, alternately, falling edge.

Stop **Digital Oscilloscope (Express)**, if it's running, and then configure it for digital edge triggering as follows. On the block diagram, double-click on **DAQ Assistant**. When the DAQ Assistant dialog window opens, click on the **Triggering** tab, and select **Trigger Type>>Digital Edge** and **Edge>>Rising**. **Trigger Source** will identify the possible DAQ device pins at which the digital signal can be applied; choose one of the available pins. In the window shown, pin **PFI0** is chosen on a PCI-6251, which from the pinout diagram for this device corresponds to pin 11. The ground for the digital signal would also need to be connected to the digital GND pin 12. Once you have made your selection, click on the **OK** button.

That's it. **Digital Oscilloscope (Express)** is now configured for digital edge triggering. Return to the front panel and save your work.

Beside producing a sine wave with frequency f at its *main* output, your function generator creates a TTL digital signal of the same frequency f, which is available at its *sync* (sometimes called *TTL* or *trigger out*) output. The digital edges of this TTL signal coincide in time with particular points on the sine-wave's cycle. For example, on some function generator models, the rising edge of the TTL signal occurs when the sine wave is at its peak; for other models, the rising TTL edge coincides with the positive-going zero-crossing of the sine wave. Thus, the sync output provides a convenient digital signal with which one can control the acquisition of sine-wave data (i.e., always begin at the same equivalent point on the cycle) via digital edge triggering.

Connect the sync output from your function generator to the digital triggering pins on your DAQ device. For example, sync's positive and negative (ground) terminals would connect to PFI0 (pin 29) and GND (pin 32), respectively, on a USB-6009 device. On a PCI-6251, sync's positive and negative terminals connect to PFI0 (pin 11) and GND (pin 12), respectively. On an NI ELVIS II, the appropriate pins are conveniently labeled PFI 0 and GROUND.

On the front panel, set **Number of Samples** and **Sampling Frequency** equal to *100* and *1000*, respectively, and set the sine-wave frequency on your function generator to approximately 50 Hz. Run **Digital Oscilloscope (Express)**. If all goes well, you will see a "stationary" plot of about five cycles of the 50 Hz sine wave.

On the function generator, set the frequency f of the input sine-wave signal to several values within the range $0 \leq f \leq 500$ Hz. You should find that frequencies of signals in this frequency range are faithfully reproduced by **Digital Oscilloscope (Express)**, but that sine-wave shape becomes distorted as the signal frequency approaches 500 Hz (because of the paucity of sampled points per sine-wave cycle at these higher input frequencies).

We have set the sampling frequency equal to $f_s = 1000$ Hz; thus, the Nyquist frequency is $f_{nyquist} = f_s/2 = 500$ Hz. For input frequencies greater than the Nyquist frequency, that is, $f > 500$ Hz, we expect the aliasing effect. Demonstrate that for $f = 1100$ Hz, the digitized sine wave appears to have a frequency of 100 (rather than 1100) Hz. Use Equation [2] to predict two other input frequencies f that will appear as a 100 Hz sine wave when digitized because of aliasing, and then demonstrate that your predictions are correct using **Digital Oscilloscope (Express)**.

4.9   ANALOG OUTPUT

Multifunction DAQ devices manufactured by National Instruments typically have two analog output (AO) channels. Each AO channel can perform n-bit digital-to-analog conversion operations at speeds up to a *maximum update rate* given in samples per second (S/s). The possible analog voltages produced fall within a range from V_{min} to V_{max}. Within this voltage span $V_{span} \equiv V_{max} - V_{min}$, the output analog voltage can be one of 2^n possible values. Thus, the voltage resolution ΔV with which a voltage can be produced (i.e., the smallest difference between one possible output voltage and the next) is $\Delta V = V_{span}/2^n$. As a typical example, if $n = 16$ and $V_{span} = 20$ V, $\Delta V = 0.3$ mV. The specifications for the PCI-6251 and USB-6009 are listed in Table 4.3.

The voltage difference associated with Analog Output channel *ao0* is produced between pins AO 0 and AO GND, while for channel *ao1* the output is between pins AO 1 and AO GND. On the PCI-6251 and USB-6009, the relevant pins are given in Table 4.4.

TABLE 4.3   Analog Output (AO) Specifications

Device	Number of AO Channels	AO Voltage Range (Maximum)*	Bits	Update Rate (Maximum)
PCI-6251	2	−10 to +10 Volts	16	2.8 MS/s
USB-6009	2	0 to +5 Volts	12	150 S/s
myDAQ	2	−10 to +10 Volts	16	200 kS/s
ELVIS II	2	−10 to +10 Volts	16	2.8 MS/s

* Other ranges are software selectable.

TABLE 4.4 Channel Assignments for Analog Output (AO)

Board	AO 0 (pin)	AO 1 (pin)	AO GND (pin)
USB-6009	14	15	13, 16
PCI-6251	22	21	54, 55

On the myDAQ and NI ELVIS II, AO 0 and AO 1 are labeled that way, and AO GND is labeled AGND and GROUND, respectively.

4.10 DC VOLTAGE SOURCE

To familiarize ourselves with the AO process, let's start by writing a program that simply outputs a requested voltage value. On a blank VI, place a **Numeric Control** labeled **Voltage** and a **Stop Button**. If you'd like, hide the Stop Button's label (by toggling it off via **Visible Items>>Label**) and make the Voltage control's **Digits of precision** appropriate for the voltage resolution of your particular DAQ device. Save this VI in the YourName\Chapter 4 folder under the name DC Voltage Source (Express).

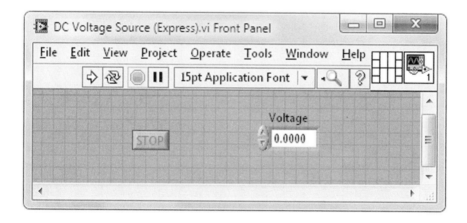

Switch to the block diagram and place **DAQ Assistant** within a While Loop there. When the **Create New Express Task...** window opens, make the following sequence of selections: **Generate Signals>>Analog Output>>Voltage>>ao0**, and then press **Finish**.

When the **DAQ Assistant** dialog window opens, input the selections shown in the following diagram, and then click on the **OK** button. Note we are programming V_{min} and V_{max} to be 0 and +5 V, respectively (values appropriate for all DAQ devices, including the USB-6009). With the **Generation Mode>>1 Sample (On Demand)** option selected, the DAQ device will update the voltage output at channel ao0 every time that DAQ Assistant is executed in your program. This method of updating the AO channel is called *software timing*.

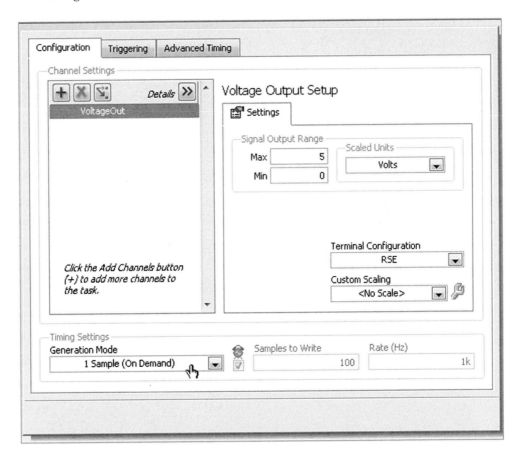

Upon return to the block diagram, use the ↖ to expand all of the terminals of DAQ Assistant when programmed for a **1 Sample (On Demand)** Analog Output operation.

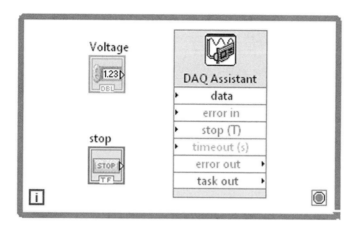

Then, resize DAQ Assistant so that only the **data**, **stop**, and **error out** terminals are expanded, and in the order shown next. Complete the block diagram, remembering the front-panel **error out** cluster indicator can be created by popping up on the **error out** terminal and selecting **Create>>Indicator**. The red coercion dot indicates that the DBL value from **Voltage** is converted to the dynamic data type required at the **data** input to DAQ Assistant. When this code runs, the While Loop will execute once every 10 ms until the **Stop Button** is pressed. With each iteration, DAQ Assistant will instruct the DAQ device to output a voltage difference between pins AO 0 and AO GND equal to the current value of the **Voltage** control.

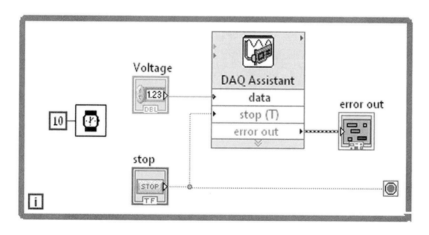

Switch to the front panel and arrange the objects as you wish.

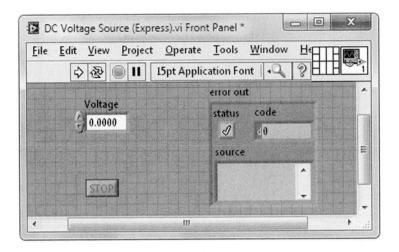

Connect the positive and negative (ground) input of a voltmeter (or oscilloscope in autotriggering mode) to the AO 0 and AO GND pins of your DAQ device, respectively.

Then, set the **Voltage** control to some value within the range 0 to +5 V and press the **Run** button. The voltmeter should inform you that the requested voltage difference is being produced by your DAQ device. With the VI still running, set **Voltage** to a few other values within the 0 to +5 V range to convince yourself that the VI is functioning properly.

Next, let's purposely cause a runtime error in this program and see what happens. Remembering that we programmed the DAQ device to output voltages only within the range from 0 to +5 V (by choosing $V_{min} = 0$ and $V_{max} = +5$ V in the DAQ Assistant dialog window), set **Voltage** equal to a value outside of this range, say, 6 V.

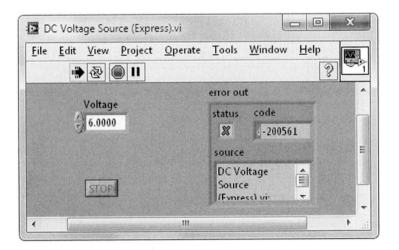

The **error out** cluster indicator informs us that an error has occurred with the ✖ in its **status** box, which corresponds to a Boolean value of TRUE. The **source** box tells us where within the program the error occurs, and **code** identifies the cause of the error via an integer code. To decipher this integer code, pop up on the code box and select **Explain Error**. A dialog window will appear, describing the source of the error, which in the present case was caused by the requested voltage being "too large or too small."

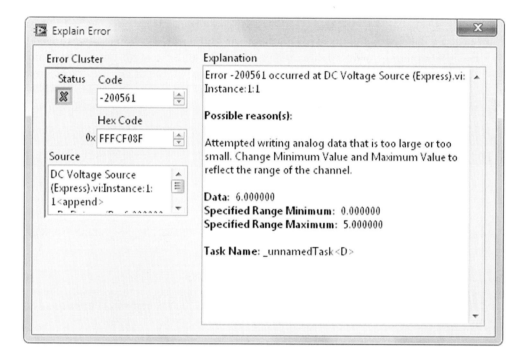

Note that your program continues to run, despite this error, and if you set **Voltage** back to within the acceptable range, the VI will function correctly as if nothing out of sorts had ever happened. In some programming situations, the "continue-to-run-as-if-nothing-is-wrong" behavior you observed here is fine, but more commonly one wants a program to respond in some appropriate way (such as to cease operation) if an error occurs. Appropriate response to an error is called *error handling*.

Let's build error handling into **DC Voltage Source (Express)** by making the VI cease operation when an error occurs. Press the **Stop Button**, if the VI is still running, and then switch to the block diagram so that we can add the desired error-handling code.

Currently, there is only one way to stop our VI. When the **Stop Button** is pressed, a TRUE Boolean value is passed to the 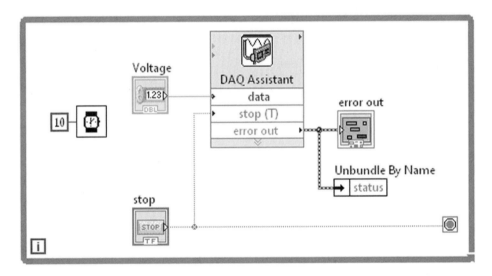, which causes the While Loop to stop iterating. Consider the following alternate way to stop the program. When an error occurs, the **status** element within the **error out** cluster becomes TRUE. By passing this value to the

, we could stop the While Loop. You can include both of these two possible program-stopping methods in your VI with the following coding.

Place an **Unbundle By Name** (found in **Functions>>Programming>>Cluster, Class, & Variant**) on the block diagram and wire it to the **error out** terminal of DAQ Assistant. Once wired to the **error out** terminal, **Unbundle By Name** can be resized to have output terminals for all three of the elements within the **error out** cluster (status, code, source), if desired. However, since we only need **status** in our present program, keep **Unbundle By Name** sized to a single terminal. The element associated with this terminal can be selected by popping up on the right (output) side of **Unbundle By Name** and using **Select Item**.

On your block diagram, allow a TRUE Boolean value from either the **Stop Button** or the error cluster's **status** element to cease While Loop iteration by including an **Or** icon (found in **Functions>>Programming>>Boolean**) as shown next.

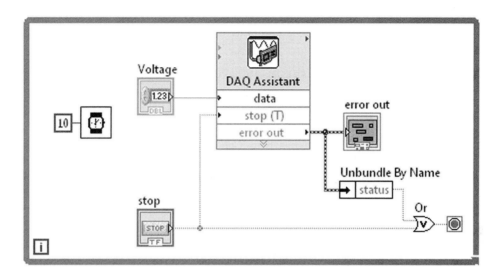

Alternately, in the latest versions of LabVIEW, the error cluster wire can be directly wired to the input of a Boolean icon such as **Or**. At its input, the **Or** icon then will monitor only the error cluster's **status** value, obviating the need for the **Unbundle By Name** icon in the previous block diagram.

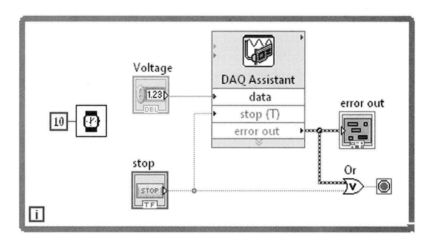

Return to the front panel and save your work. Run the VI with various values for **Voltage**. You will now find that when **Voltage** is set to a value outside of the range 0 to +5 V, the program ceases operation.

When you are finished, run **DC Voltage Source (Express)** one more time with **Voltage** set to *0*, so the channel ao0 is left in the zero-voltage state.

4.11 SOFTWARE-TIMED SINE-WAVE GENERATOR

By continually updating an Analog Output channel with appropriately chosen voltage values, one can turn a DAQ device into a waveform generator with a time-dependent voltage output of any desired shape (e.g., sine wave). To obtain an output voltage whose time dependence accurately represents that desired shape, one must assure that the AO channel changes ("updates") to particular values at precisely the right moments. The most accurate method for timing these required updates involves the use of a high-frequency hardware clock. Such clocks are contained internally on DAQ devices such as the PCI-6251, myDAQ, and ELVIS II and can be used to create high-quality analog waveforms. Unfortunately, a hardware clock is not included on the USB-6009, and so only the (much) less accurate method of software-timed waveform generation is possible on these devices. Let's first write a program that creates an analog sine wave using software timing. Owners of any DAQ device will be able to run this program on their systems. After that, you will have the option of writing a VI that implements hardware-timed AO operations if your DAQ device possesses this capability.

With a few quick edits, we can transform **DC Voltage Source (Express)** into a software-timed sine-wave generator. In this program, DAQ Assistant is configured in the **1 Sample (On Demand)** mode and so outputs a given voltage level each time the While Loop iterates. To obtain a sine-wave output, we will arrange for the sequence of voltage levels fed to DAQ Assistant over the course of several iterations to follow a sine-wave pattern. **Wait (ms)** will control the time per iteration of a While Loop and hence will determine the rate at which the sine wave is generated, that is, its frequency. We want to make this VI compatible with the capabilities of a USB-6009, which has the following two limitations: First, the USB-6009's maximum AO update rate is 150 S/s, which means that the fastest time interval over which an AO channel's voltage level can be changed is 1/150 s, or about 7 ms. Thus, we will program **Wait (ms)** to wait 10 ms. If you have a device capable of faster update, you can (slightly) improve this VI by programming **Wait (ms)** to wait 1 ms, which is its shortest allowable time to wait when using this icon. Second, AO channels on the USB-6009 can only produce voltages in the range from 0 to +5 V. Hence, we will create a sine wave with amplitude of 1 V, which oscillates about the 2 V (rather than 0 V) level.

Let's use our previous work on **DC Voltage Source (Express)** to jump start our development of this sine-wave generator program. With **DC Voltage Source (Express)** open, select **File>>Save As...** In the dialog window that appears, select **Copy>>Substitute copy for original**, and then press the **Continue** button. In the next window that appears, navigate to the **YourName\Chapter 4** folder in the **Save in:** box, name this copied VI **Sine Wave Generator (Software-Timed)** in the **File name:** box, and press the **OK** button. The original file **DC Voltage Source (Express)** will be closed (and safely saved in

YourName\Chapter 4), while the newly created file **Sine Wave Generator (Software-Timed)** is open and ready for you to modify.

On the front panel, change the Numeric Control's label from **Voltage** to **Samples per Cycle**. Then, modify the block diagram to appear as shown below. The **Mathscript Node** (found in **Functions>>Programming>>Structures**) has two inputs (*i* and *N*) and one output (data). Pop up on the **data** output and make sure that **Choose Data Type>>All Types>>Scalar>>DBL** is selected. As the iteration terminal increments, the code within the Mathscript Node evaluates the function $y\,(x) = 2 + \sin x$ at *N* equally spaced arguments *x* in each sine-wave cycle. The spacing in radians between adjacent arguments is $2\pi/N$, where *N* is the number of samples per cycle.

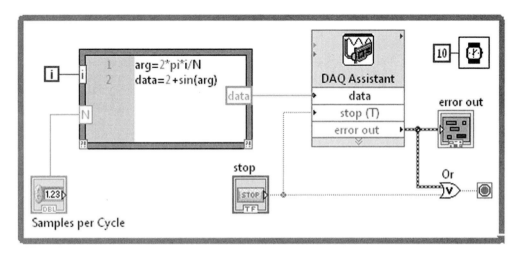

Connect the positive and negative (ground) input of an oscilloscope to the AO 0 and AO GND pins of your DAQ device, respectively.

Switch to the front panel and save your work.

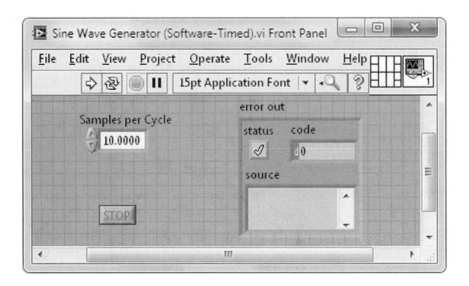

You can now observe how well a digitally produced sine wave approximates the (continuous) sine function. Run your VI with **Samples per Cycle** set to *10*, so that each sine cycle is represented by ten voltage levels. Given that each While Loop iteration takes 10 ms, the period of the observed sine wave should be (10 ms/update) × (10 updates) = 100 ms, making its frequency 10 Hz. Observe the 10 Hz sine wave on an oscilloscope. Although the underlying sine pattern is apparent, the ten voltage levels used to form this waveform are too well resolved to give the appearance of a continuous function.

Try increasing **Samples per Cycle** to *20*, then *30*, and so on. Since the While Loop iteration time remains constant, as you increase **Samples per Cycle**, the frequency of the output waveform will decrease. You will need to adjust your oscilloscope's *Time/Div* control appropriately to view a complete cycle of the waveform. At what value for **Samples per Cycle** does the output waveform start to take on a continuous appearance?

4.12 HARDWARE-TIMED WAVEFORM GENERATOR

If your DAQ device supports hardware-timed analog output operations (PCI-6251, myDAQ, and ELVIS II do; USB-6009 doesn't), try building the following high-quality waveform generator. On the front panel of a blank VI, place a **Stop Button** and an **XY Graph** with *x*- and *y*-axes labeled **Time** and **Voltage**, respectively. Then, save this VI under the name **Waveform Generator (Express)** in **YourName\Chapter 4**.

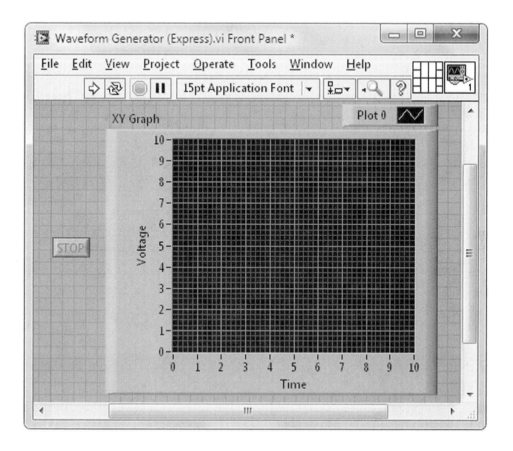

Switch to the block diagram. Place a **While Loop** there and then a **DAQ Assistant** within it. When the **Create New Express Task...** window opens, make the following sequence of selections: **Generate Signals>>Analog Output>>Voltage>>ao0**, and then press **Finish**. When the **DAQ Assistant** dialog window opens, make the selections shown in the following diagram. Be sure to uncheck the **Use Waveform Timing** box (located next to the 👆 in the illustration). Note we are programming V_{min} and V_{max} to be −10 and +10 V, respectively. In the **Generation Mode>>Continuous** mode, a 1D array consisting of a sequence of voltage values is first written into a memory buffer associated with DAQ Assistant, and then the DAQ device outputs this sequence of voltages at channel AO 0 in a circular fashion (i.e., after the last array value is output, the next output is the first array value). When all of the selections are properly made, click on the **OK** button.

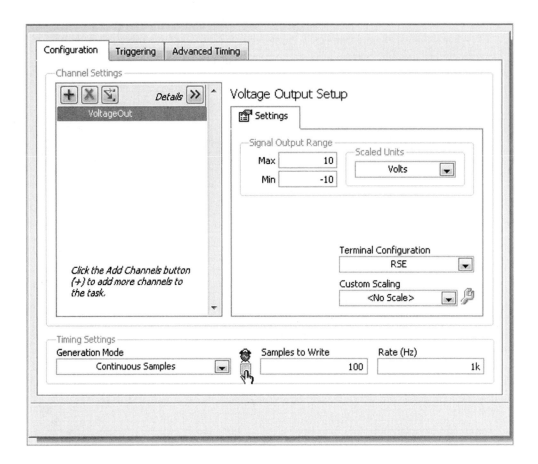

Upon return to the block diagram, expand the terminals of DAQ Assistant and order them as shown next using **Select Input/Output** in each terminal's pop-up menu.

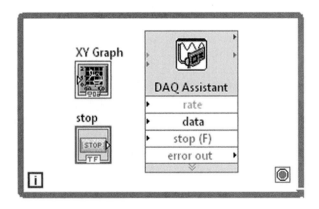

The plan for our VI is this: Use the **Waveform Simulator** program written in Chapter 3 to generate a 1D array of voltage values that describes a desired waveform. Then, pass this 1D array to the **data** input of **DAQ Assistant**, which initiates the process of outputting the voltage waveform at channel ao0.

4.13 PLACING A CUSTOM-MADE VI ON A BLOCK DIAGRAM

First, we must place **Waveform Simulator** (as a subVI) on our block diagram. Here's how to do that. On the Functions Palette, choose the **Select a VI...** subpalette.

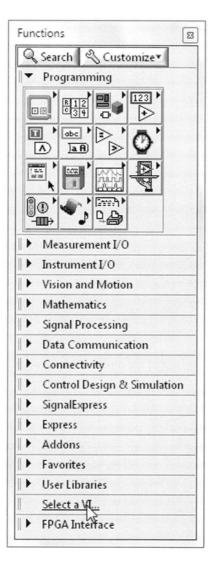

The **Select the VI to Open** dialog window will appear. In the **Look in:** box, navigate to the YourName\Chapter 3 folder. Once the file list for this folder is displayed, double-click on **Waveform Simulator** or highlight it, and then click on the **OK** button in the dialog box.

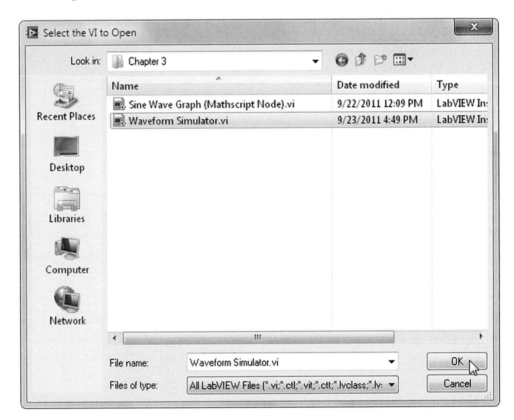

You will then be returned to the block diagram, where you can place your custom-written icon in a convenient position. Then, with the ✎, create **Digitizing Parameters** and **Waveform Parameters** control clusters by popping up on these terminals and using **Create>>Control**.

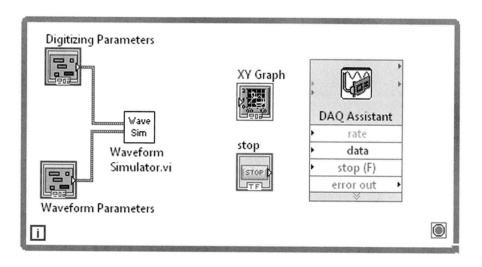

4.14 COMPLETING AND EXECUTING WAVEFORM GENERATOR (EXPRESS)

Finally, complete the block diagram as shown next. When you wire Unbundle By Name's **Displacement** output to DAQ Assistant's **data** input, the **Convert to Dynamic Data** icon will be inserted into the wire automatically. This icon converts the 1D floating-point numeric array produced by **Waveform Simulator** to the dynamic data type used by Express VIs such as DAQ Assistant. If you ever need to place **Convert to Dynamic Data** on a block diagram manually, it is found in **Functions>>Express>>Signal Manipulation**.

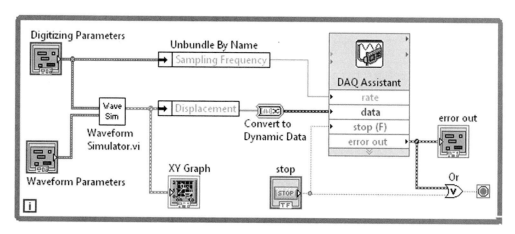

Return to the front panel, tidy it up, and then save your work.

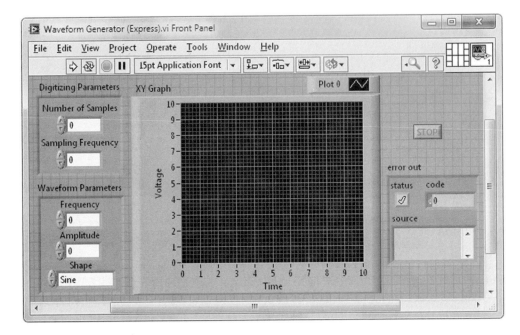

Here's how to program your VI to output a continuous waveform. Using the **Digitizing Parameters** and **Waveform Parameters** control clusters, chose values that instruct **Waveform Simulator** to produce one (or, possibly, a few) complete cycle(s) of the desired periodic waveform. This software representation of the waveform's cycle will then be written into a memory buffer by **DAQ Assistant**. The array of waveform values will then be output sequentially (in a circular fashion) by your DAQ device at a rate given by the **Sampling Frequency** until a TRUE value is sent to DAQ Assistant's **stop** input.

Test drive your VI by using it to output a 100 Hz sine wave of amplitude 5 V. First, in the **Waveform Parameters** control cluster, program **Frequency**, **Amplitude**, and **Shape** to be *100*, *5*, and *Sine*, respectively. Next, assuming 100 points are required to well define each cycle of the sine wave, the sampling frequency should be 100 times larger than the sine wave's frequency. Thus, for **Waveform Simulator** to create one complete cycle, in the **Digitizing Parameters** control cluster let **Number of Samples** and **Sampling Frequency** be *100* and *10000*, respectively.

Run your VI. The software waveform produced by **Waveform Simulator** will be displayed on the **XY Graph** for you to view. Using an oscilloscope, observe the voltage waveform output at channel ao0. Is it a (quasi-)continuous sine wave with a 100 Hz frequency and 5 V amplitude?

With your VI still running and keeping all other parameters the same, change **Frequency** (in the **Waveform Parameters** control cluster) to *150*. From the XY Graph, you will find that **Waveform Simulator** produces 1 1/2 cycles of this waveform. Using your knowledge of the circular fashion in which the DAQ device sequences through the memory buffer values, can you explain the discontinuities that you observe in the voltage waveform being output at channel ao0?

With the VI still running, try changing **Frequency** to *200, 400, 500*, and *1000*. With each of these choices, **Waveform Simulator** will produce an integer number of complete cycles of the waveform and so produce voltage waveform outputs with no discontinuities. Why does the waveform output at channel ao0 become less well defined for the higher frequencies? Also, you might enjoy changing **Shape** and **Amplitude** as the VI is running.

Finally, this VI has one constraint, which reflects a limitation the DAQ Assistant Express VI. Namely, DAQ Assistant sets the DAQ device's sampling frequency during its first execution within the program (i.e., during the first iteration of the While Loop) and this value cannot be changed as the program continues to run. You can verify this point as follows: Start the VI with **Frequency, Amplitude**, and **Shape** equal to *100, 5*, and *Sine*, respectively, and **Number of Samples** and **Sampling Frequency** equal to *100* and *10000*, respectively. **Waveform Simulator** will create one cycle of a 100 Hz sine wave, and you will observe a 100 Hz sine wave at ao0. Now, with the VI continuing to run, change **Frequency** and **Sampling Frequency** to *1000* and *100000*, respectively. **Waveform Simulator** will create 100 points that represent one cycle of a 1000 Hz sine wave if output at a rate of 100 kHz, but you will observe a 100 Hz sine-wave output at channel ao0 because the DAQ device's sampling frequency did not change from its initial value of 10 kHz.

4.15 MODIFIED WAVEFORM GENERATOR

Above, we discovered the following properties of DAQ Assistant when placed in an iterative loop: After being set during this icon's initial execution, the value of **Sampling Frequency** remains fixed. However, the array read into its **data** input can be changed during any execution of the icon, allowing the parameters **Frequency, Amplitude, Shape**, and **Number of Samples** to be adjusted as an iterative program runs. Consider the following approach for increasing the flexibility of **Waveform Generator (Express)**: Set **Sampling Frequency** to a large value. Then, during each iteration as the program runs, determine an appropriate data array based on the current **Waveform Parameters** settings and input this array to DAQ Assistant. Let's arbitrarily define the "appropriate data array" to be four cycles of the waveform described by the **Waveform Parameters** settings.

If interested, with **Waveform Generator (Express)** open, use **File>>Save As...** to create **Waveform Generator (Modified Express)** and save it in **YourName\Chapter 4**. Then delete the **Digitizing Parameters** control cluster so that the front panel is as shown next.

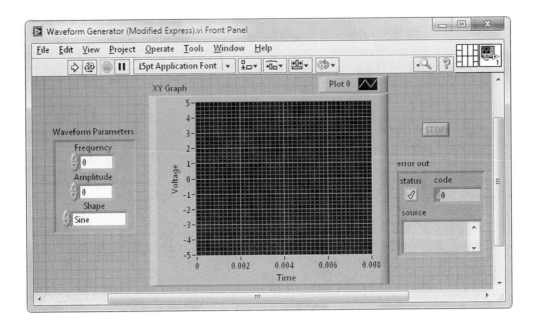

On the block diagram, hardwire **Sampling Frequency** to, say, 100000 Hz. Additionally, using the value of **Frequency** input on the front panel, the **Number of Samples** value required to produce four cycles of the waveform is determined and input to **Waveform Simulator** as shown in this block diagram.

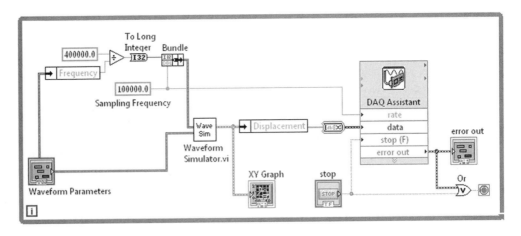

Run **Waveform Generator (Modified Express)**. It will produce high-quality voltage waveforms of frequencies up to a few kilohertz, where any of these frequencies can be chosen during runtime.

DO IT YOURSELF

In this exercise, you will write a VI that turns an LED on or off. A Digital Input/Output (DIO) pin on an NI DAQ device is called a *line*. A typical DAQ device has two or three *ports*, where each port is composed of eight *lines*. A particular line is specified as *P port:line*, so for example P0.3 is line 3 of port 0. Locate the DIO line P0.0 on your DAQ device, and then construct the hardware circuit shown in Figure 4.3. In this circuit, when P0.0 goes HIGH, it supplies (approximately) a 5 V voltage difference across the resistor–LED series combination, resulting in a current of 5 V/220 $\Omega \approx$ 20 mA that illuminates the LED.

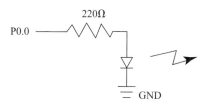

FIG. 4.3 Hardware circuit for Do It Yourself project. Digital Input/Output line P0.0 of the data acquisition device connected to (current-limiting) resistor in series with LED.

Write a VI, which can illuminate the LED as follows:

Place a Boolean switch (e.g., **Push Button**) on the front panel of a new VI. Then on the block diagram, place **DAQ Assistant** and configure it as **Generate Signals>>Digital Output>>Line Output>>port0/line0** and **Generation Mode>>1 Sample (On Demand)**. With this configuration, when a Boolean TRUE (FALSE) is passed to DAQ Assistant's **data** input, P0.0 will go HIGH (LOW). Build the following block diagram. Since a DAQ device has *N* DIO lines, **data** is formatted to accept an array of Boolean values with up to *N* elements. Hence, the **Build Array** icon (found in **Functions>>Programming>>Array**) is necessary to convert the Boolean switch's single Boolean value to a Boolean array with one element.

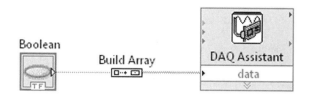

Run the VI several times and show that you can toggle the LED in your hardware circuit on and off via the setting of the Boolean switch.

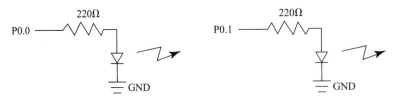

FIG. 4.4 Hardware circuits controlled by Alternating LEDs VI.

Next, construct the hardware circuit shown in Figure 4.4 with two LEDs, which we will call A and B. Then write a While Loop–based VI called **Alternating LEDs**. When this program runs, during even iterations of its While Loop, A is lit and B is dark; during odd iterations, B is LED is lit and A is dark. The LEDs alternately light until a front-panel **Stop Button** is clicked.

In constructing **Alternating LEDs**, beside **Build Array**, you may find the following icons useful: **Select** and **Equal To 0?** in **Functions>>Programming>>Comparison** as well as **Quotient & Remainder** in **Functions>>Programming>>Numeric**.

PROBLEMS

1. Writing the aliasing condition as $f = \pm f_{alias} + nf_s$, show that $\sin(2\pi f t_i) = \pm \sin(2\pi f_{alias} t_i)$ at all $t_i = i\,\Delta t$.

2. Observe the voltage resolution ΔV of an analog input signal digitized by your DAQ device and verify that your observation is consistent with the prediction of Equation [1]. Attach a slowly varying analog signal (e.g., 1 Hz triangle wave) and the sync output from a function generator to the analog input channel and digital triggering pins programmed into **Digital Oscilloscope (Express)**. Run **Digital Oscilloscope (Express)** with a fast sampling rate and fairly small number of samples so that the input signal is effectively constant over the trace that you obtain. Stop the VI, and then, by changing the y-axis scaling, zoom in on the acquired data (alternately, you can use a **Probe** to view the numerical values of the data samples). Under "high magnification," you should find that the data samples of this "effectively constant" input signal are distributed in discrete levels (because of electronic noise). Measure the spacing ΔV between two of these adjacent levels. Does your value for ΔV agree with the prediction of Equation [1]?

3. By passing an analog signal through a low-pass filter prior to inputting it to a digitizer, the aliasing effect can be suppressed. To demonstrate this procedure, attach an analog sine-wave signal and the sync output from a function generator to the analog input channel and digital triggering pins programmed into **Digital Oscilloscope (Express)**.

    On the front panel of **Digital Oscilloscope (Express)**, set **Number of Samples** and **Sampling Frequency** equal to *100* and *1000*, respectively, and set the sine-wave frequency f on your function generator to approximately 100 Hz. Run **Digital**

Oscilloscope (Express). If all goes well, you will see a "stationary" plot of about ten cycles of the 100 Hz sine wave. Turn off autoscaling on the *y*-axis.

Next, set *f* on your function generator to approximately 1900 Hz, and then fine-tune *f* until **Digital Oscilloscope (Express)** displays an aliased waveform of about 100 Hz. Use Equation [2] to explain why the input with *f* = 1900 Hz appears as a 100 Hz sine wave when digitized, as a result of aliasing.

Finally, without changing the frequency setting on your function generator, pass the *f* = 1900 Hz sine-wave voltage through the low-pass op-amp filter shown in Figure 4.5 with *R* = 15 kΩ and C = 0.1*μ*F prior to inputting it to the DAQ device. By what factor is the amplitude of the aliased signal attenuated in comparison to the unfiltered situation? What is the 3-dB frequency $f_{3dB} = 1/2\pi RC$ of the low-pass filter? If you wanted the aliased signal to be attenuated even further, what would you change? [This process can be optimized using a low-pass filter with a steeper roll-off than the simple (first-order) circuit used here.]

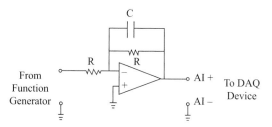

FIG. 4.5 Problem 3

4. Build a two-channel digital oscilloscope. With **Digital Oscilloscope (Express)** open, use **File>>Save As...** to create a new VI named **Dual Digital Oscilloscope (Express)**. On the block diagram, you must reprogram DAQ Assistant to read two channels rather than just one. Open DAQ Assistant and add the new channel, and then close the dialog window. Your VI is now complete.

Attach a voltage waveform and the sync out digital signal from a function generator to the two AI channels of your DAQ device programmed on the block diagram. Additionally, attach the sync out digital signal to the PFI0 channel of your DAQ device. Run **Dual Digital Oscilloscope (Express)** and verify that it simultaneously displays traces of the two inputs.

5. Thermocouples are widely used as temperature sensors. A thermocouple is constructed by joining the ends of two dissimilar metals, for example, a copper and a constantan wire for a Type T thermocouple. This junction produces a millivolt-level voltage, which has a well-documented temperature dependence, where the temperature is measured relative to a "cold junction" reference temperature. Conveniently, this cold junction can be provided by a compact electronic device called a cold junction compensator (CJC), which effectively makes the reference temperature equal to 0°C.

Connect a thermocouple to a CJC and then connect the plus and minus outputs of the CJC to the AI 0⁺ and AI 0⁻ pins of your DAQ device. Using DAQ

Assistant, write a program called **Thermocouple Thermometer (Express)** that, every 250 ms until a Stop Button is pressed, reads the thermocouple voltage, converts this value to the corresponding temperature in Celsius, and then display this temperature in a front-panel indicator. Program DAQ Assistant as follows: **Acquire Signals>>Analog Input>>Temperature>>Thermocouple>>ai0**, followed by **Scaled Units>>deg C**, **CJC Source>>Constant**, and **CJC Value>>0**, and **Acquisition Mode>>1 Sample (On Demand)**. Program **Thermocouple Type** for your particular type of thermocouple (e.g., T).

Run **Thermocouple Thermometer (Express)** and use it to measure room temperature as well as the temperature of your skin.

6. If your DAQ device supports hardware-timed analog output operations, write a program called **Bode Magnitude Plot**, which performs the following process: Apply an AC voltage with frequency f and 1 V rms amplitude to the input of a circuit, measure the resulting rms voltage amplitude V_{out} at the circuit's output, and then calculate the gain (in decibels) $G = 20 \log (V_{out}/V_{in})$, where $V_{in} = 1$ V_{rms}. Construct the VI such that this process is repeated for frequencies in the range $f = 10$ Hz to $f = 2010$ Hz in 25 Hz increments, and the resulting data are displayed on a plot of G vs. $\log (f)$. Build your VI according to the following guidelines.

Block Diagram:

The workhorse of this VI is formed by configuring three Express VIs as shown below. Here **Simulate Signal** (found in **Functions>>Express>>Input**) is used to create a sine wave of frequency f and 1 V rms amplitude, which is passed to a **DAQ Assistant** configured to generate this waveform at an analog output channel. Assuming this sine-wave signal is applied at the input of a circuit, the second **DAQ Assistant** is configured to read 500 samples of the response at the circuit's output at a sampling frequency of $50\,f$ (i.e., 10 cycles of the output will be read and made available at the **data** terminal). Finally, these 10 cycles are supplied to **AC & DC Estimator.vi** (found in **Functions>>Signal Processing>>Signal Operation**). Note in the dialog window for Simulate Signal, select **Samples per second (Hz)>>40000** and check **Integer number of samples**. For the analog output DAQ Assistant, select **Generation Mode>>Continuous** and check **Use Waveform Timing**. For the analog input DAQ Assistant, select **Acquisition Mode>>N Samples**.

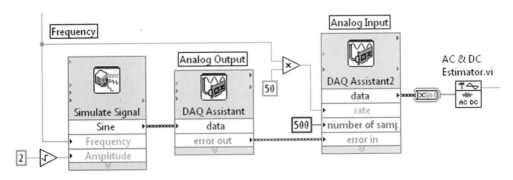

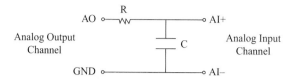

FIG. 4.6 Problem 6

Front Panel:

Place an XY Graph on the front panel and supply it with bundled arrays of f and G from the block diagram. To obtain a log scale on the x-axis, pop up on the XY Graph and select **X Scale>>Mapping>>Logarithmic**.

When completed, use **Bode Magnitude Plot** to obtain a Bode plot on the RC circuit shown in Figure 4.6, where $R = 4.7$ KΩ and $C = 0.1$ μF.

7. If your DAQ device supports hardware-timed analog output operations, write a program called **Music Box**, which plays the scale of A major with each note sounding for 500 ms. The eight notes in this scale have the following frequencies in Hertz: 440.0, 493.9, 554.4, 587.3, 659.3, 740.0, 830.6, and 880.0. These frequencies can be stored either in a front-panel **Array Control** or in a block-diagram **Array Constant** (created by placing a **Numeric Control** within an **Array Constant** shell). On the block diagram, input each frequency f one at a time into a For Loop that contains the subdiagram shown below. Here, the **Simulate Signal** Express VI (found in **Functions>>Express>>Input**) creates an array representing an integer number of cycles of the desired sine wave with frequency f, which is then output as a real voltage waveform on an AO channel by DAQ Assistant. When programming **Simulate Signal**, select **Samples per second (Hz)>>100000**, **Number of samples>>1000** (and uncheck **Automatic**). Check **Integer number of cycles**.

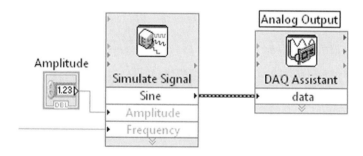

To play the scale, attach an amplifier and speaker to the AO channel such as shown in Figure 4.7, and then run **Music Box**.

8. For myDAQ owners, include digital triggering in your digital oscilloscope program using software, rather than hardware. Assume that you wish to use your scope to observe a waveform produced by a function generator and that this waveform is

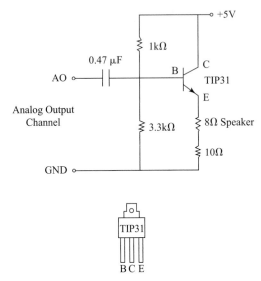

FIG. 4.7   Amplifier circuit for Problem 7

being digitized on AI channel 0 of your myDAQ. Further assume that the sync output from the function generator is simultaneously being digitized on AI channel 1 of the myDAQ.

First, create a VI called **Digital Trigger (Software)**, whose front panel is shown next. To create the 2D Array Control **Acquired Data**, first create a 1D Array Control, and then pop up on its index display and select **Add Dimension**. The representation of **Number of Samples** is **I32**.

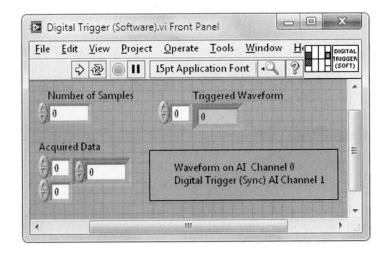

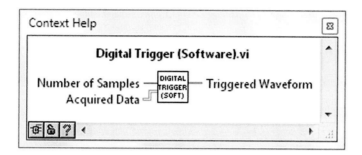

Next, code the block diagram of **Digital Trigger (Software)** as shown. **Index Array** and **Array Subset** are found in **Functions>>Programming>>Array.** For description of array operations, see Sections 6.8 and 9.6.

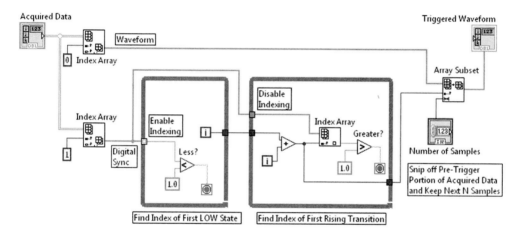

Finally, modify the block diagram of **Digital Oscilloscope (Express)** as shown below. Here, you must program **DAQ Assistant** to read both channels AI 0 and AI 1. As you configure this Express VI, it will instruct you how to select multiple AI channels as you define a task. Also, when you wire DAQ Assitant's **data** output to Digital Trigger (Software)'s **Acquired Data** input, the **Convert from Dynamic Data** icon will be inserted into the wire automatically. This icon converts the dynamic data type produced by Express VIs to the 2D floating-point numeric array used by Digital Trigger (Software). If you ever need to place **Convert from Dynamic Data** on a block diagram manually, it is found in **Functions>>Express>>Signal Manipulation.**

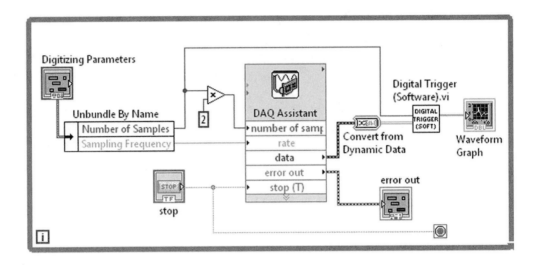

Once completed, connect a function generator's main and sync output to the AI 0 and AI 1 pins of your myDAQ. Configure the generator to produce a sine wave of frequency 50 Hz with amplitude less than 10 V, and then run **Digital Oscilloscope (Express)** with **Number of Samples** and **Sampling Frequency** set to *1000* and *10000*, respectively. Is the observed waveform on your scope properly triggered (i.e., stationary in appearance)?

CHAPTER 5

Data Files and Character Strings

5.1 ASCII TEXT AND BINARY DATA FILES

The outcome of any experiment is embodied in the data it generates. In computer-controlled experimentation, then, it is important to have the ability to save experimental data in disk files for future analysis. Hopefully, these data are stored in a convenient format so that they can be easily read by a user-written program or a commercially available software package that analyzes and displays the experimental results.

Computers commonly communicate alphanumeric data using the *American Standard Code for Information Interchange* (ASCII). In this coding scheme, seven bits of a byte are used to represent a character, while the byte's eighth "parity" bit is used for error checking when the data are received by a reader. The ASCII set of $2^7 = 128$ distinct states is used to represent all of the keyboard's alphanumeric characters as well as non-displayable control characters such as carriage return (CR) and line feed (LF). In the Extended ASCII set, all eight bits of the byte are utilized for character coding, yielding 256 distinct states.

A sequence of ASCII characters is called a *string*. Character strings, of course, can be used to represent text messages. However, strings also are useful in passing commands and data to and from the computer and stand-alone instruments. Additionally, as you will find in this chapter, strings can be used to store numerical data in files on a computer disk.

The closest approximation to a universally readable data file format is the *ASCII text file*. Storing data in this manner has the following advantages. Your files will be read accurately by computers of all manufacturers. Additionally, the files can be viewed with a word processor and, if desired, easily cut and pasted into a document for report generation. Finally, your data will be easy to import into commercially available data analysis software packages. A large number of these application programs prefer that your file is in tab-delimited ASCII text, or what is commonly called the *spreadsheet format*. In this format, tabs separate columns and end of line (EOL) characters separate rows as shown next.

$$
\begin{array}{l}
0.00 \rightarrow 0.4258\P \\
1.00 \rightarrow 0.3073\P \\
2.00 \rightarrow 0.9453\P \\
3.00 \rightarrow 0.9640\P \\
4.00 \rightarrow 0.9517\P
\end{array}
\qquad
\begin{array}{l}
\rightarrow = \text{Tab}\P \\
\P = \text{EOL}\P
\end{array}
$$

Opening this file using a spreadsheet program such as Microsoft Excel yields the following:

	A	B	C
1	0.00	0.4258	
2	1.00	0.3073	
3	2.00	0.9453	
4	3.00	0.9640	
5	4.00	0.9517	
6			

Alternately, numerical data may be stored as a *binary byte stream file*. This file format is simply a bit-for-bit image of the data that reside in your computer's memory. Thus, when data are exchanged between a binary file and computer memory, little or no data conversion is required, so you get maximum performance. Also, binary data files provide the most memory-efficient storage method for numerical data. To demonstrate this fact, consider the number of bytes necessary to store the integer *54321*. Since this number is less than $2^{16}=65636$, it will require only two bytes of memory in a binary-format file. However, in an ASCII text file, this number will occupy five bytes of memory, one byte for each of the characters *5, 4, 3, 2,* and *1*. The disadvantage of binary files is their lack of portability. They cannot be viewed by a word processor and they cannot be read by any program without a detailed knowledge of the file's format.

LabVIEW contains built-in VIs that facilitate data storage and retrieval using either the ASCII text or the binary file format. If ease of use and compatibility with a spreadsheet application program are of most concern, use the ASCII-based file storage. If, on the other hand, efficient memory use and high speed are desired, a binary file is the best choice.

In this chapter, you will learn how to store data in an ASCII-based spreadsheet file. You will write a program that generates data using your custom-made **Waveform Simulator** as a subVI and then store these data by implementing icons from **Functions>>Programming>>File I/O**.

5.2 STORING DATA IN A SPREADSHEET-FORMATTED FILE

Here's the plan for our data-storage VI: Use the **Waveform Simulator** program created in Chapter 3 to generate some data, and then store these data in a spreadsheet-formatted data file, which is saved on your computer's hard drive.

Start by opening a new VI. Using **File>>Save**, create a new folder called **Chapter 5** within the **YourName** folder, and then save this VI under the name **Spreadsheet Storage** in **YourName\Chapter 5**.

Now switch to the block diagram. First, you must place **Waveform Simulator** (as a subVI) on your block diagram. In summary, here's how to do that (you may wish to review Section 4.13, "**Placing a Custom-Made VI on a Block Diagram**," which illustrates the steps in the process). On the Functions Palette, choose the **Select a VI...** subpalette. The **Select the VI to Open** dialog window will appear. In the **Look in:** box, navigate to the **YourName\Chapter 3** folder. Once the file list for this folder is displayed, double-click on **Waveform Simulator**, or highlight it, and then click on the **OK** button in the dialog window. You will then be returned to the block diagram, where you can place the custom-written icon in a convenient position.

Next, with the ✎, create **Digitizing Parameters** and **Waveform Parameters** control clusters on the front panel by popping up on these terminals and using **Create>>Control**.

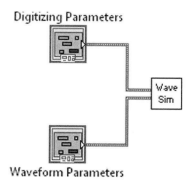

Then, place a **Write To Spreadsheet File.vi** icon (found in **Functions>> Programming>>File I/O**) on the diagram.

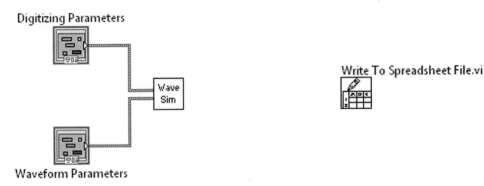

5.3   STORING A ONE-DIMENSIONAL DATA ARRAY

To understand how the **Write To Spreadsheet File.vi** icon functions, view its Help Window, as shown below. Here, you will find that this VI has the potential to function in numerous ways. As is generically true for Help Windows, default values for this icon's inputs are shown in parentheses. Also, inputs are labeled in boldface, plain, or dimmed text to denote whether each input is *required*, *recommended*, or *optional*, respectively. Required inputs (**Write To Spreadsheet File.vi** has none of these) must be wired, or else the icon will not execute. Recommended and optional inputs control features are available for your use, if desired. Optional inputs are less commonly used and so their labeling only appears when you click (as I have done) on the **Show Optional Terminals and Full Path** button in the Help Window's bottom left-hand corner.

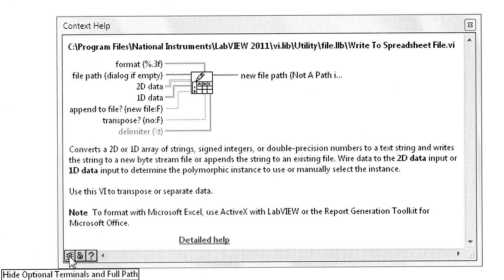

In its most basic operating mode, one simply provides a 1D array of numerical values to Write To Spreadsheet File.vi's **1D data** input. The VI then prompts the user for a file name and stores the values under this title using the spreadsheet format. By operating **Write To Spreadsheet File.vi** in this elemental way, we will gain insight into the necessity of its other available input options.

Place **Unbundle By Name** (found in **Functions>>Programming>>Cluster, Class, & Variant**) on your diagram, and then wire it to the **Waveform Output** terminal of Waveform Simulator. Using the ↳, expand **Unbundle By Name** so that output terminals for both elements of the cluster (**Time** and **Displacement**) are accessible, and then wire the **Displacement** terminal to the Write To Spreadsheet File.vi's **1D data** input.

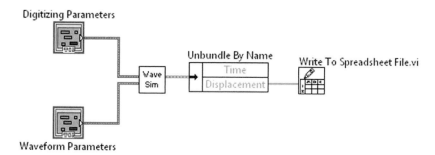

Now return to the front panel, tidy up the object arrangement, and then save your work. Program the inputs to create a small number (such as *10*) of displacement values for a 100 Hz sine wave of amplitude 4.0 with a sampling frequency of 1000 Hz. Then run your program.

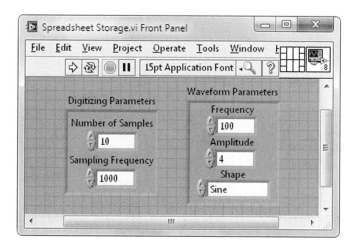

215

A dialog window called **Choose file to write** will appear. Navigate to the **YourName\ Chapter 5** folder in the **Save in:** box, name the file **Sine Wave Data.txt** in the **File name:** box, and then click on the **OK** button. The program will then complete execution by creating the requested spreadsheet file in **YourName\Chapter 5**.

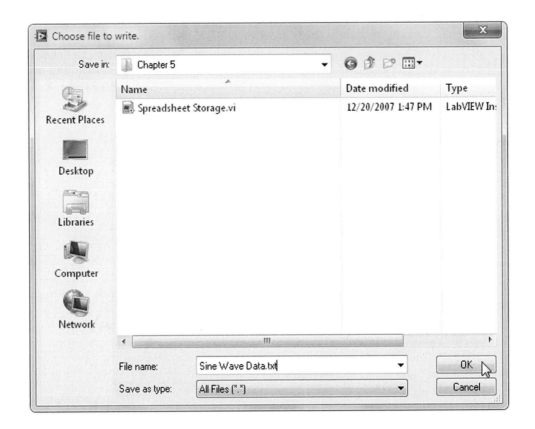

Use a word processor program to open the ASCII-based **Sine Wave Data.txt** spread-sheet file (in the Open dialog window, you may have to choose the file type as **All Files** to see **Sine Wave Data.txt** listed). Shown next is what I found by viewing this data file in Microsoft Word (the appearance of your file may vary slightly because of the particular tab settings of your word processor). Here I have activated Word's **Show ¶** command, which allows one to see the file's non-displayable characters such as Tab (→) and EOL (¶). If your word processor has a similar option, use it to view these usually hidden characters.

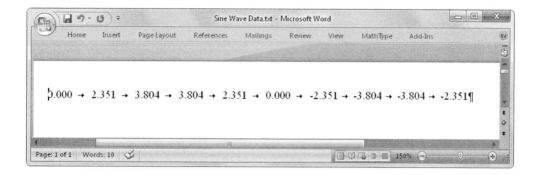

We see that the 10 sine-wave values are delimited by tabs and that the sequence concludes with an EOL. From our knowledge of the spreadsheet formatting convention, we conclude that a spreadsheet application program will interpret this string as a row of data, placing these 10 numerical values in a sequence of columns contained within a single row. Close **Sine Wave Data.txt** with the word processor, but don't save it (or else the program will possibly embed its own characteristic formatting statements within the file).

If available, use a spreadsheet application program to read your **Sine Wave Data. txt** file. To do this, your application program may give you the option of using an **Open** or an **Import** command. Either will work. You may have to tell the program that you're reading in a text file (as opposed to binary, Excel, etc.). Also, you may encounter a dialog window in which you must tell the program that your data is tab-delimited. Next, I show the result of reading the **Sine Wave Data.txt** file using Microsoft Excel. As expected, the 10 numerical values appear in 10 sequential columns of a single row.

	A	B	C	D	E	F	G	H	I	J	K
1	0	2.351	3.804	3.804	2.351	0	-2.351	-3.804	-3.804	-2.351	
2											
3											

5.4 TRANSPOSE OPTION

If you have a bit of experience with spreadsheet applications, the above illustration will cause some concern. In using such programs to generate plots, perform curve fitting, and other useful data analysis operations, the array of values for a particular quantity (e.g., a sine-wave displacement) is expected to reside in a single column, with the array's elements indexed by the row numbers. This organizational scheme for data is often called *column-major order.* We have seen that, in its default setting, the **Write To Spreadsheet**

File.vi icon does not function in a manner consistent with this convention. However, once this problem has been identified, it is trivial to remedy by consulting the icon's Help Window. Here you will find that the VI possesses a **transpose?** input, whose default value is FALSE. By wiring a TRUE **Boolean Constant** to this input, the file will be recorded in a "column-like" manner, rather than "row-like." Perform this wiring on your block diagram. An easy way to obtain the **Boolean Constant** is to pop up on the **transpose?** input and use **Create>>Constant**. The value of the **Boolean Constant** can be toggled from the FALSE (default) to the TRUE state using the 🖐.

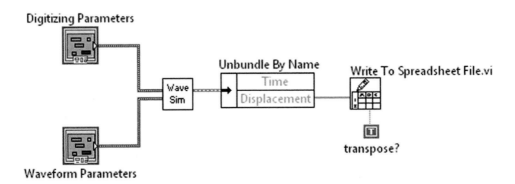

Run your program to produce a spreadsheet file of 10 sine-wave values. You can either reuse the file name **Sine Wave Data.txt** (by opening this name in the **Choose file to Write** dialog window, assuming the file is initially closed) or invent a new name. View this file using a word processing program. As shown next (using Microsoft Word), we see that each data value is separated from its neighbor by an EOL character. Thus, when read by a spreadsheet application, this array of values will be placed in a single column.

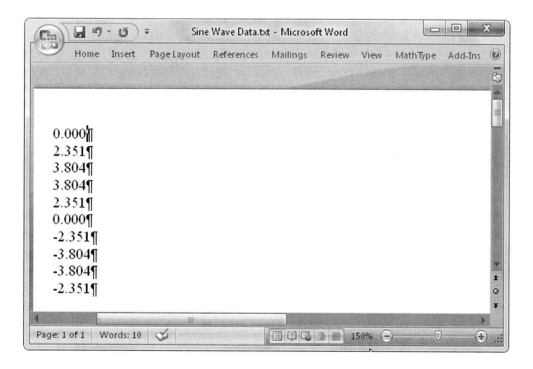

If available, open the data file in a spreadsheet application. In the next illustration (using Microsoft Excel), we see that our expectation of column-like data is fulfilled.

	A	B	C
1	0.000		
2	2.351		
3	3.804		
4	3.804		
5	2.351		
6	0.000		
7	-2.351		
8	-3.804		
9	-3.804		
10	-2.351		
11			

5.5 STORING A TWO-DIMENSIONAL DATA ARRAY

To analyze and/or plot **Waveform Simulator**'s sine-wave data using a spreadsheet application, one would need to import both its **Time** and its **Displacement** arrays into two spreadsheet columns. Let's see how, by implementing the **2D data** input of the **Write To Spreadsheet.vi**, it is easy to modify our **Spreadsheet Storage** program to produce the appropriate data file. We start with **Time** and **Displacement**, the two 1D array outputs from **Waveform Simulator**. Each of these 1D arrays has *N* elements, where *N* is determined by the value of the **Number of Samples** control. We now need to splice these two 1D arrays together to form a 2D array with the *N* values of **Time** and **Displacement** in the 2D array's first and second rows, respectively. Mathematicians would call this object a 2-by-*N* matrix; in LabVIEW it is called a 2D array.

To construct this 2D array, you will use an icon called **Build Array**. However, first you must open up some room for **Build Array** between **Unbundle By Name** and **Write To Spreadsheet File.vi**. To widen the gap between these two icons, click on the wire that connects them, and then delete it. Next, place the ↖ (it will appear as a cross in the empty diagram region) as shown below.

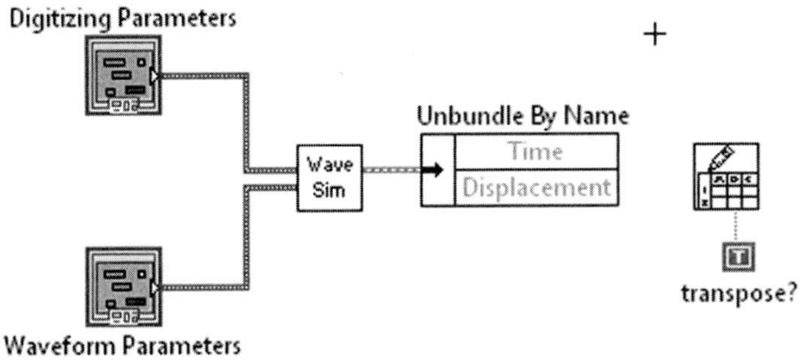

Click, and then drag cursor until a dotted rectangle frames both the **Write To Spreadsheet File.vi** and the **Boolean Constant** icon.

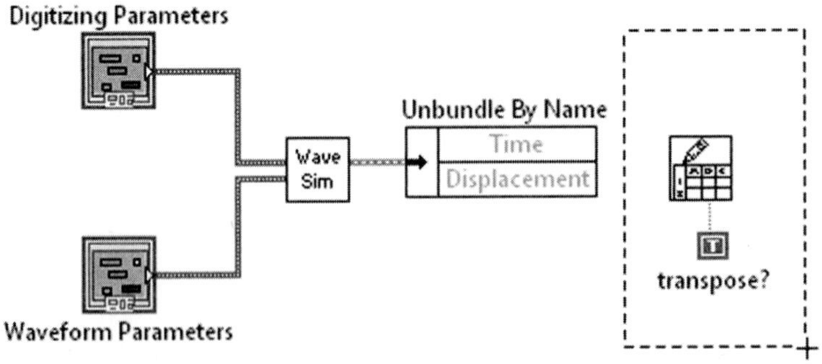

When you release the mouse button, both icons will be highlighted with a marquee. You can then move them to the right as one object by either using the keyboard's *<Right Arrow>* key, dragging with the Positioning Tool while depressing the *<Shift>* key (allowing only horizontal motion), or simply dragging with the 🖈.

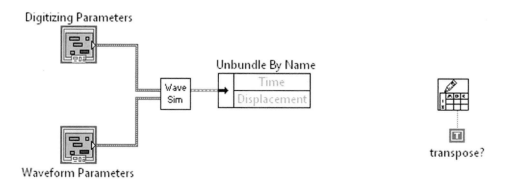

Now select the **Build Array** icon from **Functions>>Programming>>Array**. When you put this icon on the block diagram, it will initially appear with only one input.

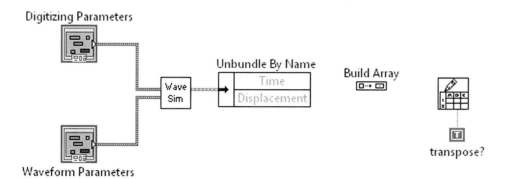

Place the 🖈 at the bottom center of the icon until it morphs into a resizing handle. Resize the icon so that it can accommodate two inputs.

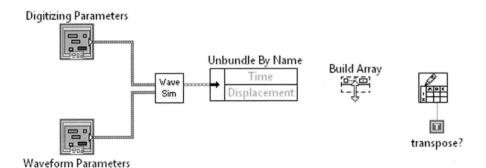

Then wire Unbundle By Name's **Time** and **Displacement** outputs to the top and bottom inputs of **Build Array**, respectively. **Build Array** will splice these inputs together to form the required two-dimensional array.

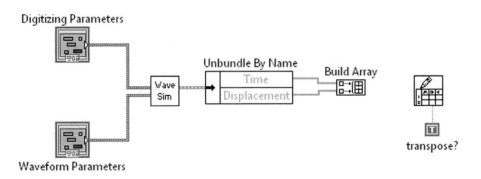

Wire the output of **Build Array** to the **2D data** input of Write To Spreadsheet File. vi. Note that this connection appears as a double wire, the LabVIEW designation for a 2D array.

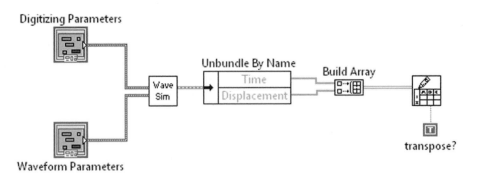

Return to the front panel and run your program with a small value such as *10* for **Number of Samples**. Then view your resulting data file using a word processor. It should appear like this.

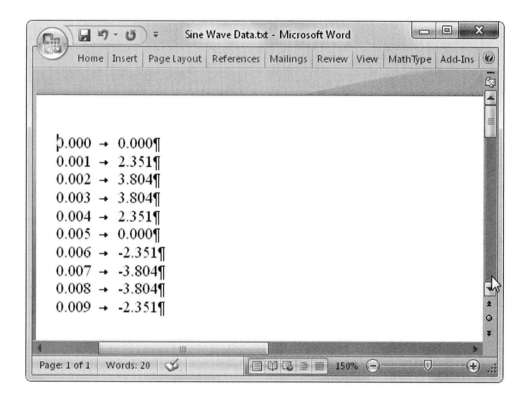

This formatting appears correct for importing the **Time** and **Displacement** data into two adjacent columns of a spreadsheet sheet. To verify this expectation, open your file with a spreadsheet application program. It should appear as below. If you're familiar with the operation of the spreadsheet application, you may enjoy plotting column *B* vs. column *A* or curve fitting the column *B* data to a sine function.

	A	B	C
1	0.000	0.000	
2	0.001	2.351	
3	0.002	3.804	
4	0.003	3.804	
5	0.004	2.351	
6	0.005	0.000	
7	0.006	-2.351	
8	0.007	-3.804	
9	0.008	-3.804	
10	0.009	-2.351	
11			

5.6   CONTROLLING THE FORMAT OF STORED DATA

Now let's explore some of **Write To Spreadsheet File.vi**'s other options. First, you may have noticed that, when viewing your data file, each of its numerical values was stored with a precision of three digits to the right of the decimal point. Is this the precision of the numerical values produced by the **Waveform Simulator** VI?

Try this test: Open **Waveform Simulator** by double-clicking on its icon within **Spreadsheet Storage**'s block diagram. Pop up on the **Time** and **Displacement** front-panel indicators, select **Display Format...**, and then increase **Digits of precision** from its default value to something large like *10* and disable the **Hide trailing zeros** option. You may then want to resize the indicators so that all of the new digits are visible. Run the VI with some appropriate choices for the **Digitizing Parameters** and **Waveform Parameters** inputs. You will then find that the array indicators display values of the precision that you requested. Now, with this modification in **Waveform Simulator**, rerun **Spreadsheet Storage**, and then view the new **Sine Wave Data.txt** file it produces. You will find that the stored values there still have a precision of three digits. Thus, we surmise that **Waveform Simulator** is not influencing the formatting of values within **Sine Wave Data.txt**. In fact, independent of its **Digits of precision** setting, the **Waveform Simulator** VI calculates highly accurate floating-point values, which it outputs to connecting wires. The **Digits of precision** parameter merely selects how many decimal places of these highly accurate values are to be displayed on the VI's front-panel indicator.

Since we have eliminated **Waveform Simulator** as a suspect, we must look elsewhere within **Spreadsheet Storage** to find where the decision is being made to truncate highly precise **Waveform Simulator**–produced values in the thousandths place upon storage in the file. A quick glance at Write To Spreadsheet.vi's Help Window identifies that the decision is made at this icon's **format** input. By leaving this input unwired, we have instructed the icon to store numerical values in the default format of *%.3f*. To understand this instruction, we need to know that the **format** string obeys the following syntax:

$$\% \text{ [WidthString] [.PrecisionString] ConversionCharacter}$$

Here, optional features are enclosed in square brackets. Each syntax element, and how it affects the form of the stored number within the data file, is explained in Table 5.1.

Thus, we see that *%.3f* instructs **Write To Spreadsheet File.vi** to store numbers in the fractional floating-point format with each number's total width auto-adjusted under the restriction of maintaining three digits to the right of the decimal point.

Place a **String Constant** on the block diagram, define it to be *%8.5f*, and then wire it to Write To Spreadsheet File.vi's **format** input. The easiest way to create the **String Constant**, of course, is to pop up on the **format** input and then select **Create>>Constant**. You can also find this icon in **Functions>>Programming>>String**.

TABLE 5.1

%	Character that denotes the beginning of a format specification.
WidthString (Optional)	Integer specifying the total number of ASCII characters to be used to represent the stored number. If your specified value of **WidthString** exceeds the number of characters actually required to represent a particular number, the excess portion of the string will be padded with spaces. If **WidthString** is not specified (or the stored number requires more characters than the specified **WidthString**), the numeric string will expand to as long as necessary to represent the stored number.
.PrecisionString (Optional)	Period (.) followed by an integer specifying the number of digits to the right of the decimal point in the stored number. If **.PrecisionString** is not specified, six digits to the right of the decimal point are stored. If only the period (.) appears and **PrecisionString** is missing or zero, no digits to the right of the decimal point are stored.
ConversionCharacter	Single character that specifies in what manner the number is to be stored with the following code: d decimal integer x hex integer o octal integer f floating point with fractional format e floating point with scientific notation

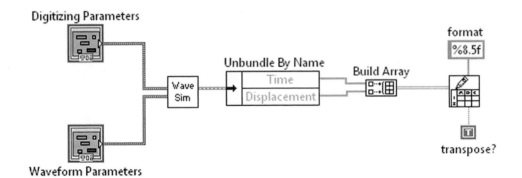

Run your program and store the data in a file. View the file using a word processor. Do the file's numerical values appear in the format that you specified? You might try exploring the effect of changing the **ConversionCharacter** to *d* (decimal integer) or *e* (floating point with scientific notation).

5.7   THE PATH CONSTANT AND PLATFORM PORTABILITY

Next, let's look more closely at the act of naming the data file. We note that, in some cases, it would be desirable to hardwire the name of a data file into our program code. Glancing back at the Write To Spreadsheet File.vi's Help Window, we find that this option is available via the **file path (dialog if empty)** input. Previously, we have left this input unwired; thus, the default **dialog if empty** (i.e., a dialog window prompting us for a file name) has been operational. Now we will take the alternate approach of programming in a specific file name.

As you probably know, the *path* to a particular file in your computer system is specified by a hierarchical directory structure. The path is typically denoted by a **drivename**, followed by **directory names** (commonly called **folders**), followed by the **filename** itself. In Windows, these various levels of the hierarchy are separated by backslashes (\); on the Macintosh, forward slashes (/) are used. Thus, on a Windows machine with a hard drive named **C:**, the file **Sine Wave Data.txt** within the directory **Users\Account\ Desktop\YourName\Chapter 5** has the path **C:\Users\Account\Desktop\YourName\ Chapter 5\Sine Wave Data.txt**. On a Mac with a hard drive named **Macintosh HD**, a similarly named file on the **Desktop** has the path **Macintosh HD/Users/Account/ Desktop/YourName/Chapter 5/Sine Wave Data.txt**. To hardwire a specific file name into a program, one might anticipate that the path appropriate to your computing platform (Windows or Macintosh) is simply inserted into a **String Constant** and then this object is wired to the **file path** input of the Write To Spreadsheet File.vi icon. However, in deference to the issue of *platform portability*, this expectation is slightly modified.

Platform portability is one of the outstanding features of LabVIEW programming. This means, for example, that you can write a LabVIEW program called **Widget** on a

Windows-based system and then send it (e.g., electronically, on a flash drive, or on a disk) to a colleague for use in a Macintosh-based laboratory. Once your colleague has copied the VI onto his or her Mac system, **Widget** can be opened from within LabVIEW running on the Macintosh system. While opening it, LabVIEW detects that **Widget** was written on another platform and recompiles it to use the correct instructions for the local processor. As seen above, one small part of this recompilation must include translating any hardwired paths that are contained within **Widget** into the form appropriate for the new system. To facilitate this translation process, a special LabVIEW object called the **Path Constant** (found in **Functions>>Programming>>File I/O>>File Constants**) is used for the specification of paths. As a consequence of dedicating the **Path Constant** to the sole use of enclosing paths, these special strings are distinguished from other character strings when the program is being ported.

By popping up on the **file path** input of Write To Spreadsheet File.vi, place a **Path Constant** on your block diagram, and then enter a path appropriate to your computing platform into it. Here, I show the resulting diagram on my Windows system. If you encounter an error when your VI tries to open the data file, carefully check that the file's path within the **Path Constant** is perfectly correct (to find the exact path for an existing Windows file, open the folder containing the file, and then right-click on the file's icon or name and select **Properties**).

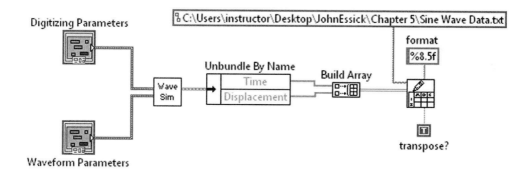

Save your work, and then run this program. Verify that it performs as expected by viewing the resulting data file with a word processor or spreadsheet application program.

5.8 FUNDAMENTAL FILE I/O VIs

Write To Spreadsheet File.vi is actually a high-level VI constructed from more fundamental File I/O functions. To gain more control of the file input/output process, let's first use the fundamental File I/O functions to write a program with the same capabilities as **Write To Spreadsheet File.vi** and then use our deepened understanding to enhance its functionality.

All data storage ("file output") operations must perform the following three steps: *open a file*, *write data to the open file*, *close the file*. Retrieving previously stored data ("file input") is performed in the same three steps, except that data are read from the open file (rather than written to it). In **Functions>>Programming>>File I/O**, the following three icons are available to perform the three required steps of data storage: **Open/Create/Replace File**, **Write To Text File**, and **Close File**. To understand how to wire these three icons together to accomplish data storage, we will first briefly describe the function of each individual icon.

The Help Window for **Open/Create/Replace File** is shown next. Given a file defined at its **file path** input, the job of this icon is either to open that existing file or to create it (if it doesn't already exist). If nothing is wired to **file path**, a dialog window will appear when the icon operates to allow one to input the file's path. Using the given file path as well as other configurational information such as the location within the file of the last I/O operation and degree of user access (e.g., read only), this icon also produces a "reference number" (called *refnum* for short). To pass this needed information to other File I/O icons, the refnum is available at the **refnum out** output terminal. Note that all of the inputs for this icon are (plain text) recommended or (dimmed text) optional inputs; if there were required inputs, they would be shown in boldface text. Thus, when **Open/Create/Replace File** executes, each input left unwired will take on its default value, which is shown in parentheses.

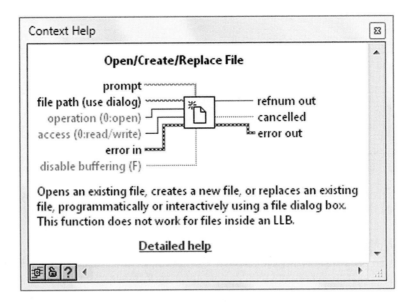

Write To Text File, whose Help Window follows here, performs the actual data storage. Once presented with the open file's *refnum* configurational information at its **file** input, this icon writes the ASCII string at its (required) **text** input into the open file. Additionally, the refnum is available at the **refnum out** output terminal.

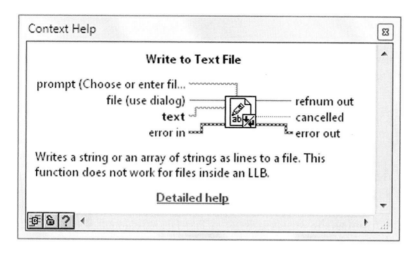

Finally, **Close File**'s Help Window is given below. This icon closes the file specified at its **refnum** input. Also, the file path, which is contained within the input refnum, is made available at the **path** output terminal.

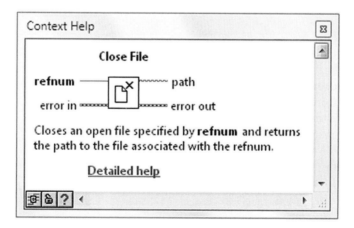

Note that all three of the above icons include error reporting via the error cluster, which appears at the **error in** and **error out** terminals.

The three-step data storage process is accomplished by wiring these three icons together as follows.

This wiring scheme takes advantage of the principle of LabVIEW programming called *data dependency*. Simply stated, data dependency means that an icon cannot execute until data are available at *all* of its inputs. In this diagram, all of the inputs to **Open/Create/Replace File** are wired (the **Path Constant** is explicitly wired to **file path**, and the other optional inputs are implicitly wired to their default values). So, when this diagram is run, **Open/Create/Replace File** executes immediately. Upon completion, **Open/Create/Replace File** outputs a refnum at its **refnum out** terminal, which is passed through the wiring to Write to Text File's **file** input. Because of data dependency, **Write to Text File** cannot execute until it receives the refnum from **Open/Create/Replace File.** Then, after **Write to Text File** completes its execution, it passes the refnum from its **refnum out** terminal to Close File's **refnum** input. Only then can **Close File** execute. Thus, through this programming scheme, we are assured that the icons will execute in the desired sequence: **Open/Create/Replace File** followed by **Write to Text File** followed by **Close File**.

Also, the correct manner of chaining together File I/O icons for error reporting is shown above. If an error does occur at one point in the chain, subsequent icons will not execute and the error message will be passed to the **General Error Handler.vi**, which will display the message in a dialog window. **General Error Handler.vi** is found in **Functions>>Programming>>Dialog & User Interface**. Rather than using this dialog window approach to error reporting, one could alternately pass error information to a front-panel error cluster for viewing (as you did in your Chapter 4 VIs).

With **Spreadsheet Storage** open, select **File>>Save As...** and then use it to create a new VI named **Spreadsheet Storage (OpenWriteClose)** in the **YourName\Chapter 5** folder. Modify the block diagram to appear as shown below. The required icons are found in **Functions>>Programming>>File I/O** and **Functions>>Programming>>Dialog & User Interface**.

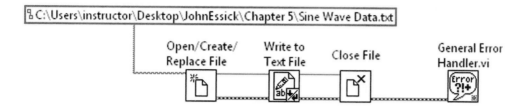

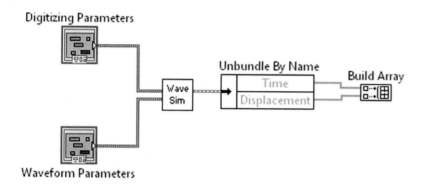

Pop up on the **operation** input of Open/Create/Replace File and select **Create>> Constant**. A block-diagram **Enumerated Constant** ("Enum") will be created. Place

the 🖑 atop the **Enum**, click the mouse, and then select the **replace or create** option. This option will be secured when you release the mouse button. With the path input at **file path**, this icon will then either create and leave open a new file (if a file with that path does not previously exist) or open the existing file with that path and instruct the subsequent icons to overwrite ("replace") its existing data with new data. This mode of operation is the same as that programmed into **Write To Spreadsheet File.vi**. Also, note that, by leaving the other inputs (**prompt, access, error in**) to **Open/Create/Replace File** unwired, these inputs will be automatically programmed to their default values (**no prompt, read/write, no error**).

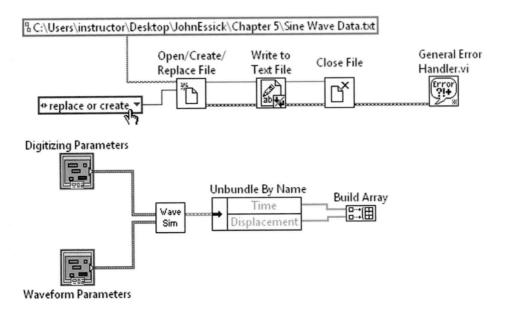

Finally, we need to supply **Write to Text File** with a string containing the 2D array of waveform values in spreadsheet formatting. This string can be created through the use of **Transpose 2D Array** (found in **Functions>>Programming>>Array**) and **Array To Spreadsheet String** (found in **Functions>>Programming>>String**). Complete the block diagram shown next, with the spreadsheet values formatted as *%8.5f.*

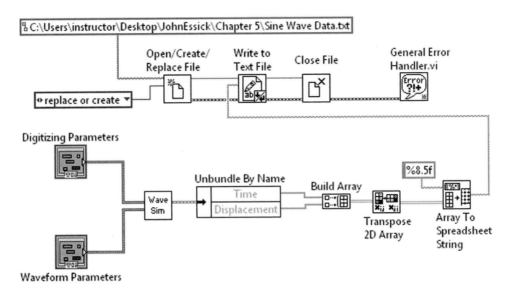

Save your work. Run **Spreadsheet Storage (OpenWriteClose)** with the same choice of values for its front-panel controls as when you ran **Spreadsheet Storage** a few minutes ago. Then use a word processor to verify that the file **Sine Wave Data.txt** appears the same as when it was created using **Spreadsheet Storage** (which implements **Write To Spreadsheet File.vi**).

If interested, place **Write To Spreadsheet File.vi** on a block diagram, and then open this VI by double-clicking on it. Look at its block diagram and open (by double-clicking) its subVI **Write Spreadsheet String.vi**. You will find that **Open/Create/Replace File**, **Write to Text File**, and **Close File** are the basic building blocks from which **Write To Spreadsheet File.vi** is constructed.

5.9 ADDING TEXT LABELS TO A SPREADSHEET FILE

In the previous exercise, you created an ASCII text file with two tab-delimited columns of (x, y) data, the first column containing a sequence of **Time** values and the next column containing the associated values of sine-wave **Displacement**. Because of the file format, you were able to use a word processor or spreadsheet application program to view the contents of this file. A desirable addition to this data file would be the ability to provide descriptive text at the top of each column, labeling each experimental quantity. More generally, one might wish to include all sorts of text statements within the file that would provide a complete record of the experimental conditions and choice of parameters under which the data were taken. Let's see how to provide labels at the top of each data column. Once this skill is in place, it's a small step to including any other desired text in the file.

In our text-based method of data storage, the entire set of (x_i, y_i) values is contained within one long ASCII character string. LabVIEW indexes these N values in the range of $i = 0$ to $N - 1$, so the structure of this spreadsheet-formatted string is as follows:

$$x_0 <Tab> y_0 <EOL> x_1 <Tab> y_1 <EOL> ...x_{N-1} <Tab> y_{N-1}<EOL>$$

One can then attach the labels **Time** and **Displacement** as the initial entries in the first and second columns, respectively, by simply adding these text characters as a spreadsheet-formatted prefix *Time <Tab> Displacement <EOL>*. The string will then appear as

$$Time <Tab> Displacement <EOL> x_0 <Tab> y_0 <EOL> x_1 <Tab> y_1 <EOL>$$
$$...x_{N-1} <Tab> y_{N-1}<EOL>$$

Thus, our task is to construct the label prefix, append the long ASCII string containing all the data values, and then store this entire string (termed a *byte stream*) in a file. Let's see how to do this.

The required byte stream is constructed using the **Concatenate Strings** icon, which is found in **Functions>>Programming>>String**. This icon splices together all its input strings into a single output string.

Delete the pink wire connecting **Array To Spreadsheet String** to **Write to Text File**, and then place **Concatenate Strings** on your block diagram as shown next. You will find that at first this icon has only two inputs. Position the ⬉ at the icon's bottom center, where ⬉ will morph into a resizing handle, and then expand the icon until it has five inputs.

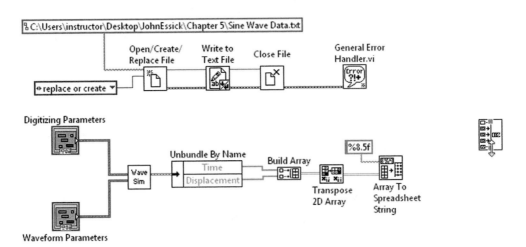

You will now construct the desired label prefix using the top four inputs of **Concatenate Strings**. Wire a **String Constant** (found in **Functions>>Programming>> String** or by popping up on the input) containing the text *Time* and *Displacement* to the first and third inputs of **Concatenate Strings**, respectively. Obtain a **Tab Constant** and an **End of Line Constant** icon from **Functions>>Programming>>String** and wire them to the second and fourth **Concatenate Strings** inputs, respectively.

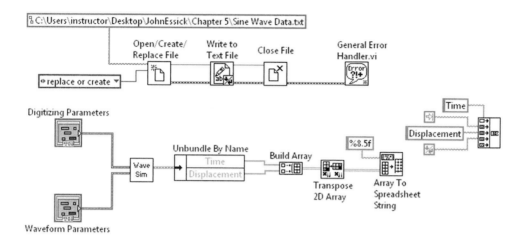

Complete construction of the desired byte stream by wiring Array To Spreadsheet String's **spreadsheet string** output to the fifth input of **Concatenate Strings**. Then, wire Concatenate Strings' **concatenated string** output to Write to Text File's **text** input.

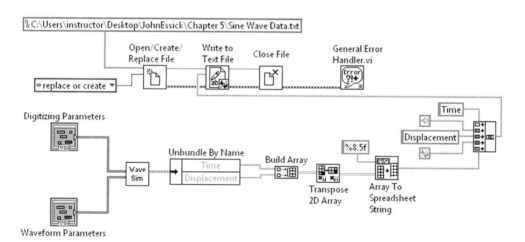

Save your work. Run **Spreadsheet Storage (OpenWriteClose)** with some choice of values for its front-panel controls. Then view the resulting file using a word processor. As shown next, with Microsoft Word I found the desired labels atop each column just as we had planned. You may have to play with your word processor's tab settings to get the labels to nicely align with the numerical data.

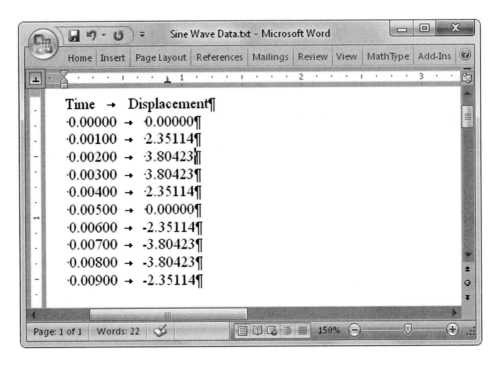

If available, try opening this text-based data file into a spreadsheet application program. In Excel, after a sequence of dialog windows within which I selected the option to begin data import at the file's first line, the data appeared as shown below. If the column labels are not needed, begin the data import at the file's second line. With the data input thusly, an experimentalist can then analyze and graph his or her experimental results in Excel.

	A	B	C
1	Time	Displacement	
2	0.000	0.00000	
3	0.001	2.35114	
4	0.002	3.80423	
5	0.003	3.80423	
6	0.004	2.35114	
7	0.005	0.00000	
8	0.006	-2.35114	
9	0.007	-3.80423	
10	0.008	-3.80423	
11	0.009	-2.35114	
12			

5.10 **BLACKSLASH CODES**

To conclude, let's explore a shortcut method for including non-displayable ASCII characters within a **String Constant**. The trick is to pop up on the **String Constant** icon and then select '\' **Codes Display** (as opposed to **Normal Display**) from the menu that appears. When this option is activated, LabVIEW will interpret characters immediately following a backslash (\) as a code for non-displayable characters in a manner listed in Table 5.2. You must use uppercase letters for the hexadecimal characters and lowercase letters for the special characters such as Tab and CR.

Let's demonstrate the convenience of backslash codes by employing them in the construction of the labeling string *Time <Tab> Displacement <EOL>*. To build this string, I must tell you that the EOL marker is platform dependent. The definitions for EOL peculiar to each computing systems are as follows:

Windows: Carriage Return, then Line Feed (\r\n)
Macintosh: Carriage Return (\r)

Thus, by enabling the '\' **Codes Display** option, the entire column-labeling text can be constructed by inserting *Time\tDisplacement\r\n* within a single **String Constant**, assuming we are working on a Windows system. However, if in the future an attempt is made to port this program to another platform, the explicit expression of the EOL character will cause problems. Because of this portability issue, it is good programming practice to take the following slightly modified approach to the string construction: Enclose *Time\tDisplacement* within a single **String Constant**, and then concatenate this object with the **End of Line** icon found in **Functions>>Programming>>String**. The **End of Line** icon is a "smart" EOL character. At compile time, it checks which platform is currently being implemented and automatically generates the appropriate EOL character(s), thus infusing the program with portability.

TABLE 5.2

Code	LabVIEW Interpretation
\00–\FF	Hexadecimal value of an eight-bit character
\b	Backspace (ASCII BS, equivalent to \08)
\f	Formfeed (ASCII FF, equivalent to \0C)
\n	Line Feed (ASCII LF, equivalent to \0A)
\r	Carriage Return (ASCII CR, equivalent to \0D)
\t	Tab (ASCII HT, equivalent to \09)
\s	Space (equivalent to \20)
\\	Backslash (ASCII \, equivalent to \5C)

Here is a step-by-step procedure for modifying **Spreadsheet Storage (OpenWrite-Close)** to accomplish the above. Start by replacing all of the labeling-string construction code by a **Concatenate Strings** icon with three inputs. Wire an empty **String Constant** to the top input (by popping up and selecting **Create>>Constant**), an **End of Line** icon (from **Functions>>Programming>>String**) to the middle input, and Array To Spreadsheet String's **spreadsheet string** output to the bottom input.

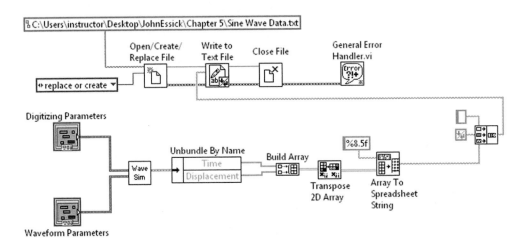

Next, pop up on the empty **String Constant** and enable backslash coding by selecting '\' **Codes Display** from the pop-up menu. Using the ✎ or 🅰, enter *Time\t Displacement* into the **String Constant**, and then press the numeric keypad's *<Enter>*.

When you are finished, you can resize and reposition this object using the ↖, if needed. Then wire the **Concatenate Strings** output to the Write Character To File.vi's **character string** input.

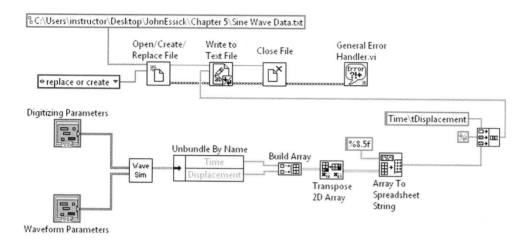

Save your work, and then run the program. Use a word processor to verify that backslash coding produces a data file with the desired column labeling.

DO IT YOURSELF

Write a VI called **Spreadsheet Read**, which reads and displays data from a two-column spreadsheet-formatted disk file. Name the first and second columns of data within the spreadsheet file *X Array* and *Y Array*, respectively. Construct **Spreadsheet Read** to do the following: Open the desired two-column spreadsheet file, read its contents, and then plot *Y Array* vs. *X Array* on an **XY Graph** as well as display these arrays in an indicator cluster labeled **output cluster**. Design an icon for your VI and assign its connector terminals consistent with the following Help Window. Save this VI in **YourName\ Chapter 5**.

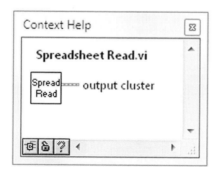

When **Spreadsheet Read** is completed, run **Spreadsheet Storage** with its front-panel controls set to the values shown below. After **Spreadsheet Storage** executes, there will be a two-column spreadsheet data file named **Sine Wave Data.txt** in the **YourName\Chapter 5** folder with its X Array and Y Array containing Time and Displacement values for five cycles of a 50 Hz sine wave of amplitude 5.

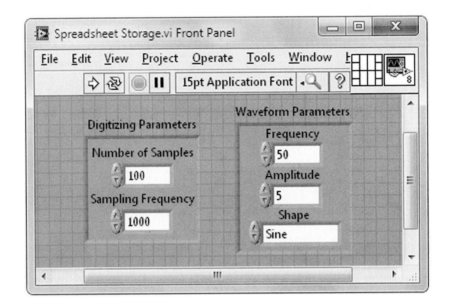

Next, use **Spreadsheet Read** to read **Sine Wave Data.txt**. After successfully running this VI, the front panel of **Spreadsheet Read** should appear as below.

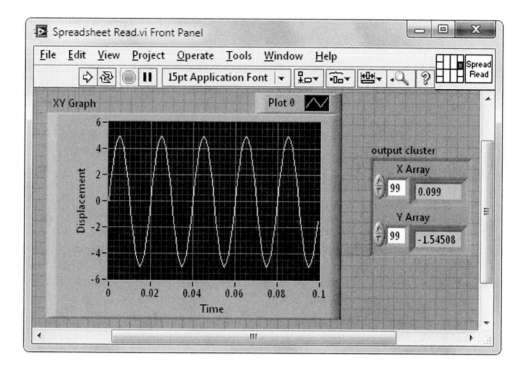

Here are three helpful tips:

- Use **Read from Spreadsheet File.vi** (found in **Functions>>Programming>>File I/O**) to open and read the contents of **Sine Wave Data.txt**. Leave **file path** unwired so that the user will be prompted for the file name via a dialog window. The data will be output as a 2D array at the **all rows** output.

- A column of a 2D array can be "sliced off" (that is, isolated as a 1D array) using **Index Array** (found in **Functions>>Programming>>Array**). When first placed on a block diagram, **Index Array** has just one **index** input (which appears as a small black box), as shown below at the left. However, when a 2D array is wired to its **n-dimensional array** input, two **index** inputs appear as shown below in the middle. The top and bottom **index** inputs are the row and column indices of the 2D array, respectively. To slice off the first column (indexed as column 0), just wire a **Numeric Constant** of value *0* to the bottom (column) **index** input, as shown below at the right. The top (row) **index** input is left unwired. The **subarray** output of Index Array will then be a 1D array containing the values of column 0 from the input 2D array.

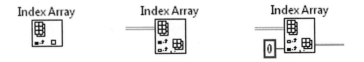

Thus, to slice off both columns of a two-column 2D array, the required subdiagram is as follows.

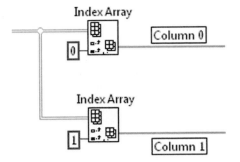

- Create **output cluster** from the block diagram by popping up on an appropriate block-diagram object and using **Create>>Indicator**. Then, on the front panel, pop up on the index display of each Array Indicator within **output cluster**, select **Visible Items>>Label**, and then enter its appropriate name.

PROBLEMS

1. Write a program called **Read Spreadsheet Text**, which uses the icons **Open/Create/Replace File**, **Read from Text File**, and **Close File** to read the contents of a spreadsheet-formatted text file and display the resulting string in a front-panel **String Indicator** as well as a decoded numeric representation in a numeric array indicator as shown in the following illustration.

   (a) Create a spreadsheet-formatted text file by running **Spreadsheet Storage** with its front-panel controls set to create 10 samples of a 50 Hz sine wave. After **Spreadsheet Storage** executes, there will be a two-column spreadsheet text file with ten rows named **Sine Wave Data.txt** in the **YourName\Chapter 5** folder.

   Next, read **Sine Wave Data.txt** by running **Read Spreadsheet Text**. Does the text displayed in the front-panel indicator appear as expected? Pop up on this indicator and select '**\' Codes Display**. Non-displayable ASCII characters will now be visible in the indicator. Explain the pattern of characters that you now see.

   (b) On the block diagram, use the icon **Spreadsheet String To Array** (found in **Functions>>Programming>>String**) to convert the text string to a 2D double-precision floating-point array with a format consistent with that used in creating **Sine Wave Data.txt**. Display this 2D array in a front-panel indicator as shown below. Run **Read Spreadsheet Text**. Is the array of data properly represented in the front-panel array indicator?

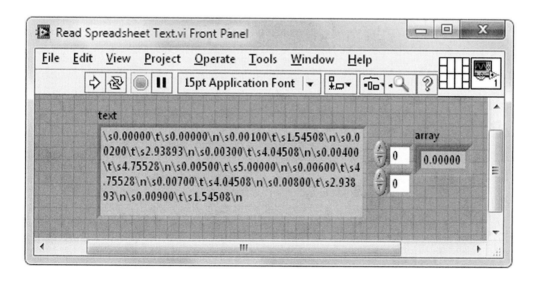

2. A LabVIEW programmer modifies **Spreadsheet Storage** as shown next in an attempt to add column labels to the stored spreadsheet data.

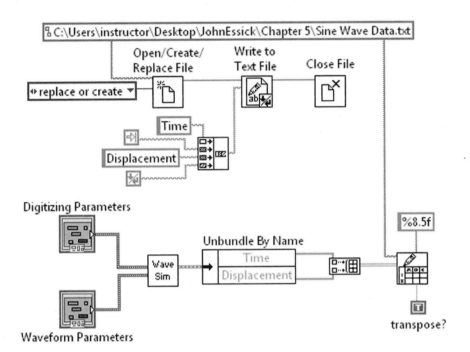

With **Spreadsheet Storage** open, use **File>>Save As...** to create a new VI called **Spreadsheet Storage with Labels** in **YourName\Chapter 5**. Modify the block diagram of this VI so that it replicates the code shown above. Run this program and demonstrate (with the help of a word processor) that it does not work as hoped.

(a) Explain why this code does not accomplish its stated goal.
(b) Fix the block diagram code so that it functions properly. You should just have to change one wire.
(c) Once the code functions properly, explain why this approach to achieving column labeling is less efficient than that used in **Spreadsheet Storage** (OpenWriteClose).

3. Write a For Loop–based program called **Four Text Files**, which (with no run-time input from a user) creates four separate files named **Text0.txt**, **Text1.txt**, **Text2.txt**, and **Text3.txt**, each containing a unique text message, and stores these files in the **YourName\Chapter 5** folder. Create one file during each iteration of the For Loop. The file's path can be constructed using **Build Path** (found in

Functions>>Programming>>File I/O) as well as **Format Into String** (found in **Functions>>Programming>>String**). This latter icon can be expanded so it has multiple inputs, for example, one for an **I32** integer and another for the string **.txt**. Make the text messages in **Text0, Text1, Text2,** and **Text3** be *This is file 0*, *This is file 1*, *This is file 2*, and *This is file 3*, respectively.

4. Write a VI called **Binary Storage**, which stores the 2D array of **Time** and **Displacement** values produced by **Waveform Simulator** in a binary-formatted file named **Sine Wave Data.txt**. Much of the program you will write is similar to **Spreadsheet Storage (OpenWriteClose)**, so you may want to begin there and create **Binary Storage** using **File>>Save As…** and save it in **YourName\Chapter 5**. The front panel of **Binary Storage** should look as follows.

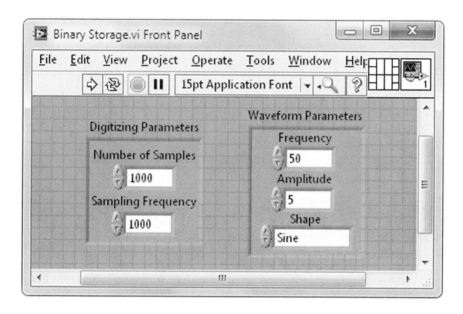

On the block diagram, implement the required three-step data storage process using **Open/Create/Replace File**, **Write to Binary File**, and **Close File** found in **Functions>>Programming>>File I/O**. Wire a TRUE **Boolean Constant** to Write to Binary File's **prepend array or string size?** input. This selection instructs LabVIEW to begin the binary file with an eight-byte header, which records the number of rows and columns into which the data are arranged.

(a) Run **Binary Storage** with its front panel controls programmed as shown above. Find the icon for the binary-formatted **Sine Wave Data.txt** file in the **YourName\Chapter 5** folder and determine its size (in kilobytes). The file's exact size is found by right-clicking on the icon and selecting **Properties**. Given the front-panel input values and binary formatting, explain why **Sine Wave Data.txt** has the size it does.

(b) Open **Sine Wave Data.txt** with a word processor. Offer a qualitative explanation for what you observe. Close **Sine Wave Data.txt**.

(c) Run **Spreadsheet Storage** with its front-panel controls programmed as shown in the previous illustration. Determine the size (in kilobytes) of the spreadsheet-formatted **Sine Wave Data.txt**. You will find that this file is now slightly larger than its previous binary-formatted size. Given the front-panel input values, explain why **Sine Wave Data.txt** has the size it does under spreadsheet formatting.

(d) By changing a particular parameter on the block diagram of **Spreadsheet Storage**, you can cause spreadsheet-formatted **Sine Wave Data.txt** to be approximately twice as large as binary-formatted **Sine Wave Data.txt**. Predict the block-diagram change needed to result in this file-size doubling. Make this change and verify your prediction.

5. Create a binary-formatted file called **Sine Wave Data.txt** by running **Binary Storage** (the VI written in Problem 4) with the choice of front-panel control values shown next. In **Binary Storage**, make sure that the Write to Binary File's **prepend array or string size?** input is wired TRUE, so that an eight-byte informational header is inserted at the beginning of the **Sine Wave Data.txt** file.

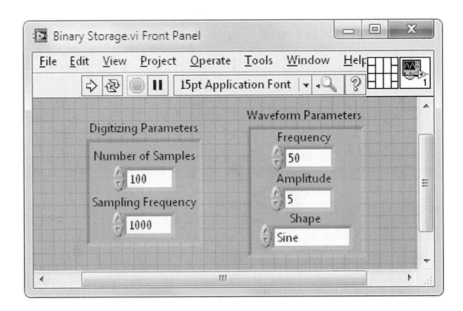

Your assignment is to write a VI called **Binary Plot**, which reads the binary-encoded **Time** and **Displacement** data within **Sine Wave Data.txt** and then plots these data on an **XY Graph**. When the completed VI is run, its front panel should look as follows.

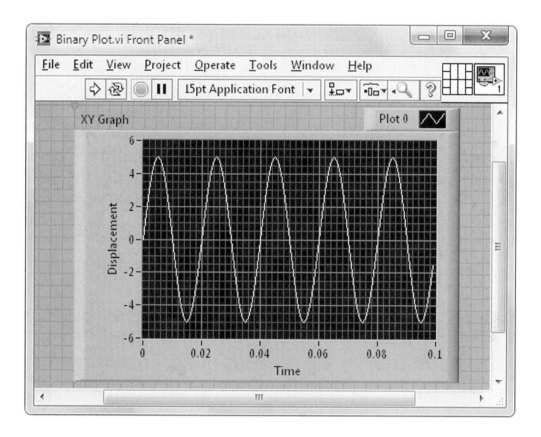

Here are some guidelines to follow when writing your program:

- Retrieval of data from a file is a three-step (*open*, *read*, *close*) process. These three steps are accomplished by configuring the appropriate File I/O icons (found in **Functions>>Programming>>File I/O**) as shown next. The eight-byte header at the beginning of the **Sine Wave Data.txt** file informs **Read from Binary File** about the size of the data file to be read as well as the required form (i.e., the number of rows and columns) of the data array output.
- The fact that the data consist of a 2D array of double-precision floating-point numerics is communicated to **Read from Binary File** by wiring a 2D DBL

Array Constant to its **data type** input as shown in the next diagram. This constant inputs no data to **Read from Binary File**, but rather simply specifies the desired data type for the output data. To form this Array Constant, first place an **Array Constant** shell from **Functions>>Programming>>Array** on the block diagram. Then, place a **Numeric Constant** (from **Functions>>Progra mming>>Numeric**) within the Array Constant shell, pop up on the **Numeric Constant**, and select **Representation>>Double Precision (DBL)**. Finally, pop up on the Array Constant's **index display** and select **Add Dimension**. The icon is now a 2D Array Constant, where the top and bottom index displays show the row and column index, respectively.

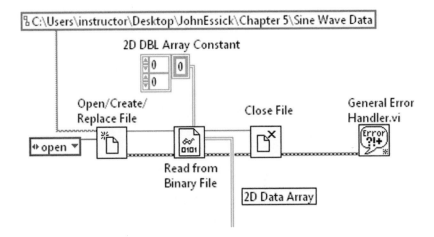

- The 2D data array emanating from Read from Binary File's **data** output has two rows, where the first and second rows contain the **Time** and **Displacement** values, respectively (note that this 2D array has never been transposed, in contrast to the 2D array involved in the spreadsheet VIs that we wrote). Each of these rows must be sliced off in preparation for plotting on an **XY Graph**. To accomplish this slicing operation, use **Index Array** (from **Functions>>Programming>>Array**) with a **Numeric Constant** of appropriate value wired to its top (row) **index** input. Leave the bottom (column) index input unwired (see the helpful tips section at the end of the Do It Yourself project in this chapter).

6. Write a program called **Palindrome Detector**, which determines whether an input word, phrase, or sentence reads the same backward as forward. As shown below, a **Round LED** indicator is lit on the front panel when the input string is determined to be a palindrome.

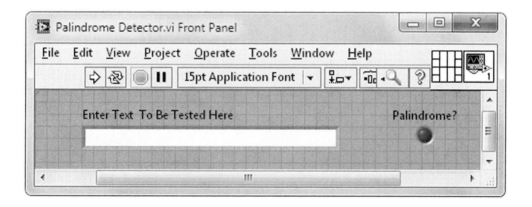

Construct your block diagram using icons from **Functions>>Programming >> String** and its subpalette **Additional String Functions**. Before comparing the equality of the input string with its reverse, first change the input to all lowercase characters and also remove all spaces. The spaces can be removed by replacing each space character with an **Empty String Constant**.

When completed, test **Palindrome Detector** with the following inputs: *radar*, *Able was I ere I saw Elba*.

7. The day of the week for a given (Gregorian calendar) date can be found using the following formula called Zeller's congruence:

$$day = \mod (I, 7)$$

where

$$I = D + floor\left[(M+1)2.6\right] + Y + floor\left[\frac{Y}{4}\right] + floor\left[\frac{C}{4}\right] + 5C$$

Here, the *floor* [*x*] function truncates its argument *x* to the closest lower integer and the modulo function mod (*I*, 7) equals the remainder when the integer *I* is divided by 7. The quantities appearing in Zeller's convergence are defined as follows:

day = day of the week (0 = Saturday, 1 = Sunday, ..., 6 = Friday)
  D = day of the month
  M = month (3 = March, 4 = April, 5 = May, ..., 10 = December)
  C = century given by *floor* (*year*/100)
  Y = year within century given by mod (*year*, 100)

For the months of January and February, $M = 13$ and $M = 14$ for the previous year, i.e., *year* → *year* − 1.

Write a program called **Day of the Week**, which implements Zeller's convergence to find the day of the week for a given date. The front panel of this VI should appear as shown below. The date is input in a **String Control** in the format *mm/dd/yyyy*, and the result is displayed as text (e.g., *Thursday*) in a **String Indicator**.

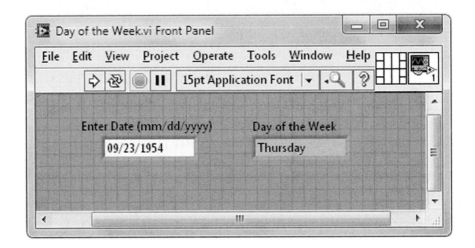

On the block diagram, use icons from **Functions>>Programming>>String** and its subpalettes (e.g. **Match Pattern, Decimal String To Number**) to parse the input string and convert it into numeric representations of *M*, *D*, and *year*. Also, create a seven-element array of strings (e.g., using **Build Array** or a block-diagram **Array** shell containing a **String Constant**) where the index-zero, index-one,..., index-six elements are the strings *Saturday*, *Sunday*,..., *Friday*, respectively. Then, use **Index Array** to select the correct element from this array to output to **Day of the Week**. Finally, the icon **Select** from **Functions>>Programming>> Comparison** may come in handy.

Verify your VI works correctly by finding the day of the week for July 4, 1776 (Thursday), and January 1, 2000 (Saturday).

8. Computer-controlled scientific instruments often report their measurement results in the form of an ASCII string. As an example, assume a multimeter sends the string *VOLTS DC +1.345E+02* in which *VOLTS DC* identifies the type of measurement being reported and *+1.345E+02* is the actual measured value. For this measured value to be useful within a program (e.g., plotted on a chart, input to a calculation), the string received from the multimeter must be "parsed" (i.e., the identifier portion separated for the measured value portion) and the measured value portion converted from its ASCII string representation to a numeric format.

Write a program called **Parse and Convert Multimeter String**, in which *VOLTS DC +1.345E+02* is input in a front-panel **String Control**. On the block diagram, this string is then parsed appropriately and the measured value portion is converted to the double-precision floating-point format. The needed icons for these operations are found in **Functions>>Programming>>String** and its subpalette, **String/Number Conversion**.

When properly functioning, the front panel of **Parse and Convert Multimeter String** should appear as shown below. In the String Control, '\' **Codes Display** is selected so that non-displayable characters are made visible.

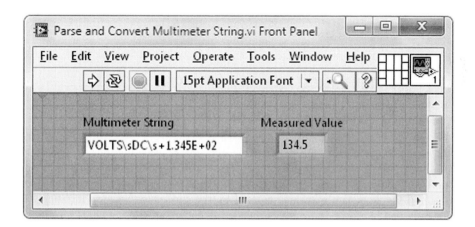

CHAPTER 6

Shift Registers

6.1 SHIFT REGISTER BASICS

A looping structure allows one to implement a common requirement in computer programming—repeating the same operation many times. In earlier chapters, we found that LabVIEW provides two such looping structures, the For Loop and the While Loop. We discovered that, through the creation of a tunnel at its border, each of these loop structures possesses a form of memory. By activating the tunnel's auto-indexing feature, the loop (once it has completed its full execution) will output an array in which is stored the sequence of values created by the succession of loop iterations. Often, however, another form of memory is needed—that which interconnects successive loop iterations. In this guise, a value created by the previous iteration is transferred for use within the calculations of the present iteration. This form of memory is called a *local variable* and can be accomplished in LabVIEW's looping structures via the creation of a *shift register*.

Shift registers are created through the use of a pop-up menu at a loop boundary. For example, you may create a shift register by popping up on a For Loop's border and then selecting **Add Shift Register**.

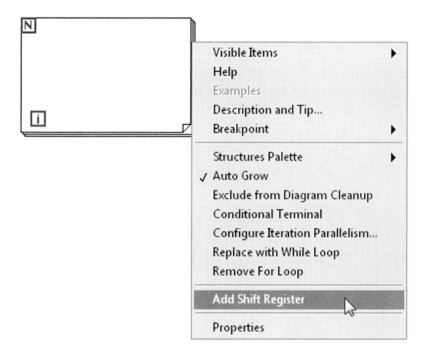

When you release the mouse button, the shift register will appear. It is comprised of a pair of terminals directly opposite each other on the vertical sides of the loop border.

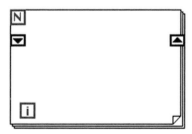

The right terminal stores a value upon completion of an iteration. This stored value is then shifted to the left terminal for use in calculations during the next iteration. Such a feature is useful, for example, in summing the components of a quantity calculated over the span of the entire set of loop iterations.

You can configure the shift register to remember values from more than one previous iteration. To accomplish this feat, pop up on the shift register's left terminal and select **Add Element** from the menu.

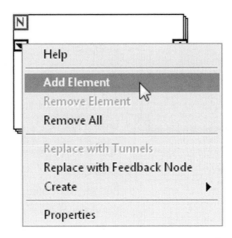

Upon this selection, a second left terminal will appear.

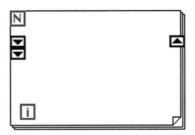

By repeating this operation, you can create as many left terminals as desired.

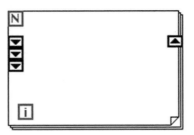

In the previous illustration, we have created three left terminals that will function as follows. When a subdiagram within the For Loop (not shown) is calculating some quantity during the *i*th loop iteration, the top left terminal contains the value of this quantity calculated during the $(i - 1)$th iteration, the middle left terminal contains the value from the $(i - 2)$th iteration, and the bottom left terminal that from the $(i - 3)$th iteration. Over the course of the complete For Loop execution, this set of left terminals will behave as a First In, First Out (FIFO) digital shift register, if that is a familiar concept to you.

6.2 QUICK SHIFT REGISTER EXAMPLE: INTEGER SUM

We'll first explore the use of local variables in LabVIEW programming with the following quick example. A famous story begins like this: In the late 1700s, a German elementary-school teacher wanted to keep the kids in his unruly classroom occupied for 30 minutes so he gave them the following "busy work" assignment: Find the sum of all the integers from 1 to 100, that is, determine $S = 1 + 2 + 3 + \ldots + 98 + 99 + 100$.

Let's show that this assignment would have taken the kids much less than 30 minutes if they had had LabVIEW available. To write this summing program, first open a blank VI, use **File>>Save** to create a folder named **Chapter 6** within the **YourName** folder, and then save this new program under the name **Sum to 100** in **YourName\Chapter 6**. As shown next, on the front panel place a single **Numeric Indicator** labeled **Value of Sum**, whose representation is **I32**.

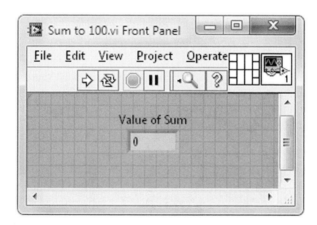

Now switch to the block diagram. Add a **For Loop** to your diagram and place a shift register on its boundary. To add the shift register, pop up on either of the For Loop's vertical borders and select **Add Shift Register**. You can then reposition the terminal pair using the ↖, if desired.

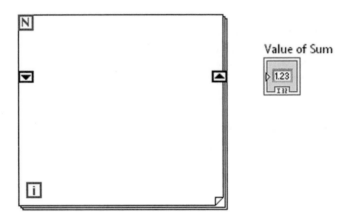

Value of Sum

We are going to use the For Loop to compute the sum S of the integers I where $I = 1, 2, 3, \ldots, 100$. In a text-based language, we would write something along the lines of the following:

$$S = 0.0$$
$$\text{For } i = 0, 99$$
$$I = i + 1$$
$$S = I + S$$

To implement this code in LabVIEW, initialize the shift register to zero by wiring a **Numeric Constant** (defined as 0 with representation **I32**) to its left terminal. Place an **Add** icon within the loop and wire one of its inputs and its output to the shift register's left and right terminals, respectively.

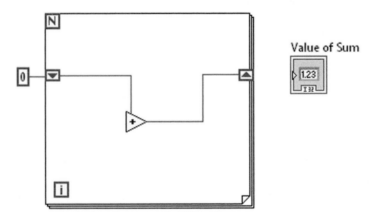

Value of Sum

Complete the block diagram as shown next. The convenient **Increment** icon, which takes the input i and then outputs $i + 1$, is found in **Functions>>Programming>> Numeric**. Save your work.

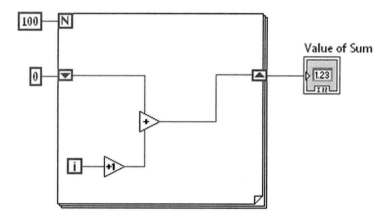

The above diagram works like this. The For Loop steps through the set of integers from *1* to *100*, accumulating the sum of these numeric values as it goes. During a particular iteration, the shift register's left terminal provides the sum S accumulated up through the last iteration. The new integer I is added to this sum and the result is stored in the shift register's right terminal for use in the next loop iteration. That is, on the first loop iteration when equals 0, the integer $I = 1$ is constructed by **Increment** and added to the accumulated sum S, which comes from the left shift register and has the initialized value of zero. The new value of $S = 1$ is stored in the right shift register. On the second iteration when equals 1, the new integer $I = 2$ from Increment is added to the accumulated sum stored in the left shift register and the new sum value of $S = 3$ is stored in the right shift register, and so on until this process has been repeated 100 times. When the For Loop completes its operation, the accumulated sum of the integers, which is contained in the shift register's right terminal, is output to the **Value of Sum** indicator for display on the front panel.

Return to the front panel. Run your VI to find the integer sum. Did writing this program take you less than 30 minutes?

The end of our famous story is this: Unfortunately for the elementary-school teacher, the young mathematical genus Karl Gauss was one of his students. In less than a minute after receiving the integer-summing assignment, Gauss raised his hand and announced the correct answer. The flabbergasted teacher asked how he had arrived at the answer so quickly and Gauss explained his reasoning as follows. First, he noted that the series from 1 to 100 can be divided into two 50-element sets: 1 to 50 and 51 to 100. Next, starting at the beginning of the first set and the ending of the second, 50 pairs of integers can be formed, where the sum of each pair equals 101. That is, $1 + 100 = 101$, $2 + 99 = 101$, $3 + 98 = 101$, and so on. Thus, given 50 pairs of integers, where each pair sums to 101, the sum of all of the pairs is

$$S = 50 \times 101 = 5050$$

Hopefully, this is the same answer your **Sum to 100** VI gave.

By the way, Gauss's trick works for summing any series of integers $I = 1, 2, 3, \ldots, N$ as long as N is even (so that all the integers in the series can be paired). Then, one has $N/2$ pairs of integers, with each pair summing to $N + 1$, so the sum S of all of the integers is

$$S = \sum_{I=1}^{N} I = \left(\frac{N}{2} \right)(N+1) = \frac{N(N+1)}{2}$$

Even better than that, because of a little bit of mathematical luck, $S = \sum_{I=1}^{N} I = N(N+1)/2$ even if N is odd. You might have fun proving that fact.

6.3 NUMERICAL INTEGRATION AND DIFFERENTIATION USING SHIFT REGISTERS

In the rest of this chapter, we will demonstrate the use of local variables while writing two of our most ambitious programs to date. In the first program, a shift register will be implemented to numerically integrate a given discrete data set, and then, in the second program, a shift register that recalls values from two past iterations will be used to numerically differentiate this same data set. Each of these programs will be carried out through the coordination of several smaller component programs. We will use this opportunity to illustrate one of the important "best practices" of LabVIEW programming—modularity.

In the following sections, we will first write a VI called **Power Function Simulator**, which will be used as a subVI in our subsequent programs to provide the needed discretely sampled data set. Then, after a brief review of numerical integration via the Trapezoidal Rule, a VI will be written to carry out and evaluate the convergence property of this theoretical procedure. Finally, the theory and LabVIEW implementation of numerical differentiation will be investigated, with an eye toward compartmentalizing distinct component tasks as subVIs within a higher-level program.

6.4 POWER FUNCTION SIMULATOR VI

Imagine some physical system for which one of its measurable quantities y (for example, temperature or pressure) is a function of the continuous variable x (e.g., position or time). That is, y is perfectly described by $y = f(x)$, where f is an analytic function. If an experimenter sets out to determine the function f, he or she will have to confront this reality: It is impossible to design an experiment that samples the quantity x and its associated y-values in a continuous manner. One must instead settle for discretely sampling x a total of N times and recording each associated y-value. Thus, the experiment will result in a set of (x, y) data samples, where $x = x_0, x_1, x_2, \ldots, x_{N-1}$ and $y = f(x_0), f(x_1), f(x_2), \ldots, f(x_{N-1})$.

In the next exercise, we will study the use of shift registers by writing a program that numerically integrates a discretely sampled data set. In an actual computer-controlled experiment, the discretely sampled data set might be obtained through the use of LabVIEW's built-in data acquisition (DAQ) VIs. Given this set, an integration VI might be used to perform some appropriate analysis on the data. Here, we will use software to generate a simulated set of discretely sampled (x, y) data that follows the power function $y = ax^b$, where a and b are the *prefactor* and *power*, respectively. For our present purposes, it is actually beneficial to use simulated data. In contrast to a real experiment, we will know the exact functional form by which the data were generated and this power-function form is easy to integrate analytically. Thus, we will be able to judge the accuracy of our numerical integration methods by comparison with an analytical result.

We desire a VI that, given a lower limit x_0 and upper limit x_{N-1}, creates N equally spaced points along an x-axis and then produces an associated array of $y = ax^b$ values. Open a blank VI and use **File>>Save** to store this new program under the name **Power Function Simulator** in **YourName\Chapter 6**. As shown next, on the front panel place two **Cluster** shells (found in **Controls>>Modern>Array, Matrix & Cluster**), and then label them **Digitizing Parameters** and **Function Parameters**. Within the **Digitizing Parameters** cluster, place three **Numeric Controls** labeled **Number of Samples**, **Lower Limit**, and **Upper Limit**; within the **Function Parameters** cluster, place two **Numeric Controls** labeled **Prefactor** and **Power**, respectively. Set the representation of **Number of Samples** to **I32**; the representation of all other Numeric Constants should be the default **DBL**. Finally, place an **XY Graph** on the front panel with its x- and y-axes labeled **x** and **y = a*x^b**, respectively.

As an aid in arranging these front-panel objects into a pleasing pattern, you might enjoy experimenting with the **Alignment Tool** and **Distribution Tool** in the toolbar. To use these tools, you must first use the to highlight the group of objects on which you wish to operate.

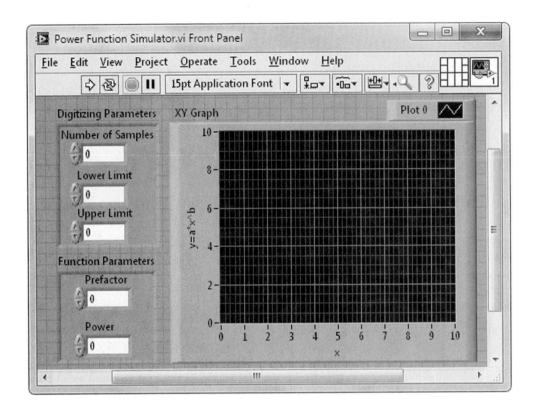

Pop up on the icon pane, select **Edit Icon...**, and then design a new icon using the Icon Editor. Save your design by pressing the **OK** button.

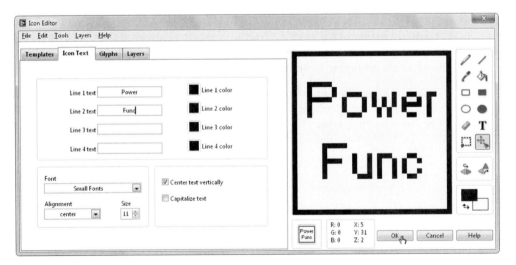

N equally spaced sampled values of x, ranging from x_0 (lower limit) to x_{N-1} (upper limit), can be created by determining the necessary spacing between neighboring points Δx:

$$\Delta x = \frac{x_{N-1} - x_0}{N-1}$$

Then

$$x_i = x_0 + i\,\Delta x \qquad i = 0, 1, 2, \dots, N-1$$

Once the array of x-values is known, then $y_i = f(x_i)$ can be calculated, where (for simplicity) we are taking f to be the power function.

Switch to your block diagram and write code shown next that produces the x and y arrays via the following Mathscript code:

$$delta\_x = (up - lo)/(N - 1)$$
$$x = lo : delta\_x : up$$
$$y = a * x.\wedge b$$

Note that the Mathscript operator .$^\wedge$ (period followed by a caret) performs the element-wise power operation. If you create the Mathscript Node outputs "manually," pop up on the x and y outputs and select **Choose Data Type>>1D-Array>>DBL 1D**; the *delta_x* output should be formatted via **Choose Data Type>> Scalar>>DBL**. Those with the latest versions of LabVIEW may wish to write the code within the Mathscript Node first and then implement LabVIEW's automatic data-type assignment feature to create the three x, y, and *delta_x* outputs. However, by popping up on the y output, you may find it has been assigned the **1D CDL** (one-dimensional array of complex double-precision numbers) data type. If this is the case, then manually change the data type of y to **1D DBL** (we'll discuss why in a moment). This change will result in a red coercion dot at the y output as a block-diagram reminder that you have gone against LabVIEW's automated data-type choice. Finally, **output cluster** is easily created by popping up on the output of the associated **Bundle** icon and selecting **Create>>Indicator**.

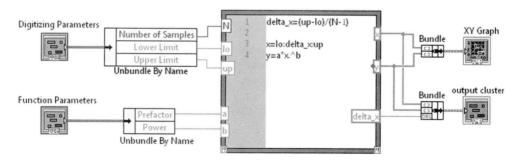

Why did LabVIEW's automated data-type assignment feature choose **1D CBL** for the y-array output? In making its assignment, LabVIEW will account for all possible values of this variable. Since $y = ax^b$, if the argument x is negative and the power is not an integer (for example, $b = 1/2$ so that $y = a\sqrt{x}$), y will be a complex number. Hence, to provide for this possibility, LabVIEW chose the data type for y to be **1D CBL**. In all uses of **Power Function Simulator** in our work ahead, we will be careful to make x positive and b an integer, so y will always be real. Thus, to avoid the complication of working with complex data types, for this program it is safe to override LabVIEW's automatic assignment and format y as a double-precision floating-point (real number) 1D array as shown next.

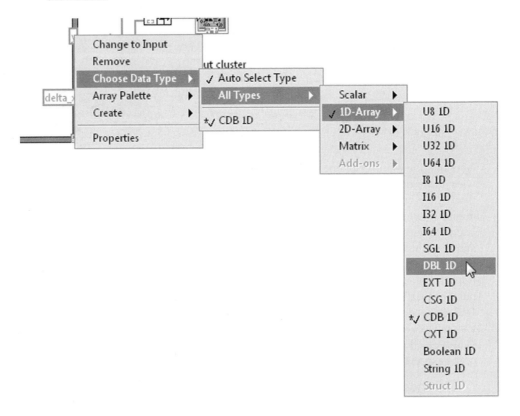

Return to the front panel and use the $\mathbb{k}$ to position all objects nicely. In **output cluster**, label the top, middle, and bottom Array Indicators as **x array**, **y array**, and **delta_x**, respectively (an Array Indicator's owned label is named by popping up on its index display and selecting **Visible Items>>Label**; do not carry out the labeling using the $\boxed{A}$ as this will create free labels). It may be helpful to activate **AutoSizing>>Size to Fit** in the cluster's pop-up menu so that it automatically resizes as you add labeling. If you make mistakes while doing this operation, **Edit>>Undo**, or its keyboard short cut *<Ctrl+Z>*, can come in handy.

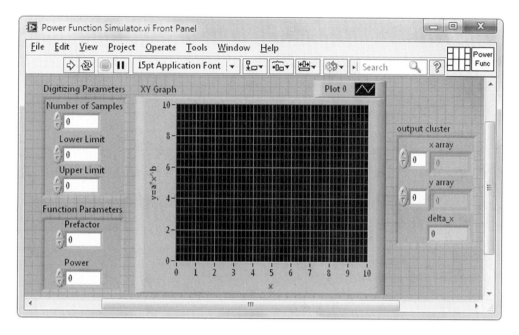

Finally, assign the terminals in the connector pane according to the scheme shown in the next Help Window (in older LabVIEW versions, you need to pop up on the icon pane and select **Show Connector** to make the connector pane visible). You can either follow the recommended approach of using the default 12-terminal pattern or else pop up on the connector pane again and select another (e.g., two-input, one-output) arrangement from the **Patterns** palette.

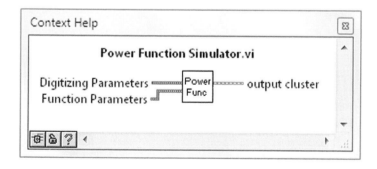

Input values for the five controls, and then run your program to verify that it works. An example of a successful run is shown below. Using **File>>Save**, store your completed program in **YourName\Chapter 6**, and then close its window.

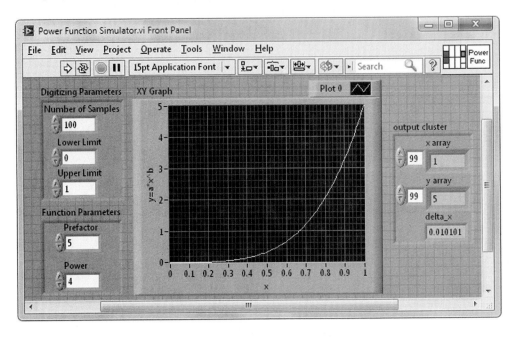

6.5 NUMERICAL INTEGRATION VIA THE TRAPEZOIDAL RULE

In the following illustration (Figure 6.1), we have a sequence of x_i values that are spaced apart by a constant step Δx. The solid curve represents a function $f(x)$ that has known values (shown as solid dots) at each x_i. Let $f(x_i) \equiv y_i$.

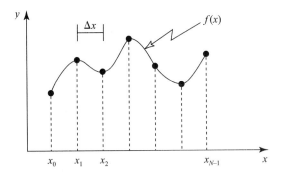

FIG. 6.1

We wish to integrate the function $f(x)$ between x_0 and x_{N-1}. This integral can be interpreted geometrically as the total area enclosed under the $f(x)$ curve between the two endpoints. In Figure 6.2, two adjacent dotted lines at the abscissas x_1 and x_2 define a columnar-shaped area bounded on top by the curve $f(x)$. We note that the sum of the areas of all such columnar shapes in the region between x_0 and x_{N-1} is equal to the integral we want to evaluate. Thus, if we have a general method for determining the area of each columnar shape, we can evaluate the integral.

The obstacle, of course, to writing a general formula for the area of a particular columnar shape is that each area has a curved top. In the *Trapezoidal Rule* for numerical integration, one assumes that the curve $f(x)$ at the top of each columnar shape can be approximated by a straight line. This approximation becomes better and better as Δx is made smaller and smaller.

Then, for example, the shaded columnar area in Figure 6.2 between x_1 and x_2 is approximated as the trapezoidal area given by

$$y_2 \, \Delta x + \frac{1}{2}\left(y_1 - y_2\right)\Delta x = \frac{1}{2}\left(y_1 + y_2\right)\Delta x$$

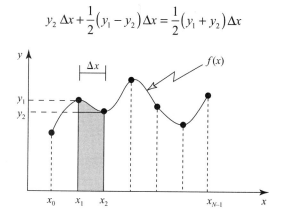

FIG. 6.2

From this equation, we find that the Trapezoidal Rule is equivalent to assuming that the columnar shape defined by x_1 and x_2 is approximately a rectangle with height equal to the average of $f(x_1)$ and $f(x_2)$. So,

$$\int_{x_1}^{x_2} f(x)\, dx \approx \left[\frac{y_1 + y_2}{2}\right]\Delta x$$

where the graphical interpretation of this relation is shown in Figure 6.3.

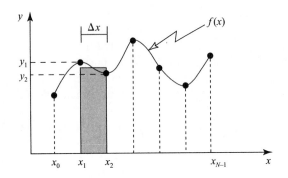

FIG. 6.3

Using the Trapezoidal Rule, the approximation for the entire integral then is given in Figure 6.4.

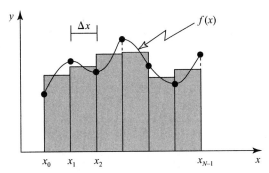

FIG. 6.4

Thus,

$$\int_{x_0}^{x_{N-1}} f(x)\, dx \approx \left[\frac{y_0 + y_1}{2}\right]\Delta x + \left[\frac{y_1 + y_2}{2}\right]\Delta x + \ldots + \left[\frac{y_{N-2} + y_{N-1}}{2}\right]\Delta x\ .$$

or

$$\int_{x_0}^{x_{N-1}} f(x)\, dx \approx \left[\frac{1}{2}y_0 + y_1 + y_2 + \ldots + y_{N-2} + \frac{1}{2}y_{N-1}\right]\Delta x \qquad [1]$$

In our program, we will find it most convenient to write this expression in the following equivalent way:

$$\int_{x_0}^{x_{N-1}} f(x)\, dx \approx \left[\{y_0 + y_1 + y_2 + \ldots + y_{N-2} + y_{N-1}\} - \frac{1}{2}\{y_0 + y_{N-1}\}\right]\Delta x \qquad [2]$$

On the right-hand side (RHS) of Equation [2], the first and second terms in the square brackets are the sum of y-array values and one-half the sum of the y-array endpoints, respectively.

6.6 TRAPEZOIDAL RULE VI USING SINGLE SHIFT REGISTER

Assume that an experiment has produced a set of N equally spaced (x, y) data points, where the x-axis spacing between points is Δx and the variation of y can be described by some function $f(x)$. Let's write a VI that, given the y array and Δx, numerically integrates $f(x)$ between the extremal x-axis values of x_0 and x_{N-1} using the Trapezoidal Rule.

Construct the front panel below and save this VI under the name **Trapezoidal Rule** in the **YourName\Chapter 6** folder. Provide plenty of **Digits of precision**, say *10*, for the **Value of Integral** indicator. Remember that the **y array** input is formed by placing a **Numeric Control** inside of an **Array** shell. Design an icon and assign the connector's terminals consistent with the Help Window shown.

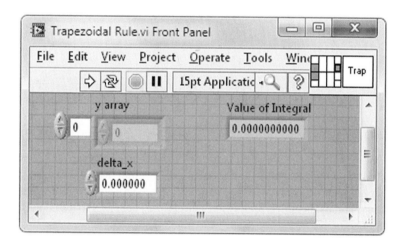

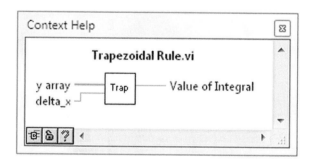

Now switch to the block diagram. We are going to use a For Loop to compute the sum of the *y*-array values of our data set. In a text-based language, we might write

$$S = 0.0$$
$$\text{For } i = 0, N - 1$$
$$S = y(i) + S$$

To implement this code in LabVIEW, write the following diagram, which is similar to the code you constructed in Section 6.2 when you began writing **Sum to 100**. Here, since we are summing double-precision floating-point numbers, initialize the shift register to zero by wiring a **Numeric Constant** defined as *0.0* with representation of **DBL** to its left terminal.

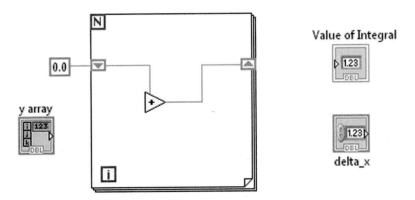

Next, we take advantage of a handy LabVIEW convenience: In addition to its (previously studied) ability to build an array at a loop's output, auto-indexing can be used to sequence through an array at a loop's input. First, complete the following wiring, and then I'll explain its meaning. Simply wire the **y array** control's terminal to the remaining **Add** input.

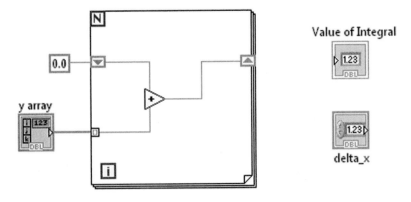

Note that the just-completed wire is thick (denoting an array) outside the loop and then becomes thin (denoting a scalar) inside the loop. Why is this so? Auto-indexing is activated in a For Loop by default, so upon execution, the loop will sequentially input one element from **y array** each time it iterates. That is, on the first loop iteration when equals 0, the element of **y array** with index 0 will be input; on the second iteration when equals 1, the element of **y array** with index 1 will be input; and so on until the end of the array is reached. As an added convenience, when auto-indexing is enabled on an *N*-element array entering a For Loop, LabVIEW automatically sets the loop's count terminal to *N*, thus eliminating the need to wire a value to .

The above block diagram functions in an analogous manner to that of your **Sum to 100** program. The For Loop steps through the complete set of **y array**'s elements, accumulating the sum of these numeric values as it goes. During a particular iteration, the shift register's left terminal provides the sum accumulated up through the last iteration. The newly indexed array element is added to this sum and the result is stored in the shift register's right terminal for use in the next loop iteration. When the For Loop completes its operation, the sum of all of **y array**'s elements is contained in the shift register's right terminal. Thus, the value output from this terminal is the first term within the square brackets on the RHS of Equation [2].

We might flirt with the idea of neglecting the second term within the square brackets on the RHS of Equation [2]. This term is an endpoint correction to the integral expression. Since it involves the sum of just two array elements, it may be small compared with the first term calculated above. If we were to neglect this small correction term, the integral would be determined by simply multiplying the For Loop output by Δx, as shown below.

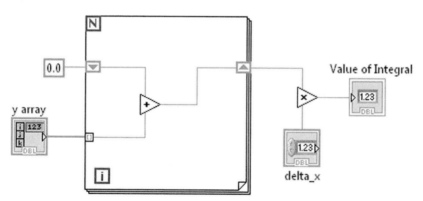

With not too much more work, though, we'll be able to write the code necessary to determine the endpoint correction, which from Equation [2] is given by one-half the sum of the first and last *y*-array elements. We can explicitly include this code on the present diagram, but let's take advantage of the modular nature of LabVIEW programming by tucking this code away within its own VI called **Endpoints**. We will then include **Endpoints** as a subVI on the **Trapezoidal Rule** diagram.

Open a new front panel and save it in YourName\Chapter 6 under the name Endpoints. Place a **Numeric Control** within an **Array** shell on the panel and label it Array. Also label a **Numeric Indicator** as Half of Endpoints Sum. Pop up on the icon pane, select **Edit Icon...**, and then design an appropriate icon. Also assign the connector pane's terminals in a manner consistent with that shown here.

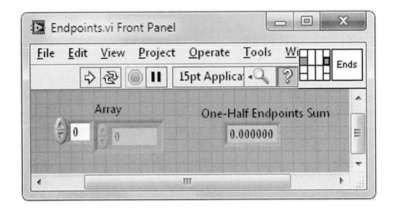

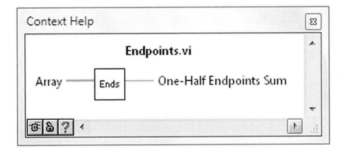

Switch to the block diagram. Here, given the *N*-element **Array**, we need to extract its first (index 0) and last (index *N* − 1) elements. The **Index Array** icon, found in **Functions>>Programming>>Array**, performs this function. Its Help Window is shown next.

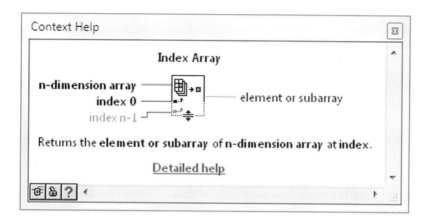

When operating on a two-dimensional (or higher) array, proper configuration of this icon can require some thought. But for our present needs, involving the one-dimensional **Array**, **Index Array** is exceedingly easy to use. One simply wires the 1D array and an integer-containing **Numeric Constant** to the **n-dimensional array** and **index** inputs, respectively. The integer denotes the index of the array element to be extracted. Once extracted, the numeric value of this element appears at the icon's **element** output terminal.

Another related icon that will prove useful is **Array Size**, also found in **Functions>>Programming>>Array**. As shown in its Help Window, this icon determines the number of elements present within an array. In the one-dimensional case, given an N-element 1D array at its input, the icon returns the integer (**I32**) N at its output.

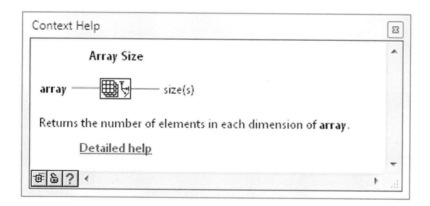

Complete the following block diagram. The left half of this diagram extracts the index 0 and index $N - 1$ elements from **Array**, while the right half adds these two numeric values together and then divides by two. Save your work and then close **Endpoints**. The convenient **Decrement** icon is found in **Functions>>Programming>>Numeric** and can be used to construct the integer $N - 1$.

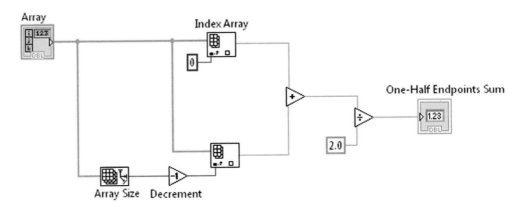

Alternately, with slightly more work, you can find $N - 1$ using the **Subtract** icon.

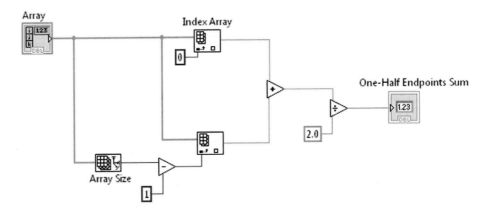

Return to the **Trapezoidal Rule** block diagram. Using **Functions>>Select a VI...**, include **Endpoints** as a subVI on this diagram in the manner shown next. The program now fully manifests all aspects of Equation [2]. Save your work as you close this VI.

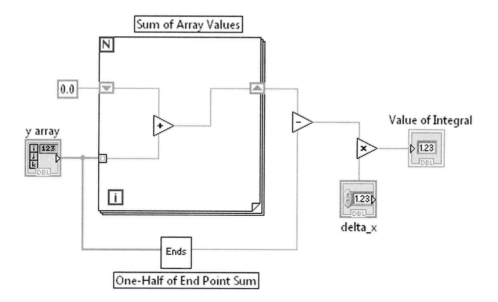

Let's see if the **Trapezoidal Rule** VI really works by supplying it with some known data from **Power Function Simulator**. Construct the following front panel and block diagram called **Trapezoidal Test** and save it in **YourName\Chapter 6**. It's easiest to first code the block diagram and then use the icons' automatic creation feature to make the front-panel objects. Remember to design an icon and assign the connector pane's terminals.

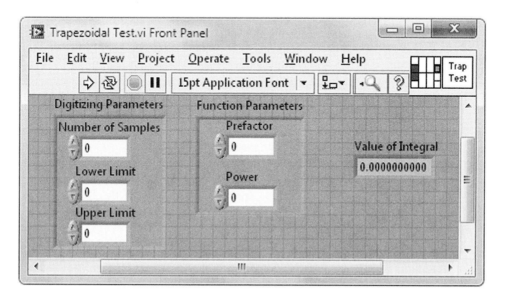

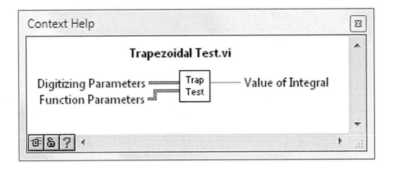

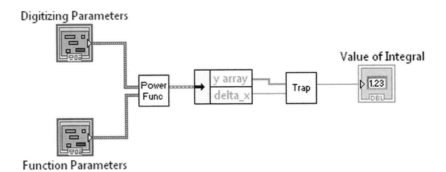

When completed, use **Trapezoidal Test** to evaluate the integral $\int_0^1 5x^4 \, dx$ numerically. It is easy to show analytically that this integral is equal to exactly one. Knowing this true value, we'll be able to check the precision of our approximate numerical method.

Enter *0* and *1* on Trapezoidal Test's front-panel **Lower Limit** and **Upper Limit** controls, respectively, and choose **Prefactor** and **Power** to be *5* and *4*, respectively, so that **Power Function Simulator** is programmed to calculate $f(x) = 5x^4$.

The appropriate value for **Number of Samples** is more problematic. A larger value for this parameter makes Δx smaller. The smaller you make Δx, the less the function $f(x)$ will vary across the top of each columnar shape whose area is needed in the calculation of the integral. Thus, the less bent the top of the columnar shape, the more closely it resembles the straight line assumed by the Trapezoidal Rule approximation. A large value for **Number of Samples** then results in more accuracy for the Trapezoidal Rule approximation, with the drawback that the program will take longer to run. Try running your program with a variety of choices for **Number of Samples** and note the resulting **Value of Integral** in each case. Can you find a good compromise between high precision and acceptable runtime? Remember, when evaluated analytically, the value of the integral is exactly 1.0000000 . . . .

6.7 CONVERGENCE PROPERTY OF THE TRAPEZOIDAL RULE

To explore the effect of **Number of Samples** in determining the precision of the resulting integral value, write the following program called **Convergence Study (Trap)**. Place an **XY Graph** on a new front panel and label its *x*- and *y*-axes **Number of Samples** and **Value of Integral**, respectively. Save this VI in **YourName\Chapter 6**.

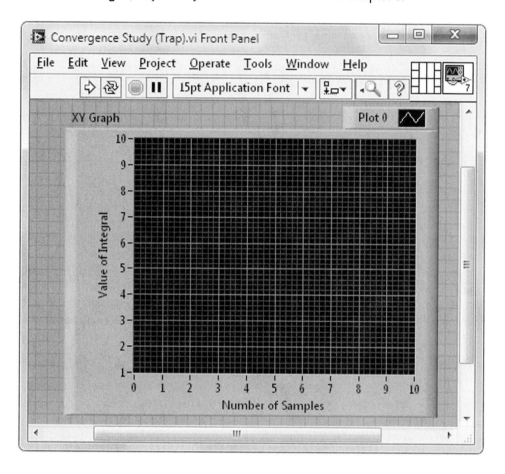

Switch to the block diagram. Our plan is to write code that increments **Number of Samples** from a small to large value, calculates (using **Trapezoidal Test**) and stores the integral value at each step, and then displays the resulting array of values on the **XY Graph**. Produce the diagram shown below. Here, the **For Loop** and **Add** function will produce **Number of Samples** values ranging from *10* to *200* as the loop iterates. The two control clusters are created by popping up on the inputs of **Trapezoidal Test** and selecting **Create>>Control**.

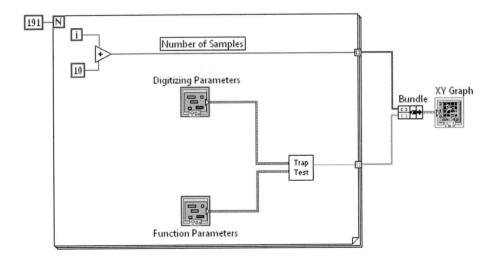

There is one last problem to solve, namely, achieving block-diagram control over the value of **Number of Samples** that is input to **Trapezoidal Test**. As the VI is currently written, the value for **Number of Samples** can only be controlled on the front panel and this value is transferred to **Trapezoidal Test** within the cluster wire emanating from the **Digitizing Parameters** terminal. However, the icon **Bundle By Name** will allow us to intercept the **Number of Samples** value "locked" within the block-diagram cluster wire and change its value before being passed on to **Trapezoidal Test**'s input.

Here's the procedure for accomplishing this feat. First, delete the cluster wire connecting the **Digitizing Parameters** terminal and **Trapezoidal Test**. Then place **Bundle By Name** (found in **Functions>>Programming>>Cluster, Class, & Variant**) as shown.

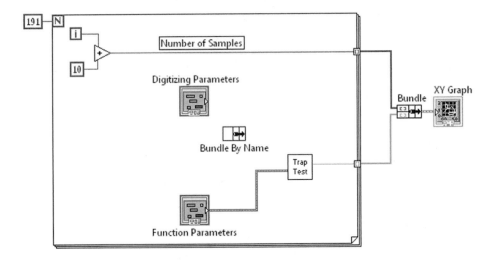

Next, wire from the **Digitizing Parameters** terminal to the central terminal of **Bundle By Name**. When this connection is made, the input (left) terminal of **Bundle By Name** will provide access to the first (index-zero) element in this cluster by default. If you want access to a different element, simply pop up on this input and use **Select Item**, or expand **Bundle By Name** to gain access to all the cluster's elements.

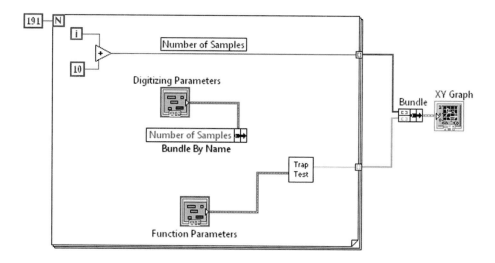

Finally, complete the block diagram coding as shown below. The cluster input to **Trapezoidal Test** will now have the value of **Number of Samples** generated by the **Add** icon (rather than the value from the front-panel control).

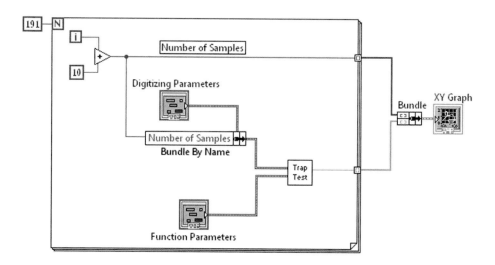

Switch back to the front panel and tidy it up using the 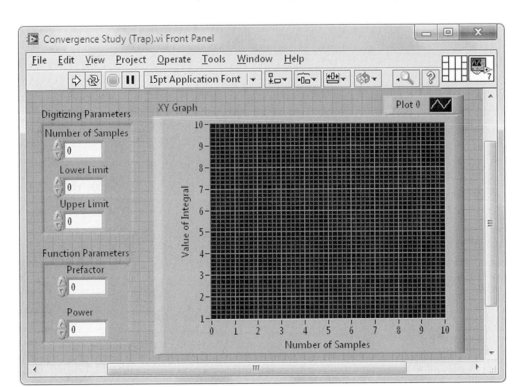. Save your work.

Run **Convergence Study (Trap)** with it programmed to calculate the integral $\int_0^1 5x^4 \, dx$. The value of **Number of Samples** on the front panel will not matter; it is controlled by the block diagram now. You should find that a (diffusely defined) optimal value for **Number of Samples** exists. Increasing **Number of Samples** to values greater than this optimal value only produces incremental enhancement in the accuracy of the integral value. Such a study of the convergence properties of your calculational method would be quite useful in making an intelligent choice for the parameter **Number of Samples**, if you were to use this VI to evaluate a complicated integral numerically.

6.8   NUMERICAL DIFFERENTIATION USING MULTIPLE SHIFT REGISTERS

In the last section, a shift register was used to recall a value created in the loop iteration immediately preceding the current one. In this section, we will explore how this register can be configured to remember values from not just one, but several previous loop iterations. We will explore this feature by writing a program that evaluates the derivative at each data point in a discretely sampled data set.

Assume we are given an equally spaced N-element set of (x, y) data such that

$$x_i = x_0 + i\,\Delta x \qquad i = 0, 1, 2, \ldots, N-1$$
$$y_i = f(x_i)$$

where

$$\Delta x = \frac{x_{N-1} - x_0}{N-1}$$

and f is an analytic function. At the ith data point, the derivative can be evaluated numerically using the following expression that involves knowledge of the two neighboring data points:

$$\frac{dy_i}{dx} = \frac{y_{i+1} - y_{i-1}}{2\,\Delta x} \qquad\qquad [3]$$

In writing a program to evaluate Equation [3], one encounters the following problem: As the elements of an array are sequentially read into a For Loop, during the loop's ith iteration, previously read y-values (such as y_{i-1}) can be made available in shift registers, but to-be-read values (such as y_{i+1}) cannot. To solve this problem, we simply rewrite Equation [3] by letting $i \to i-1$. Then, our expression for the derivative becomes

$$\frac{dy_{i-1}}{dx} = \frac{y_i - y_{i-2}}{2\,\Delta x} \qquad i = 2, 3, 4, \ldots, N-1 \qquad\qquad [4]$$

In Equation [4], the range of i begins at 2 (i.e., does not include 0 and 1) because y_{-2} and y_{-1} are unknown.

Given a For Loop with the configuration of shift registers shown below, all of the quantities needed to calculate dy_{i-1}/dx via the prescription of Equation [4] will be available during the ith iteration of the loop. Labels, which denote the indices of the array elements appearing in each shift register terminal during this particular loop iteration, are included in the illustration.

Thus, Equation [4] is coded by the following subdiagram:

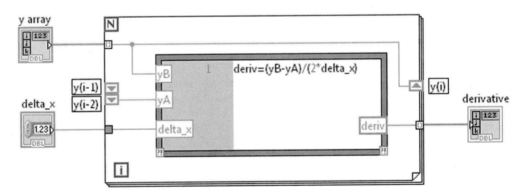

Initiating the execution of this diagram in compliance with Equation [4], however, presents us with a challenge. Since y_{-2} and y_{-1} do not exist, the derivative of lowest index that can be calculated using Equation [4] is dy_1/dx, which corresponds to $i = 2$. To calculate this derivative during the For Loop's first iteration when $\boxed{i}$ equals zero, the top and bottom left terminals of the shift register must contain y_1 and y_0, respectively, while auto-indexing should place y_2 at the y-array input to the loop. To initialize the left terminals, we simply wire the desired values to them from outside the loop. In the case of the y array, we must somehow slice off its first two values so that the y_2 value in the original N-element array becomes the index-zero element in the newly formed array. The new array then will have $N - 2$ elements.

Let's write a VI called **Extract First Two** that manipulates the y array as desired above. Construct the following front panel, which accepts **Array** as an input and outputs its $i = 0$ and $i = 1$ elements to the **Index-Zero Element** and **Index-One Element** Numeric Indicators, respectively. In addition, **Extract First Two** outputs a new **Sliced-Off Array** that is formed by deleting the index-zero and index-one elements from the input **Array**. Package all three outputs in a cluster as shown. Save the VI in **YourName\Chapter 6**.

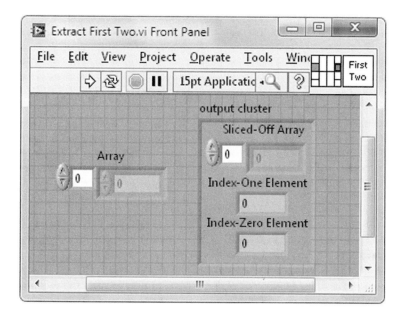

Design an icon for the icon pane and assign the connector pane's terminals consistent with the following Help Window.

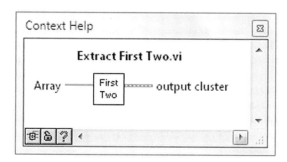

Then write the following block diagram.

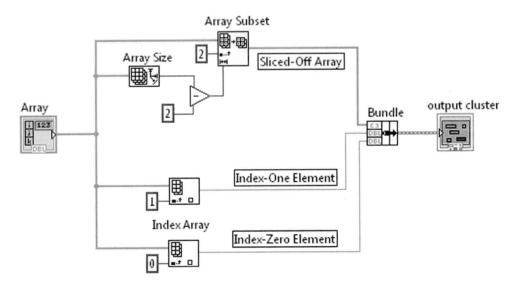

Here, we have used the **Array Subset** icon, found in **Functions>>Programming>> Array**. Its Help Window is reproduced below. This VI receives an array as input and sets aside a given number (= **length**) of its sequential elements, starting at a prescribed index (= **index**). The VI then outputs this subset of elements as a new array at **subarray**. When using this icon, always keep in mind that arrays (as all other LabVIEW structures) are *zero-indexed*. That is, the first element has an index of zero, the second has index of one, and so on.

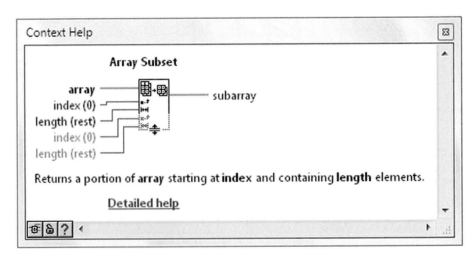

Now we are prepared to write a VI that performs differentiation. This VI will evaluate the derivative numerically at each point, except at the endpoints, when given an N-element discretely sampled data set supplied by **Power Function Simulator**. The calculated derivative will be displayed on an **XY Graph** as well as in an Array Indicator labeled **derivative**.

On a blank VI, place an **XY Graph** with its x- and y-axes labeled as **x** and **Derivative**, respectively. Then use **File>>Save** to store this VI under the name **Power Function Derivative** in **YourName\Chapter 6**. Switch to the block diagram and complete the following code. This diagram will compute the derivative as an array of $N - 2$ elements whose first and last elements are dy_1/dx and dy_{N-2}/dx, respectively, where we are using the indexing appropriate to the original y array input from **Power Function Simulator**. The **Digitizing Parameters** and **Function Parameters** control clusters are made by first placing **Power Function Simulator** on the block diagram and implementing the pop-up menu automatic creation feature.

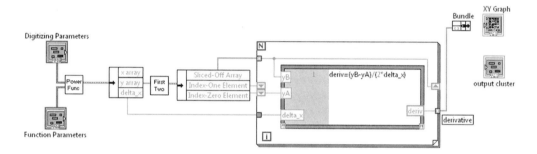

Finally, we need to slice the first and last elements off of the **x array** output from **Power Function Simulator**, so that its indexing properly matches an x-value with the correct **derivative** array element. Use **Array Subset** to perform this slicing operation, as shown below. Then, complete bundling the resulting **x** and **derivative** arrays and wire this cluster to the **XY Graph** terminal. Pop up on the cluster wire and create the **output cluster**.

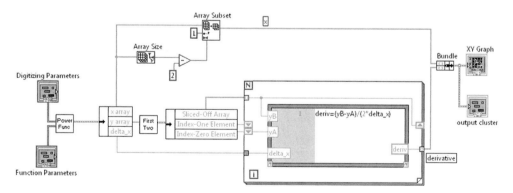

Return to the front panel and tidy up. Within **output cluster**, label the top and bottom array indicators as **x** and **derivative**, respectively, and then save your work.

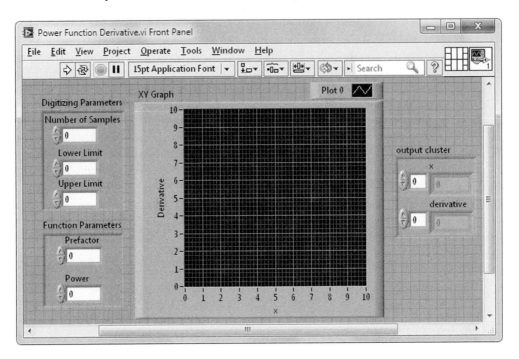

To judge whether your VI is coded properly, program its front panel so that **Power Function Simulator** will create the function $f(x) = 5x^2$. Take **Number of Samples, Lower Limit**, and **Upper Limit** to be *100*, *0*, and *1*, respectively. Then run **Power Function Derivative**. Is the slope of the curve on your **XY Graph** as anticipated?

Next, use your VI to calculate the derivative of $f(x) = 5x$. Is the result as expected? You may want to scale the *y*-axis manually here.

6.9 MODULARITY AND AUTOMATIC SUBVI CREATION

One of the hallmarks of a good LabVIEW program is its *modularity*. In such a program, the top-level VI acts as a supervisor, coordinating the execution of its collection of sub-VIs. In this approach to programming, a subVI is written to carry out each component task required for the successful completion of the greater program's overall mission. Modularity means that each subVI can be executed as a "stand-alone" program (i.e., independent of the top-level VI), simplifying the programmer's inevitable job of program debugging. Additionally, since subVIs are designed to perform well-defined tasks, many times they can be reused as components in other top-level programs.

In the exercise we have just finished, **Power Function Derivative** is the top-level VI, while **Power Function Simulator** and **Extract First Two** are its subVIs. When viewed from this perspective, a programmer might suddenly wish that **Power Function Derivative** had been written with an increased level of modularity. For example, in the upper left region of the block diagram, several icons accomplish the subtask of snipping off the **x array**'s two endpoints. Thus, this group of icons performs a well-defined task and is an excellent candidate for becoming a subVI.

When such a realization is made, here's an editing trick that is most helpful. First, use the ⤵ to select the group of block-diagram objects that you wish to include within a subVI.

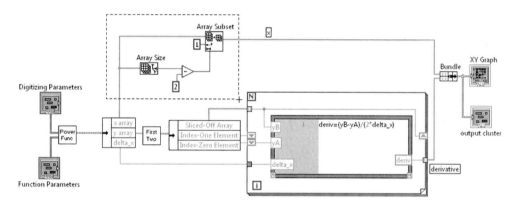

When you release the mouse button, this group will be highlighted with a marquee. Then, in the **Edit** pull-down window, select **Edit>>Create SubVI**. The highlighted group will be automatically converted into a subVI with a default icon as shown below.

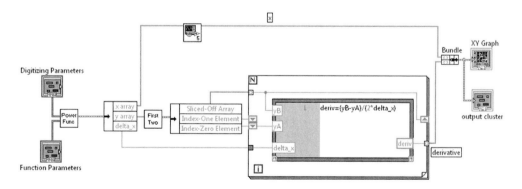

To view this subVI, double-click on its icon. The front panel will open and you will find the new subVI's automatically created controls and indicators with default labels. The subVI itself will also have a default name.

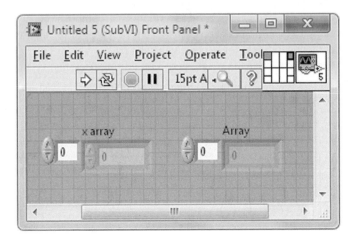

You can then use the to create more accurate labels. Also, a name can be given to the subVI using **File>>Save** and the subVI's icon can be designed by popping up on the icon pane and selecting **Edit Icon...** You will find that the terminals have been automatically associated with the controls and indicators, but these associations can be changed on the connector pane, if you wish. Below, I have kept LabVIEW's automatic terminal assignments.

Carry out these suggested procedures to create the front panel of a subVI called **End-Less Array** as shown next.

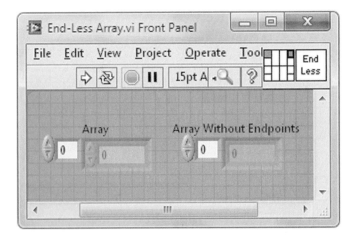

When you switch to the block diagram, you will find code that you selected. Upon closing this subVI, **Power Function Derivative** will appear as shown below.

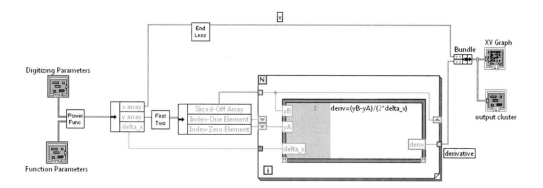

One then might suddenly realize that the entire central region of **Power Function Derivative** is the generic code required to calculate the numerical derivative of any function and so might be of future use when working with functions other than the power function. This central region can be easily morphed into a VI called **Numerical Derivative** by first selecting it as shown next.

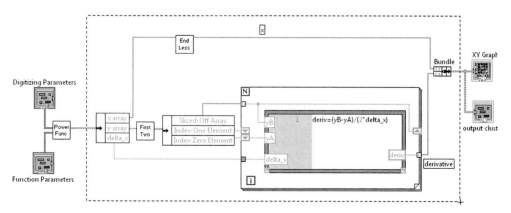

Then, using **Edit>>Create SubVI**, the **Numerical Derivative** VI shown below results after some work with the 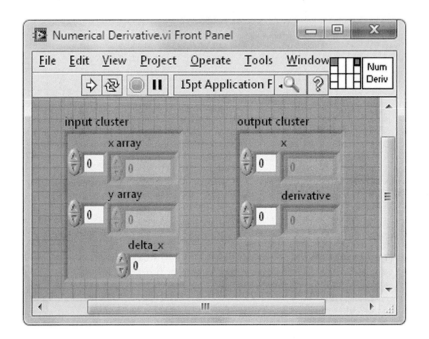.

The block diagram of **Power Function Derivative** then is a model of modularity, as shown here.

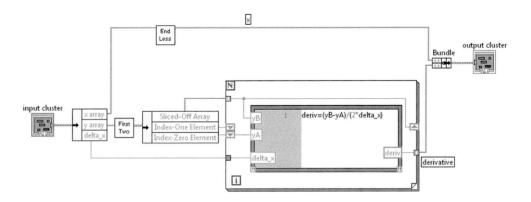

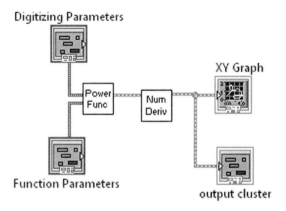

Digitizing Parameters

XY Graph

Power Func

Num Deriv

Function Parameters

output cluster

DO IT YOURSELF

The factorial of integer N is defined to be $N! \equiv 1 \times 2 \times 3 \times \ldots \times (N-1) \times N$ with $0! \equiv 1$.

   (a) Write a VI called **Factorial** that, given N, calculates $N!$. To check that your VI works, show that $10! = 3,628,800$ and $15! = 1,307,674,368,000$. If your program does not give the correct answer for both of these cases, figure out the problem within your code and fix it. **Factorial** should be able to produce correct values up to about $170!$.

   (b) Stirling's approximation states that, for large N,

$$\ln(N!) \approx N \ln(N) - N \qquad [5]$$

Using **Factorial** as a subVI, write a program called **Stirling Test** that plots the percentage deviation of Stirling's approximation from the actual value of $\ln(N!)$ over a range of N-values. Use **Stirling Test** to determine the minimum value of N so that the approximate $\ln(N!)$ deviates from the actual value by less than 1%.

PROBLEMS

   1. I have agreed to hire you for a 30-day month. I pay you one penny on the first day, two pennies on the second day, and continue to double your daily pay on each subsequent day up to (and including) day 30. How many total dollars do you earn by the end of the month? Write a For Loop–based program with a shift register called **Month's Total Pay (Shift Register)** that determines the answer to this question. You may find the **Functions>>Programming>>Mathematics>>Elementary & Special Functions>>Exponential Functions** subpalette useful.

   2. In this problem, you will explore how LabVIEW handles an uninitialized shift register.

(a) Write the following VI called **Sum To Five**.

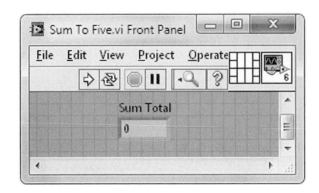

With **Sum To Five** open, run it three times in a row. What is the value of **Sum Total** after the first, second, and third runs?

Now close **Sum To Five** (removing it from your computer's RAM), and then open it again (returning it to your computer's RAM). Run the VI three times in a row. What is the value of **Sum Total** after the first, second, and third runs?

(b) Modify **Sum To Five** by deleting the **Numeric Constant** connected to the left shift register. The shift register will now be *uninitialized* when the program is run. After making this modification, save **Sum To Five**, close it, and then reopen it.

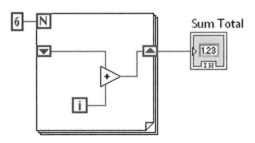

With **Sum To Five** open, run it three times in a row. What is the value of **Sum Total** after the first, second, and third runs?

Now close Sum To Five, and then open it again. Run the VI three times in a row. What is the value of **Sum Total** after the first, second, and third runs?

(c) Based on your observations, what is the default value of an uninitialized shift register when the program is first loaded into RAM (i.e., before the program has been run the first time)?

(d) Based on your observations, what is the value for an uninitialized shift register at the beginning of a program's second run? How about at the beginning of the program's third run? [Uninitialized shift registers are occasionally used in programs as a form of memory.]

3. Let x be a given positive number. Newton's iterative method for determining the square root $y = \sqrt{x}$ is as follows: As an initial guess for y, take $y_0 = x/2$. Then, for the first iteration $(i = 0)$, calculate the value for y to be $y_1 = \dfrac{1}{2}\left[y_0 + \dfrac{x}{y_0}\right]$. Iterate this process, where on the ith iteration $y_{i+1} = \dfrac{1}{2}\left[y_i + \dfrac{x}{y_i}\right]$, until the desired accuracy for y is obtained. Write a VI called **Newton's Square Root**, which implements this method.

Noting that $\sqrt{2} = 1.414213562...$, how many iterations are required to obtain a value $y = \sqrt{2}$ that is accurate to the fourth decimal place (i.e., $\sqrt{2} = 1.4142$)?

4. The icon **Reverse 1D Array**, found in **Functions>>Programming>>Array**, produces an output array whose elements are in reverse order to that of the array supplied at the icon's input. That is, if the array (0, 1, 2, 3) is input, then the array (3, 2, 1, 0) is output. Using any of the array-related icons available in **Functions>>Programming>>Array**, except for **Reverse 1D Array**, write a program called **Reverse Array Elements**, which accomplishes the same task as **Reverse 1D Array**. The front panel of **Reverse Array Elements** should contain an **Array Control** and an **Array Indicator** labeled **Input Array** and **Output Array**, respectively. Program **Input Array** with the array (0, 1, 2, 3), and then run your VI to demonstrate that it performs correctly.

5. Write a program called **Running Average of Noisy Sine**, which generates and plots samples of a noisy sine wave along with an average of the four most recent samples of this waveform. To generate a sample of a noisy sine wave, use the following code within a While Loop that iterates once every 50 ms until a Stop Button is pressed.

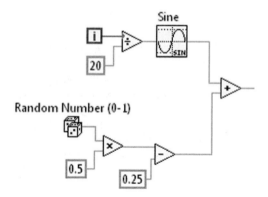

As each sample y_i of the noisy sine wave is generated, plot it on a Waveform Chart, along with $y = (y_i + y_{i-1} + y_{i-2} + y_{i-3})/4$, that is, the "running average" of the current sample and the three most recent previous samples. Refer to the Waveform Chart's Help Window to determine how to plot both y_i and $\bar{y}$ on a single chart. Run **Running Average of Noisy Sine** and demonstrate that a running average effectively "filters out" random noise.

6. Write a program called **Scale of A Major**, which plays one octave of the A major scale. In equal temperament tuning, the ratio of frequencies of neighboring notes is $2^{1/12}$. First, use a For Loop with shift register to generate an array of the frequencies f_n (where $n = 0, 1, 2, \dots, 12$) for the 13 notes within the octave from the notes A4 ($f_0 = 440$ Hz) to A5 ($f_{12} = 880$ Hz). Next, input this array to another For Loop and use it to play the eight notes of the A Major scale, which correspond to frequencies $f_0, f_2, f_4, f_5, f_7, f_9, f_{11}$, and f_{12}. Let each note sound for 500 ms. To play a note, use **Beep.vi**, which is found in **Functions>>Programming>>Graphics & Sound**, with its **use system alert?** input set to FALSE. The indices of the frequencies in the A Major scale can be stored either in a front-panel **Array Control** or in a block-diagram **Array Constant** (created by placing a **Numeric Control** within an **Array Constant** shell).

7. Fourier analysis tells us that a square wave $y(x)$ with period and amplitude of one is given by the following sum of sine waves:

$$y(x) = \frac{4}{\pi} \sum_n \frac{\sin(2\pi nx)}{n} \qquad n = 1, 3, 5, 7, \dots, \infty \qquad [6]$$

Write a VI called **Sum of Sines** that approximates the above sum by evaluating only its first N terms (i.e., if $N = 3$, then only the $n = 1$, $n = 3$, and $n = 5$ terms are included in the sum). The front panel of your VI should appear as shown next with a **Numeric Control** called **Number of Terms** to input the desired value of N and an **XY Graph** to plot y vs. x. To observe three high-resolution cycles of the square wave, evaluate the sum over the range from $x = 0$ to $x = 3$ with $\Delta x = 0.001$.

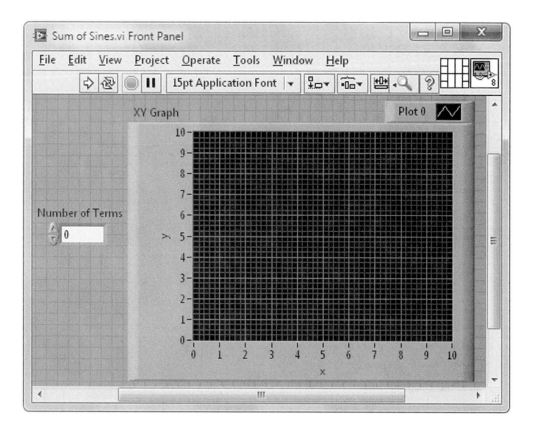

In designing your program, consider the following:

- All positive odd integers can be determined by $n = 2i + 1$, where $i = 0, 1, 2, 3, \ldots$
- If individual terms in a sum can be calculated one at a time, these terms can be summed using a For Loop with shift register. For the term associated with the odd integer n, form a 3001-element 1D array of values consisting of the quantity $\frac{4}{\pi} \sin(2\pi n x)$ evaluated over the range from $x = 0$ to $x = 3$ with $\Delta x = 0.001$. This 1D array can be summed with another similar-sized 1D array using the **Add** icon. Because of the polymorphic property of **Add**, if 1D arrays y and y' are input, each element of y is added to the corresponding element of y' and the resulting 1D array is output. That is, the ith element of the output array is equal to $y_i + y_i'$.
- To initialize an N-element array with each element equal to zero, use **Initialize Array** (found in **Functions>>Programming>>Array**) with **element** and **dimension size** wired to 0 and N, respectively.

Run **Sum of Sines** with a fairly small value of **Number of Terms**. As shown below, near the transitions between $y = +1$ and $y = -1$, you will observe "overshoot" peaks (also known as ringing artifacts), an effect called the *Gibbs phenomenon*. As **Number of Terms** is increased, you should be able to minimize this effect. For what value of **Number of Terms** does the Gibbs phenomenon become negligible (e.g., overshoot peak is only 5% of square wave's amplitude)? (With our choice of Δx, the maximum allowed value for **Number of Terms** is 500. Why?)

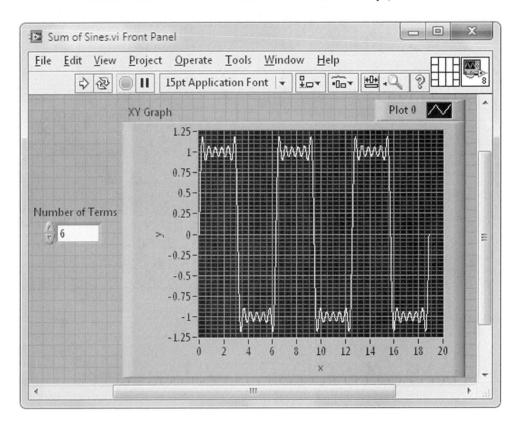

8. In the presence of air, a falling object's acceleration a (= change in velocity per unit time) can be described by the following relation

$$a = g - \alpha v^2 \qquad\qquad [7]$$

where $g = 9.8$ m/s^2 is the acceleration caused by gravity in a vacuum, v is the object's instantaneous speed in m/s, and α is a constant with units of m^{-1}. The velocity-squared term describes the effect of air resistance, and it becomes more and more significant as the object attains greater speeds. When the object achieves terminal velocity v_T, the object no longer accelerates because, at that speed, $g = \alpha v_T^2$.

Assume an elevated object is released from rest at time $t = 0$ and that the subsequent time is divided into N small intervals, each of duration Δt. Then the object's fall can be sampled at N times given by $t_n = n\Delta t$, where $n = 0, 1, 2, \ldots,$ $N - 1$. Let v_n be the object's instantaneous speed at time t_n. Then, if the object's speed v_{n-1} at time t_{n-1} is known and Δt is small, Equation [7] can be used to determine v_n as follows:

$$v_n = v_{n-1} + a\,\Delta t = v_{n-1} + \left(g - \alpha v_{n-1}^{\,2}\right)\Delta t \qquad [8]$$

(a) Write a program called **Falling In Air**, which implements Equation [8] to determine a falling object's speeds v_n at the N times t_n and then plots *Speed* vs. *Time* on a Waveform Graph with calibrated x-axis. Take $\Delta t = 0.001$ s. The front panel of **Falling In Air** should appear as follows.

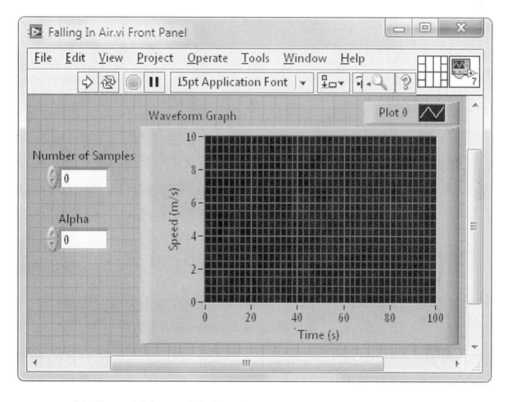

(b) For a 1.0-kg steel ball, a human skydiver, and a raindrop, the numerical value of α is about 0.00092, 0.0035, and 0.11, respectively. Run **Falling In Air** with each value of α. Determine the resulting terminal velocity v_T for each object. After being released, about how long does it take each object to attain v_T?

CHAPTER 7

The Case Structure

7.1 CASE STRUCTURE BASICS

In a LabVIEW program, conditional branching is accomplished through the use of the *Case Structure*. This structure is analogous to an "if–else" statement in a text-based programming language. The Case Structure is found in **Functions>>Programming>> Structures** and is Boolean by default. That is, by wiring a TRUE or FALSE Boolean value to its *selector terminal* ⍰, the structure will execute either the code within its TRUE window or that within its FALSE window, respectively. Using the ⍟, the selector terminal can be placed anywhere along the Case Structure's left border.

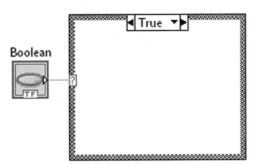

You may view only one of the Case Structure windows at a time. In the above illustration, the TRUE window is visible. To view the FALSE window, simply click the mouse cursor on the decrement (left) or increment (right) button in the *case selector label* ◀ True ▼▶ at the top of the structure. The FALSE window will then appear as shown here.

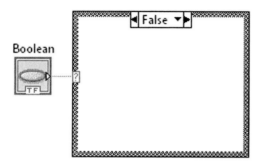

The Case Structure will automatically change its character from Boolean to numeric when you wire a numeric quantity to its selector terminal, such as the **I32** integer control shown next.

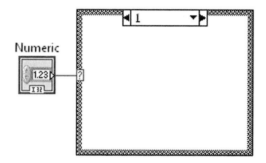

Initially, only two Case windows (**0** and **1**) are available. Above, the case selector label indicates that the **Case 1** window is currently visible. You can easily add another case by popping up anywhere on the Case Structure's border and then selecting **Add Case After** in the pop-up menu.

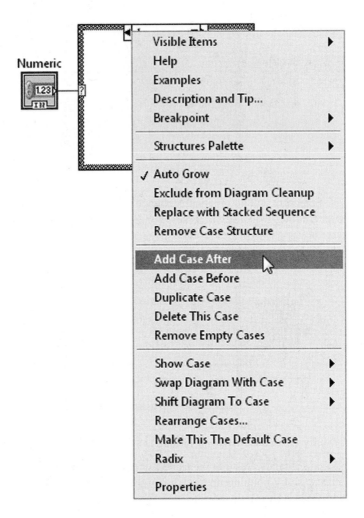

You will then find that the structure now has three (**0, 1,** and **2**) cases.

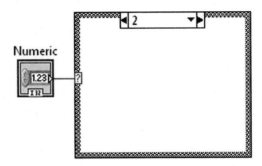

By repeating the above procedure, you may add as many (positive-integer) cases as you desire, unless you desire more than the maximally allowed 2,147,483,648 (= 2^{31}) cases! If you wire a floating-point number to the selector terminal, LabVIEW will round that number to the nearest integer value. If the number supplied to the selector terminal is negative or if the number is larger than the highest-numbered case, the case designated (using the pop-up menu) as *Default* will be selected.

Finally, when wired to a selector terminal, an *Enumerated Type Control* can be used to self-document each of a Case Structure's cases, often making it the best choice for the case-selecting control. An Enumerated Type Control (called an *Enum*, for short) is a ring-style control, which associates a unique integer value with each item on a list of text descriptors. When an Enum is wired to the selector terminal of a Case Structure, its text descriptors (rather than the associated integers) appear in the case selector label. If you define these text descriptors wisely, they will intuitively describe the purpose of each Case Structure diagram, obviating the need to document every case with, say, free labels. Such documentation is invaluable to other programmers (and your future self!) attempting to decipher your code.

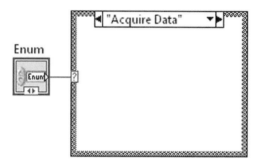

7.2 QUICK CASE STRUCTURE EXAMPLE: RUNTIME OPTIONS USING PROPERTY NODES

As an initial example of the use of the Case Structure, let's add two runtime options to the sine-wave generating program **Sine Wave Chart (While Loop)** that you wrote in Chapter 1. Here is our plan: The features of a front-panel object such as the Waveform Chart can be controlled on the block diagram by means of *Property Nodes*. First, with a **Boolean Case Structure** and **History Data** Property Node, we will enable a user to clear **Sine Wave Chart (While Loop)**'s Waveform Chart while the VI is running. Second, using an **Enum Case Structure** and **Plot Area** Property Node, we'll allow an operator to change the Waveform Chart's background color during runtime.

Start by creating a clone of **Sine Wave Chart (While Loop)** through the following procedure. Open the **YourName\Chapter 1\Sine Wave Chart (While Loop)**, and then select **File>>Save As...** In the dialog window that appears, select **Copy>>Substitute copy for original**, and then press the **Continue...** button. In the next window, first create a folder called **Chapter 7** within the **YourName** folder, and then save this copied VI under the name **Sine Wave Chart (While Loop with Runtime Options)** in **YourName\ Chapter 7**. The original file **Sine Wave Chart (While Loop)** will be closed (and safely saved in **YourName\Chapter 1**), while the newly created file **Sine Wave Chart (While Loop with Runtime Options)** is open and ready for you to modify.

On the open VI's front panel, add a **Push Button** (found in **Controls>> Modern>>Boolean**) and label it **Clear Chart?**. Pop up on this control and select **Mechanical Action>>Latch When Pressed**. By default, **Push Button** has a Boolean value of FALSE. Its mechanical action is now programmed so that when pressed, Push Button's value changes to TRUE. Immediately after this value is next read on the block diagram, Push Button will return to its default FALSE value.

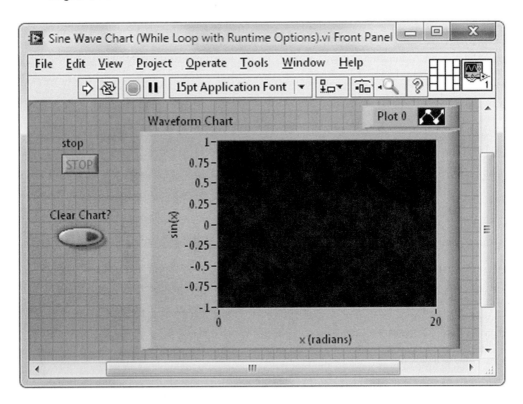

Switch to the block diagram. There, add a **Case Structure** (from **Functions>> Programming>>Structures**) and wire Push Button's Boolean terminal to its selector terminal as shown. The Case Structure is now Boolean in nature; to toggle between its TRUE and FALSE windows, click the mouse cursor on the ◀ True ▼▶ at the top of the structure.

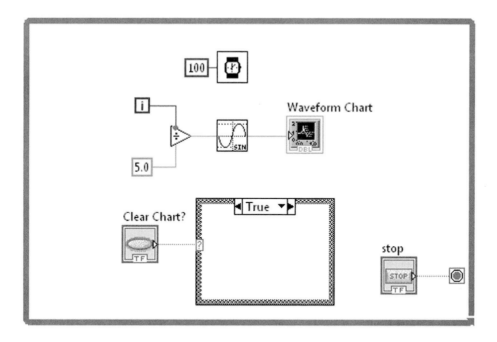

Next, we will place the appropriate Property Node that clears the Waveform Chart within the Case Structure's TRUE window. Then, when Push Button is pressed on the front panel, this chart-clearing Property Node will be executed. The appropriate Property Node is called **History Data**, which is created by popping up on the Waveform Chart's icon terminal and selecting **Create>>Property Node>>History Data** as shown next. While making this selection, you might take a moment to peruse the pop-up menu's list of Waveform Chart properties that can potentially be controlled using a Property Node. It's an impressive list.

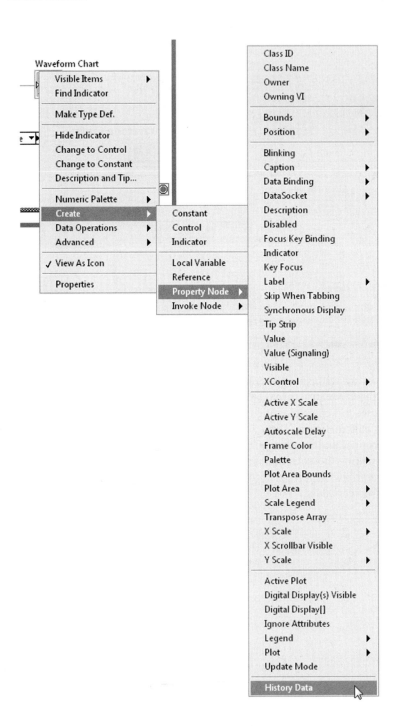

After selecting **History Data**, place this Property Node within the Case Structure's TRUE window. As with most Property Nodes, **History Data** is created as an indicator, hence the small outward-directed black arrow at its right side.

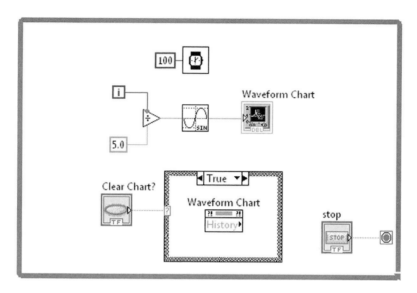

Change this icon to a control by popping up on its lower section and choosing **Change To Write**.

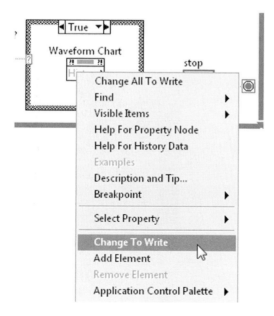

The black arrow will become inward directed on the icon's left side, indicating that **History Data** is now a control. The needed input to clear the Waveform Chart is an **Empty Array**. Pop up on the region containing the black arrow, and then, using **Create>>Constant**, produce the **Empty Array** input as shown next.

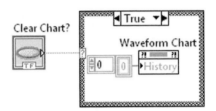

Leave the Case Structure's FALSE window empty so that, when Push Button is unpressed and in its default FALSE state, the Waveform Chart is not cleared. Save your work.

Return to the front panel and run your VI. When you press the **Clear Chart?** Push Button, the Waveform Chart should clear, with the sine-wave plot resuming on the next While Loop iteration. If plotting doesn't resume, check that that the **Mechanical Action** option of Push Button has been correctly selected as **Latch When Pressed**.

Next, we will control the plot's background color programmatically. Start by placing an **Enum** control (found in **Controls>>Modern>>Ring & Enum**) on the front panel. Label this control **Background Color** as shown.

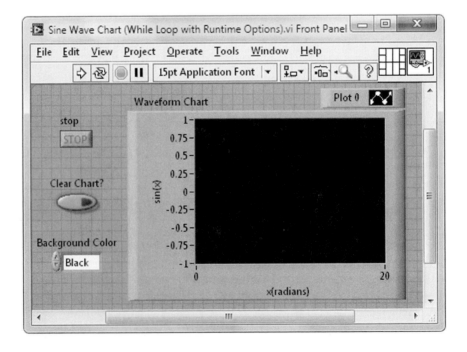

Pop up on the **Enum** and select **Edit Items...** In the dialog window that appears, program this control with the following four items—*Black*, *Red, Green, Blue*—and then click the **OK** button. For a review of this programming procedure, see Section 3.9, "**The Enumerated Type Control**."

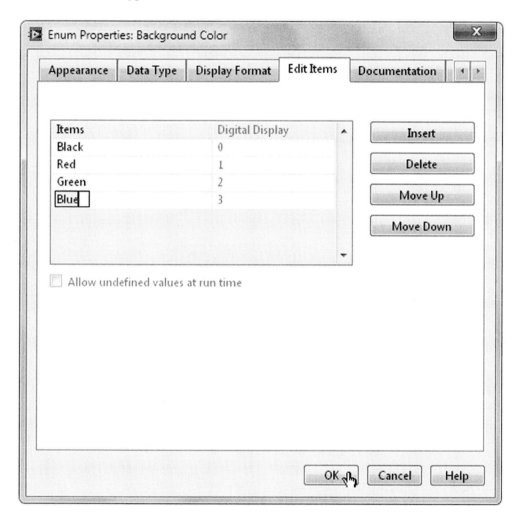

Switch to the block diagram. Add a **Case Structure** and wire **Background Color**'s Enum terminal to its selector terminal as shown.

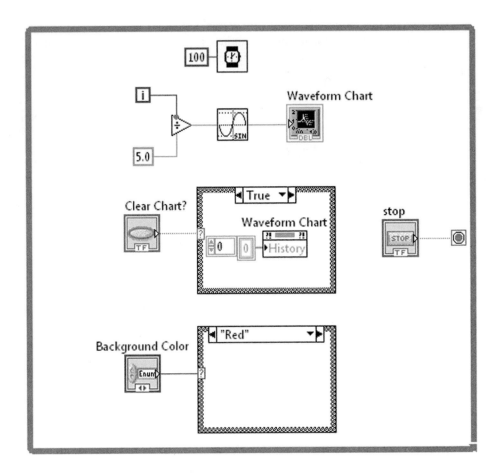

When the four-item Enum is initially wired to the selector terminal, the Case Structure will only contain two windows corresponding to the first two items in the Enum's list of items. To create the other required windows, pop up on the *case selector label*, and choose **Add Case for Every Value**.

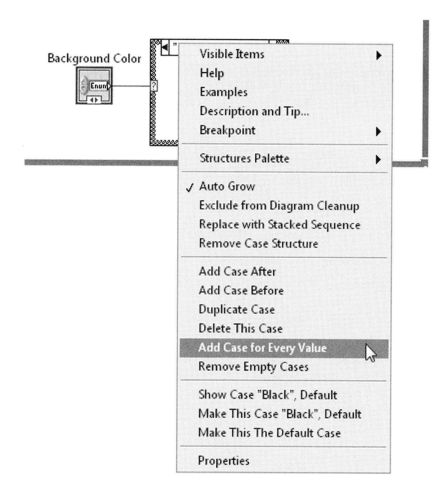

Using the *case selector label*, check to see that the Case Structure now contains four case windows labeled *Black*, *Red, Green,* and *Blue*.

The Waveform Chart's background color is generated by the RGB color method. In this approach, varying intensities of the three additive colors red, green, and blue are combined to produce a broad array of colors. For the LabVIEW implementation of RGB, eight bits are used to denote the desired intensity of each additive color, which corresponds to a decimal number in the range from 0 to 255. Each of these three decimal numbers is converted to a two-digit hexidecimal equivalents in the range from 00 to FF. The red, green, and blue hexidecimals—called *RR, GG,* and *BB*—are then packaged as a single six-digit hexidecimal number *RRGGBB*. As an example, in this scheme, *FF0000* is pure red, and *0000FF* is pure blue. Thankfully, LabVIEW supplies an icon called **RGB to Color.vi**, which carries out the details of the RGB color method. **RGB** to **Color.vi** is found in **Functions>>Programming>>Numeric>>Conversion** and its Help window is shown next. At the icon's **R**, **G**, and **B** inputs, you supply three decimal numbers, each in the range from 0 to 255 with unsigned eight-bit integer representation **U8**. These numbers denote the relative intensities of the three additive colors needed to generate the desired color. The icon then does the mathematical manipulations required to produce the *RRGGBB* hexidecimal, which is supplied at the **Color** output.

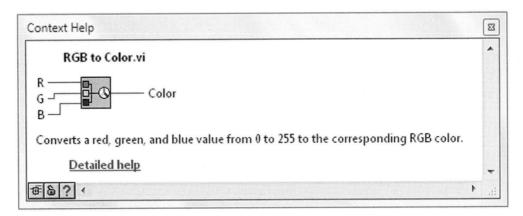

The Property Node that controls the background color of a Waveform Chart is called **BG Color**. Pop up on the Waveform Chart's terminal and create this Property Node using **Create>>Property Node>>Plot Area>>Colors>>BG Color**. Place **BG Color** on the block diagram, and then change it from its default indicator mode to a control by selecting **Change To Write** in its pop-up menu. Finally, place the appropriate three **U8** integers within each of the four Case Structure windows and wire them as shown next. For black, red, green, and blue, (**R, G, B**) are (0, 0, 0), (255, 0, 0), (0, 255, 0), and (0, 0, 255), respectively. The next block diagram shows the *Black* Enum case.

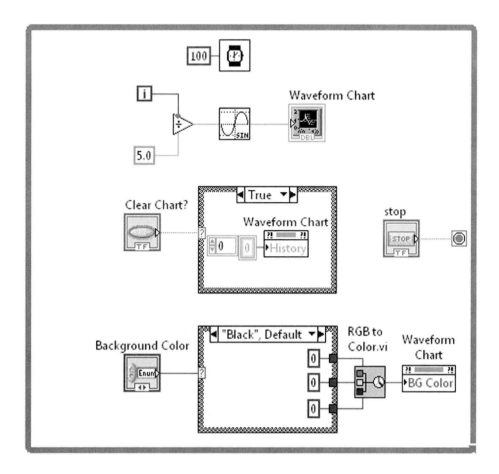

The three other Case Structure windows should be programmed as follows.

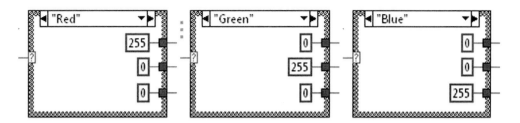

Save your work, and then return to the front panel. Run **Sine Wave Chart (While Loop with Runtime Options)** and try out each of the plot area's four possible background colors. My favorite is blue—how about you?

7.3   NUMERICAL INTEGRATION USING CASE STRUCTURES

In the rest of this chapter, we will gain further experience in using the Case Structure while writing a program that implements numerical integration via *Simpson's Rule*. A brief review of the theory behind Simpson's Rule will be given first, revealing that numerical integration of a given discrete data set via this technique divides neatly into three distinct tasks. This realization will then lead us to write three VIs, one to carry out each of these tasks. We will then package these programs together as subVIs within a top-level numerical-integration program called **Simpson's Rule**. After verifying that this VI performs correctly, we will finish our work by writing a program that compares the convergence property of the Simpson's Rule method with that of the Trapezoidal Rule technique studied in the last chapter. In this concluding VI we'll learn a new LabVIEW skill—placing multiple plots on an XY Graph.

7.4   NUMERICAL INTEGRATION VIA SIMPSON'S RULE

Assume that a curve $y = f(x)$ has known values y_1, y_2, and y_3 at x_1, x_2, and x_3, respectively, where the succession of three x-values is equally spaced by a constant step Δx. In Simpson's Rule, one assumes that $f(x)$ in the region from x_1 to x_3 can be approximated by the quadratic $y = Ax^2 + Bx + C$, as shown in Figure 7.1.

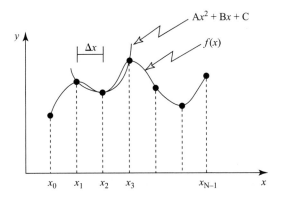

FIG. 7.1

Evaluating the posited quadratic at the three known data points yields the following three equations involving the three unknown constants A, B, and C:

$$y_1 = Ax_1^2 + Bx_1 + C$$
$$y_2 = Ax_2^2 + Bx_2 + C$$
$$y_3 = Ax_3^2 + Bx_3 + C$$

Using the fact that $x_2 = x_1 + \Delta x$ and $x_3 = x_1 + 2\,\Delta x$, linear algebra can be used to solve this set of equations for the three unknowns A, B, and C (try it yourself), yielding

$$A = \frac{1}{2(\Delta x)^2}\left[y_1 - 2y_2 + y_3\right]$$

$$B = -\frac{1}{2(\Delta x)^2}\left[y_1\left(x_2 + x_3\right) - 2y_2\left(x_1 + x_3\right) + y_3\left(x_1 + x_2\right)\right] \qquad [1]$$

$$C = \frac{1}{2(\Delta x)^2}\left[y_1 x_2 x_3 - 2y_2 x_1 x_3 + y_3 x_1 x_2\right]$$

The area under the curve $f(x)$ can then be approximated as follows:

$$\int_{x_1}^{x_3} f(x)\,dx \approx \int_{x_1}^{x_3}\left(Ax^2 + Bx + C\right)dx$$

$$= \frac{1}{3}A\left(x_3^{\ 3} - x_1^{\ 3}\right) + \frac{1}{2}B\left(x_3^{\ 2} - x_1^{\ 2}\right) + C\left(x_3 - x_1\right) \qquad [2]$$

From Equation [1], we know the values of A, B, and C. Plugging these expressions into Equation [2], the right-hand side (RHS) of [2] becomes

$$\frac{y_1}{2(\Delta x)^2}\left\{\frac{1}{3}\left(x_3^{\ 3} - x_1^{\ 3}\right) - \frac{1}{2}\left(x_2 + x_3\right)\left(x_3^{\ 2} - x_1^{\ 2}\right) + x_2 x_3\left(x_3 - x_1\right)\right\}$$

$$-\frac{y_2}{(\Delta x)^2}\left\{\frac{1}{3}\left(x_3^{\ 3} - x_1^{\ 3}\right) - \frac{1}{2}\left(x_1 + x_3\right)\left(x_3^{\ 2} - x_1^{\ 2}\right) + x_1 x_3\left(x_3 - x_1\right)\right\}$$

$$+\frac{y_3}{2(\Delta x)^2}\left\{\frac{1}{3}\left(x_3^{\ 3} - x_1^{\ 3}\right) - \frac{1}{2}\left(x_1 + x_2\right)\left(x_3^{\ 2} - x_1^{\ 2}\right) + x_1 x_2\left(x_3 - x_1\right)\right\}$$

Then, using $x_2 = x_1 + \Delta x$ and $x_3 = x_1 + 2\Delta x$, this expression greatly simplifies after some (brutal) algebra, yielding the following famous result known as the *Simpson's "Three-Point" Rule*

$$\int_{x_1}^{x_3} f(x)\,dx \approx \left[\frac{1}{3}y_1 + \frac{4}{3}y_2 + \frac{1}{3}y_3\right]\Delta x \qquad [3]$$

This relation provides an improved numerical integration method over the "two-point" Trapezoidal Rule. Note that the above formula gives the integral over an interval of size $2\,\Delta x$, so the coefficients add up to 2. Also note that, based on our derivation, we expect Simpson's "Three-Point" Rule will give exact results when the function $f(x)$ is a polynomial of order 2 or less. Surprisingly, it can be shown that symmetries in the above equations lead to fortuitous cancellations, making Equation [3] exact even for polynomials of third order.

For a set of N data points then, where

$$x_1 = x_0 + i\,\Delta x \qquad\qquad i = 0, 1, 2, \ldots, N-1$$

$$y_i = f(x_i)$$

we can group the N x-values into a succession of "three-point" units, so that

$$\int_{x_0}^{x_{N-1}} f(x)\,dx \approx \left[\frac{1}{3}y_0 + \frac{4}{3}y_1 + \frac{1}{3}y_2\right]\Delta x + \left[\frac{1}{3}y_2 + \frac{4}{3}y_3 + \frac{1}{3}y_4\right]\Delta x$$

$$+ \ldots + \left[\frac{1}{3}y_{N-3} + \frac{4}{3}y_{N-2} + \frac{1}{3}y_{N-1}\right]\Delta x$$

Let's call each term within square brackets a *"partial sum."* Then the above expression can be written as the following summation over partial sums,

$$\int_{x_0}^{x_{N-1}} f(x)\,dx \approx \sum_i \left[\frac{1}{3}y_{2i} + \frac{4}{3}y_{2i+1} + \frac{1}{3}y_{2i+2}\right]\Delta x \qquad i = 0, 1, 2, \ldots, \frac{N-3}{2}\,(N\text{ odd}) \qquad\qquad [4]$$

where the summation is over $(N-1)/2$ partial sums. We see then that this prescription for calculating the integral assumes that N is an odd integer, where N is the number of points in our data set. This, of course, is because our N data points can only be successfully grouped into a sequence of three-point units if N is odd.

If given an even number of data points, one may calculate the integral numerically with the following approach: Use Simpson's Rule to find the contribution resulting from the first $N-1$ (an odd integer) points and then the Trapezoidal Rule to calculate the contribution resulting from the region between the two last points.

$$\int_{x_0}^{x_{N-1}} f(x)\,dx \approx \sum_i \left[\frac{1}{3}y_{2i} + \frac{4}{3}y_{2i+1} + \frac{1}{3}y_{2i+2}\right]\Delta x + \left[\frac{1}{2}y_{N-2} + \frac{1}{2}y_{N-1}\right]\Delta x$$

$$i = 0, 1, 2, \ldots, \frac{N-4}{2}\,(N\text{ even}) \qquad\qquad [5]$$

The first term on the RHS of Equation [5] is the Simpson's Rule sum, which contains $(N-2)/2$ partial sums, and the second term is the Trapezoidal Rule applied to the last two elements of the y array.

In applying the above theory to integrate a discretely sampled set of N data samples, three subtasks must be performed: (1) Given the N-element data array, determine whether N is even or odd and then use the appropriate integral formula (Equation [4] or [5]) to find the number of partial sums to be calculated, (2) evaluate the Simpson's Rule term in the appropriate integral formula, and (3) calculate the Trapezoidal Rule term in Equation [5], if N is even. Thus, to write a top-level program called **Simpson's Rule**, which carries out the above theory, we will first write a VI for each of the three identified subtasks. We will then use these three programs as subVIs in the top-level **Simpson's Rule** program.

7.5 PARITY DETERMINER USING A BOOLEAN CASE STRUCTURE

First, let's write a VI called **Parity Determiner**. Given an N-element **Array** as input, this program will determine whether N is even or odd and then calculate the number of partial sums in the Simpson's Rule term of the appropriate integral formula. From Equations [4] and [5], the number of partial sums is $(N-1)/2$ and $(N-2)/2$ when N is odd and even, respectively.

Construct the following front panel with your own custom-designed icon and connector pane assigned according to the illustrated scheme. Create the **output cluster** by first placing a **Numeric Indicator** labeled **Number of Partial Sums** and then a Boolean indicator **Round LED** (found in **Controls>>Modern>>Boolean**) labeled **Odd?** within a **Cluster** shell. Use the cluster's pop-up menu option **Reorder Controls In Cluster...** to verify that **Number of Partial Sums** and **Odd?** are the cluster's index-zero and index-one elements, respectively. Format the data type of the **Array** control and **Number of Partial Sums** indicator as **DBL** and **I32**, respectively. Use **File>>Save** to create a folder called **Chapter 7** within the **YourName** folder, and then save your VI under the name **Parity Determiner** in YourName\Chapter 7.

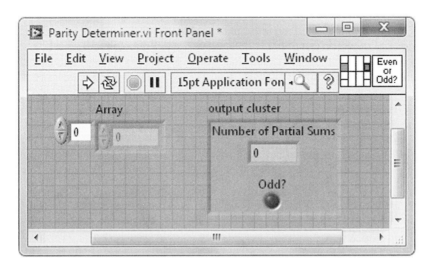

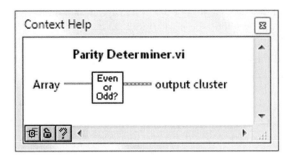

Switch to the block diagram. Here, we will determine whether the number of elements in **Array** is odd or even using **Quotient & Remainder** (found in **Functions>> Programming>>Numeric**). The Help Window for this icon is shown next.

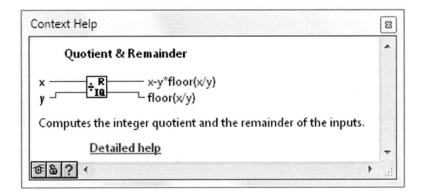

The **R** output of Quotient & Remainder, which corresponds to the modulo function *mod(x,y)* of other programming languages, calculates the remainder R when x is divided by y using the following algorithm: $R(x, y) = x - y * floor(x/y)$, where *floor* truncates its argument to the next lowest integer. Thus, $R(N, 2)$ equals *0* and *1* for N even and odd, respectively. Code the next diagram, where the Boolean output of **Equal?** (found in **Functions>>Programming>>Comparison**) will be TRUE (FALSE) when the number of elements in **Array** is odd (even). Wire this output to the second (index-one) input of a **Bundle** icon.

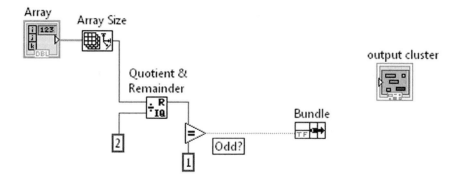

Now add a **Case Structure** (from **Functions>>Programming>>Structures**) to your block diagram and wire the **Equal?** output to its selector terminal.

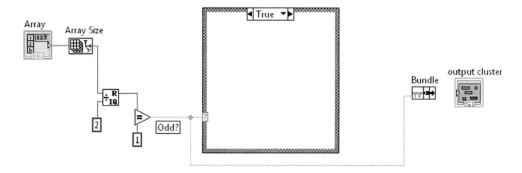

When N is odd, the number of partial sums to be calculated is $(N-1)/2$. Program this calculation into the Case Structure's TRUE window, and then wire its output to the first (index-zero) input of a **Bundle** icon. The **To Long Integer** icon is found in **Functions>> Programming>>Numeric>>Conversion.** Next, wire Bundle's output to the **output cluster** terminal (if you get a broken wire, then the ordering of **Number of Partial Sums** and **Odd?** is incorrect within the front-panel cluster; to correct this problem, see Section 3.12).

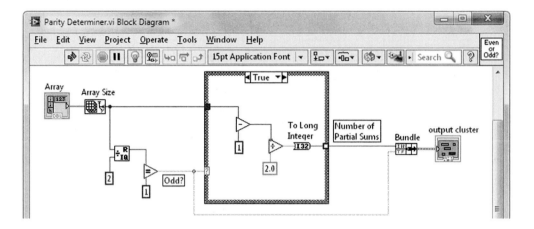

Note that the Case Structure's output tunnel is white and that the **Run** button is broken, indicating that the VI is not yet ready to run. Switch to the FALSE window and you'll find out why.

In a Case Structure, the data at all inputs (tunnels and the selector terminal) are available to all cases. Thus, the **Array Size** data tunnel, which was wired as an input in the TRUE window, will also make its data available in the FALSE window. However, a particular Case Structure window is not required to use all available inputs. So even though nothing is connected yet, the input tunnel appears black, indicating a legal wiring configuration.

Output tunnels obey a different rule: If any one Case Structure window supplies data to an output tunnel, all other windows must do so too. At present, the FALSE window lacks this requisite output-tunnel wiring and this unlawful situation is signaled by the white appearance of the tunnel. The broken **Run** button generically signals this diagram error.

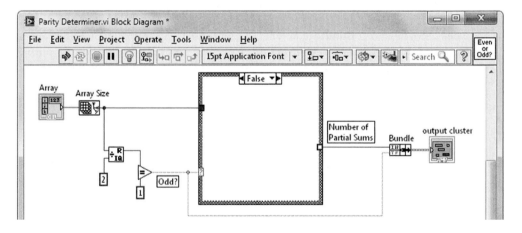

When N is even, the number of partial sums to be calculated is $(N-2)/2$. Program this calculation into the Case Structure's FALSE window as shown next. As soon as you wire to the output tunnel, it will possess connections in all the Case Structure windows (TRUE and FALSE). It will then turn black to denote it is now a legal output tunnel. Note that the **Run** button is now whole.

As an alternate method of coding the FALSE window, you might explore the following method, if interested. First, select the FALSE window, pop up on the Case Structure's border, and select **Delete This Case**. You will be automatically switched to the TRUE window, which you have previously coded. Then, since a very similar FALSE-window code is desired, pop up on the Case Structure's border and select **Duplicate Case**. A clone diagram will appear in a newly created FALSE window, which you can then modify.

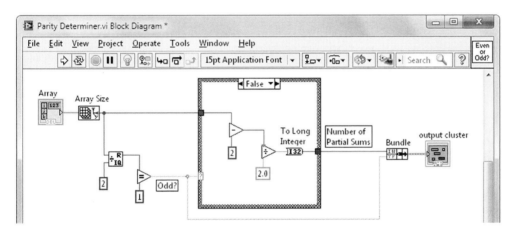

Return to the front panel and verify that the VI functions as expected by running it with **Array** programmed to have two, three, four, and then five elements (if you need to clear **Array** of all elements, pop up on its index display and select **Data Operations>>Reinitialize to Default Value**). Then save the final version of this VI in **YourName\Chapter 7** as you close it.

Parenthetically, as we found in Chapter 3, the Mathscript Node contains a conditional branching function that follows the if–else statement syntax. Above, we wrote Case Structure–based code to accomplish the following logic:

$$R = \mathrm{mod}(N, 2)$$

$$if\ \ R = 1$$

$$\qquad B = TRUE$$

$$else$$

$$\qquad B = FALSE$$

$$end$$

$$if\ \ B = TRUE$$

$$\qquad\qquad Number\ of\ Partial\ Sums = (N-1)/2$$

$$else$$

$$\qquad\qquad Number\ of\ Partial\ Sums = (N-2)/2$$

$$end$$

Such conditional branching can alternately be written using a Mathscript Node in the following way. For the Mathscript Node's **NumSum** and **B** outputs, **Choose Data Type>>All Types>>Scalar>>I32** and **Choose Data Type>>All Types>>Scalar>>Boolean** are selected.

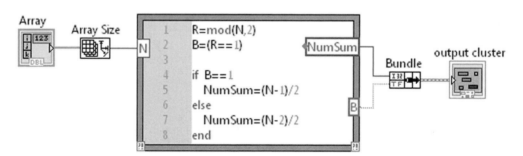

7.6   SUMMATION OF PARTIAL SUMS USING A NUMERIC CASE STRUCTURE

Next, we will write a VI called **Sum of Partial Sums**, which calculates the Simpson's Rule term in Equations [4] and [5]. Construct the following front panel. Format the **Array** and **delta_x** controls as well as the **Simp Value** indicator as **DBL** and the **Number of Partial Sums** control as **I32**. You may pop up on the DBL formatted objects and select **Display Format**...and then uncheck **Hide trailing zeros** and make **Significant digits** something on the order of 6. Save this VI under the name **Sum of Partial Sums** in YourName\Chapter 7.

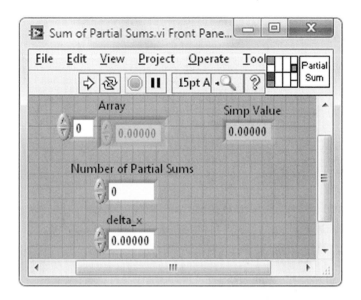

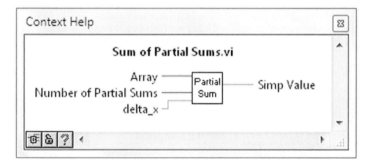

Now begin coding the block diagram as shown next. Here is our programming strategy: Use a pair of nested For Loops to implement Simpson's Rule. The outer For Loop successively picks out three-point units from the original input array using the **Array Subset** icon. The inner For Loop is used to calculate the partial sum due to each three-point unit. The summation of all the partial sums is accumulated in the outer For Loop's shift register; the number of For Loop iterations required to complete this summation is supplied by the **Number of Partial Sums** control. Since the entire input array must be passed to the **Array Subset** icon (as opposed to just one of its elements per loop iteration), you must pop up on the outer For Loop's tunnel and select **Disable Indexing**. A few free labels have been added to this block diagram to document how it functions.

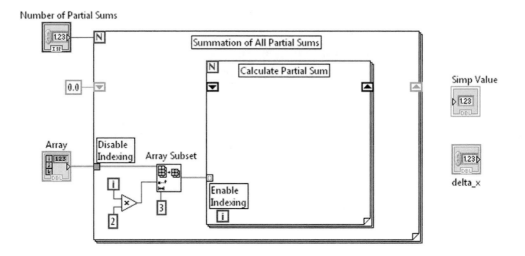

Now add the following code, which calculates each partial sum.

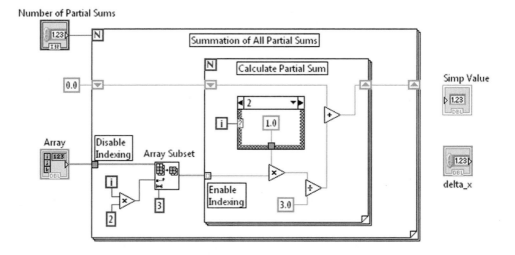

By enabling the indexing option on the inner For Loop, the three elements of the subarray created by **Array Subset** will be passed into the loop one at a time. There is no need to wire the inner For Loop's count terminal; it will automatically be set to *3*. Within the loop, each of the three values is weighted by the proper factor consistent with Simpson's Rule. The inner loop's shift register accumulates the current partial sum. By initializing this shift register with the left terminal of the outer loop's shift register, the current partial sum is added to the accumulation of all past partial sums.

In the given diagram, a numeric **Case Structure** is used to provide the correct 1–4–1 sequence of weighting factors dictated by Simpson's Three-Point Rule. Remembering that the iteration terminal begins counting at *0*, the three Case Structure windows should appear as illustrated.

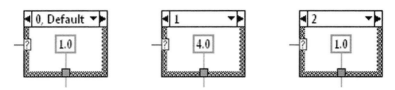

Complete the diagram to include the multiplicative factor of Δx. Be sure to save your final work as you close this VI.

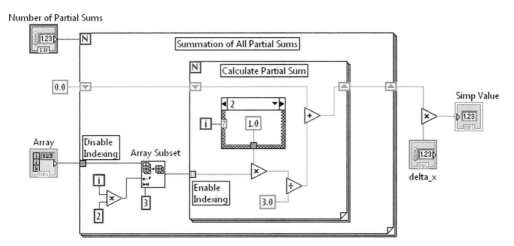

7.7 TRAPEZOIDAL RULE CONTRIBUTION USING A BOOLEAN CASE STRUCTURE

Finally, we write a VI that calculates the second term on the RHS of Equation [5], that is, the Trapezoidal Rule contribution to the integral for an even-numbered data array. Build the following front panel and design an icon in the icon pane. The Boolean control is a **Round Button**. Save the VI under the name **Even Ends** in **YourName\Chapter 7**. Assign the connector pane's terminals in a manner consistent with the Help Window shown next.

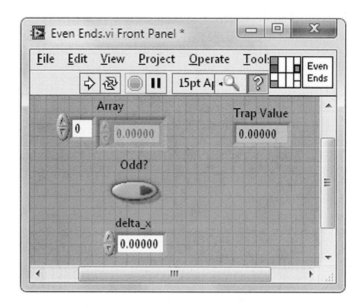

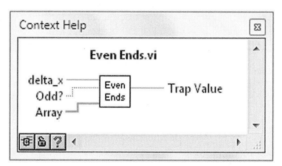

Now write the following diagram. If the array possesses an even number N of elements (**Odd?** is FALSE), the Case Structure's FALSE window determines the index of the second-to-last element ($i = N - 2$), forms a subarray consisting of solely of the original array's last two elements ($i = N - 2$ and $i = N - 1$), and then numerically integrates this subarray using the **Trapezoidal Rule** VI.

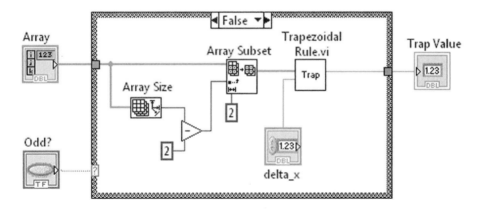

Code the TRUE window as follows, reflecting the fact that for odd-numbered arrays, the Trapezoidal Rule contribution is zero.

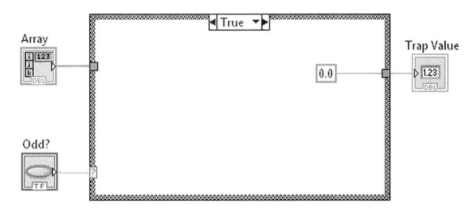

Be sure to save your final work on this VI as you close it.

7.8 TOP-LEVEL SIMPSON'S RULE VI

Now we are ready to write the top-level **Simpson's Rule** VI. Assume that a set of N equally spaced (x, y) data samples exists, where the x-axis spacing between samples is Δx and the variation of y can be described by some function $f(x)$. With the y array and Δx as input, we wish to write a program that performs a Simpson's Rule integration of $f(x)$ between the extremal x-axis values of x_0 and x_{N-1}.

Construct the following front panel. Using the **Display Format...** option in its pop-up menu, set the **Value of Integral** indicator's **Significant digits** to *10* (or so). Save this VI under the name **Simpson's Rule** in YourName\Chapter 7.

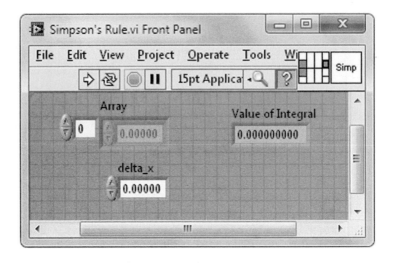

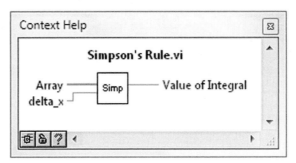

Code the block diagram as shown below, and then save your work.

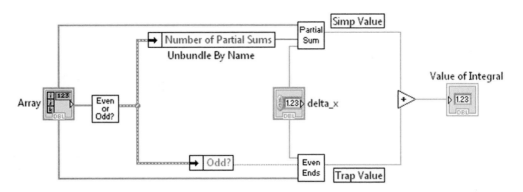

Let's see whether the **Simpson's Rule** VI really works by supplying it with some known data from **Power Function Simulator**. Construct the following front panel and block diagram called **Simpson Test**. Save this VI in **YourName\Chapter 7**. To select the desired cluster item within **Unbundle By Name**'s output terminal, either click on the terminal using the ⤢ or pop up on the terminal and use the **Select Item** option.

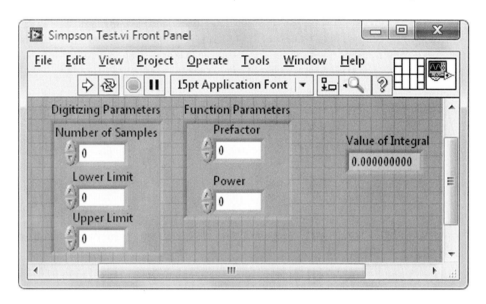

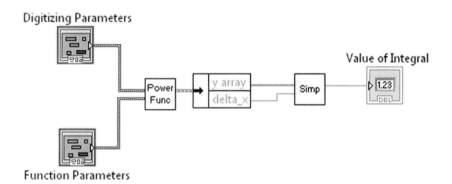

With appropriate front-panel settings, use **Simpson Test** to evaluate the integral $\int_{0}^{1} 5x^{4}dx$ numerically (i.e., **Power Function Simulator** should be programmed to calculate $f(x) = 5x^4$). Try various values for **Number of Samples** and see how this affects the precision of **Value of Integral**.

7.9   COMPARISON OF THE TRAPEZOIDAL RULE AND SIMPSON'S RULE

Finally, let's compare the convergence properties of the Trapezoidal and Simpson's Rule methods and see whether one proves to be the superior technique. To display the results of this study, place an **XY Graph** on a new front panel and save this VI under the name **Convergence Study (Trap vs. Simp)** in **YourName\Chapter 7**. Label the XY Graph's *x*- and *y*-axes **Number of Samples** and **Value of Integral**, respectively. Move the Plot Legend to the open region to the right of the XY Graph as shown.

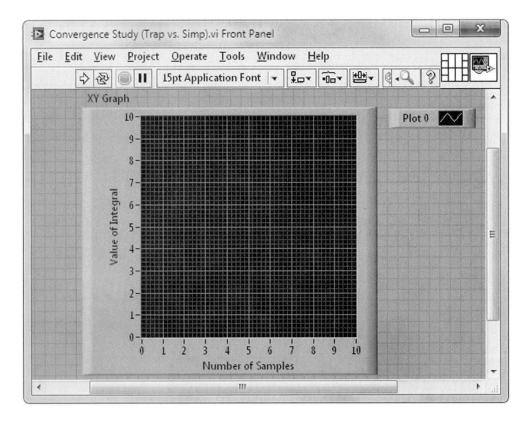

As shown next, start coding the block diagram, which will numerically integrate an *N*-element array using both methods. Here, as *N* is varied from *10* to *200*, the integral is calculated by the two methods, and the results are recorded in arrays at the For Loop boundary. To plot all of the results on a single **XY Graph**, an *xy* cluster is formed for the Trapezoidal Rule as well as the Simpson's Rule result. Then, a two-element array is built with the Trapezoidal Rule and the Simpson's Rule *xy* cluster as its first (index-0) and second (index-1) elements, respectively. Such an *array of clusters* is the required input for multiplots on an XY Graph.

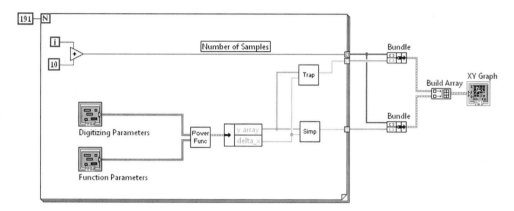

To finish this diagram, we need to intercept the cluster emanating from the **Digitizing Parameters** terminal and replace the front-panel **Number of Samples** value with the block-diagram value for that quantity. Delete this cluster wire, and then place a **Bundle By Name** icon and wire it as shown. The value for **Number of Samples** that is passed to **Power Function Simulator** is that created by the **Add** icon (rather than that passed from the front-panel control).

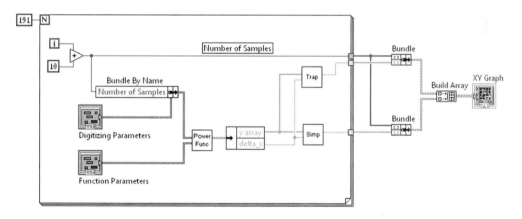

Return to the front panel. Using the ⬉, neaten the arrangement of objects. Then resize the **Plot Legend** downward to accommodate information for two plots.

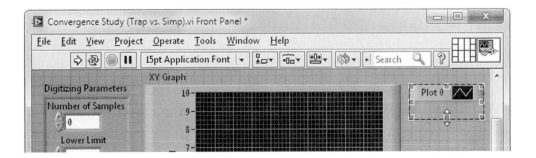

Using the 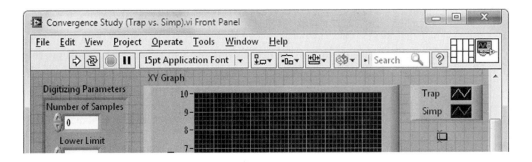 (or), highlight the text **Plot 0** in the Plot Legend, and then enter the text *Trap*. In a similar way, replace **Plot 1** by *Simp*.

The plots can be distinguished from each other by choosing unique plot characteristics. For example, to choose a particular plot's color, simply pop up on its icon within the Plot Legend, select **Color**, and click on the hue you like (LabVIEW may have already chosen different colors for the two plots automatically).

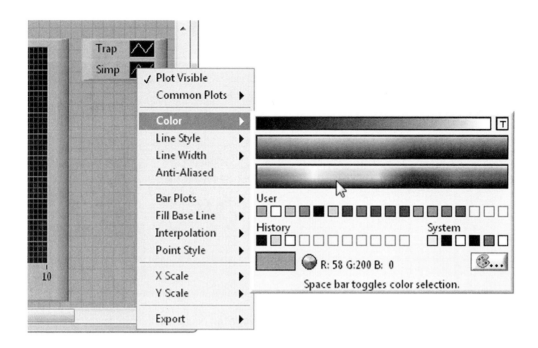

Save your work. Then run **Convergence Study (Trap vs. Simp)** with **Power Function Simulator** programmed to calculate the function $f(x) = 5x^4$. Does one of the integration methods converge to an accurate value for the integral more quickly (that is, with a smaller **Number of Samples** value) than the other?

Now that you understand the inner workings of these two numerical integration methods, check out LabVIEW's built-in icon **Numeric Integration.vi** found in **Functions>>Mathematics>>Integration & Differentiation**.

DO IT YOURSELF

Write a VI called **Five Blinking Lights** whose front panel has five **Round LED** Boolean indicators labeled *0* through *4*, as shown next. Using a **Case Structure**, develop a program that lights these LEDs one at a time in the order 0-1-2-3-4 and keeps repeating this sequence until the **Stop Button** is pressed. Make each LED stays lit for 0.2 second, so that an entire 0-1-2-3-4 sequence occurs once every second. Note that the value of a

Boolean Constant can be changed using the 🖑.

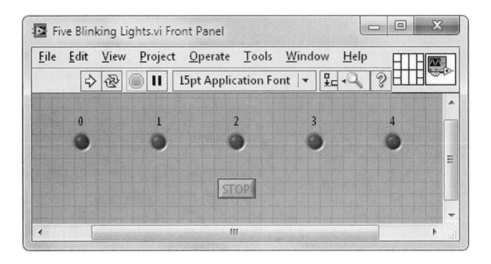

PROBLEMS

1. Construct a VI called **Seven-Segment Counter**. As shown below, use seven **Square LED** Boolean indicators, which have been appropriately resized and labeled **A** through **G**, to form a seven-segment display. Then, code the block diagram so that, when the Run button is pressed, this front-panel display counts from 0 to 9, with each digit illuminated for 1.0 second.

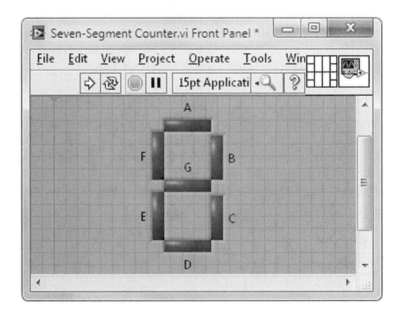

2. The Riemann zeta function ζ evaluated at the integers $n = 1, 2, 3, 4, \ldots$ is defined by

$$\zeta(n+1) = \frac{1}{n!} \int_0^\infty \frac{x^n}{e^x - 1}\, dx$$

This function appears in the theory of blackbody radiation, Bose–Einstein condensation, and the heat capacity of lattice vibrations. Taking $n = 2$ and using the fact that $2! = 2$,

$$\zeta(3) = \int_0^\infty \frac{1}{2} \frac{x^2}{e^x - 1}\, dx$$

Program **Power Function Simulator** so that it calculates the integrand of $\zeta(3)$, and then run **Simpson Test** with appropriate choices for **Number of Samples**, **Lower Limit**, and **Upper Limit**. Remember to use Mathscript element-wise operators. You'll encounter problems if you choose zero for **Lower Limit**. Find an acceptable solution to this problem. Also, you can't, of course, make **Upper Limit** equal to infinity. Because of the nature of the integrand, however, it's acceptable to take **Upper Limit** as something like *100*. Why? Optimize your choices to obtain as accurate a value as possible for $\zeta(3)$. This integral cannot be done analytically. However, high-precision numerical calculations have determined that to a good approximation $\zeta(3) \approx 1.2020569031595942853997$. By comparison with this "correct" answer, how many decimal places of accuracy does **Simpson's Rule** deliver?

3. As an alternative to implementing a loop's auto-indexing capability, an array can be built using the following subdiagram. Here, **Build Array** constructs an array by appending one element each iteration and the ever-increasing array is stored from one iteration to the next in a shift register. The shift register is initialized with the **Empty Array Constant** shown, which can be created by popping up on the left shift register (after the code within the For Loop has been completed) and selecting **Create>>Constant**.

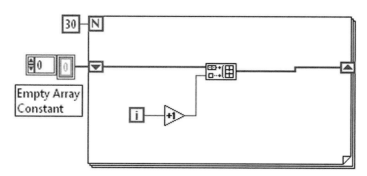

Using a modified form of the above diagram, write a program called **Multiples of 3, 4, or 5** that forms a single array containing all of the integers up to 30 that are evenly divisible by either 3, 4, or 5. You may find the **Compound Arithmetic** icon in **Functions>Programming>>Boolean** useful.

4. Write a Case Structure–based program called **Prime Detector**, which determines whether an input integer is a prime number. The front panel of this VI should appear as shown next, where the integer to be tested is input at **Integer** and, if it is determined to be prime, the **Round LED** labeled **Prime?** is lit.

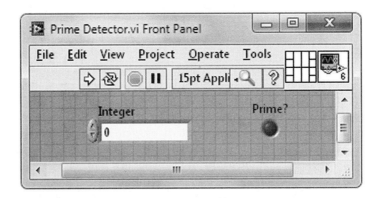

To construct your VI, consider the following properties of prime numbers. By definition, a prime number is evenly divisible by 1 and itself (i.e., only these two divisors result in a remainder of zero). Since even integers are evenly divisible by 2, even integers are not prime. An odd integer N is not prime if it can be written as the product of two integers, where one integer is less than or equal to $\sqrt{N}$ and the other is greater than or equal to $\sqrt{N}$. Thus, to test whether an odd integer is prime, it is sufficient to show that it is not evenly divisible by all of the odd integers less than or equal to $\sqrt{N}$, with the exception of 1. Note that a sequence of odd integers, starting at 3, can be constructed from $N = 2i + 3$, where $i = 0, 1, 2, \ldots$

To test your VI, try inputting 104717 (prime) and 99763 (not prime). Which of the following integers are prime: 101467, 102703, 97861?

5. Open **Simpson Test** and, using **File>>Save As...**, create a new VI called **Simpson Test (Built-In VI)**. On the block diagram, replace **Simpson's Rule** with **Numeric Integration.vi** (found in **Functions>>Mathematics>>Integration & Differentiation**) with its integration method input wired to **Simpson's Rule**. Then, with **Number of Samples** equal to *101*, run **Simpson Test (Built-In VI)** to evaluate the integral $\int_0^1 5x^4 dx$. Do you get the same result for **Value of Integral** as when running **Simpson's Rule** with the same front-panel inputs?

6. Write a program called **Temperature Scale Converter**, which converts given temperatures between the Celsius and Fahrenheit scales continuously, until a **Stop Button** is pressed.

   (a) Place an **Enum** control on the front panel and program it with the following two items: *C to F* and *F to C*. On the block diagram, wire the Enum's terminal to the selector terminal of a Case Structure, and then program the **C to F** and **F to C** case to execute the calculations $F = 1.8C + 32$ and $C = (F - 32)/1.8$, respectively. The completed front panel should appear as shown next. Run **Temperature Scale Converter** with several inputs and verify that it functions correctly, for example, $68°\,F \Leftrightarrow 20°\,C$, $100°\,C \Leftrightarrow 212°\,F$.

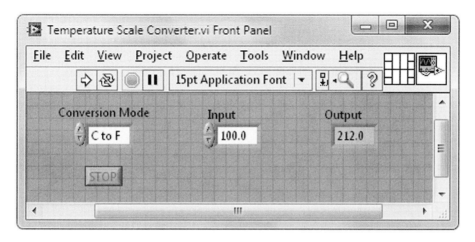

   (b) Add code to your VI so that the text above the **Input** control and **Output** indicator appropriately adapt to the **Conversion Mode** selection as the program runs. To accomplish this feat, note that the owned label of a control or an indicator cannot be changed during runtime, but its caption can be changed. Thus, in the front-panel pop-up menus of **Input** and **Output**, deactivate **Label** and activate **Caption**. Then, on the block diagram, pop up on the **Input** terminal and select **Create>>Property Node>>Caption>>Text**. Place the resulting **Property Node** within the **C to F** Case, pop up on it, and select **Change To Write** to switch it from an indicator to a control. Pop up on the Property Node again and select **Create>>Constant**, and then enter the text *Input deg C* into the resulting **String Constant**. Repeat this procedure to make the caption for **Output** be *Output deg F*. Then, program the **F to C** Case so that the captions for **Input** and **Output** are *Input deg F* and *Output deg C*, respectively. Run your VI and verify that the captioning performs properly, for example, as shown below.

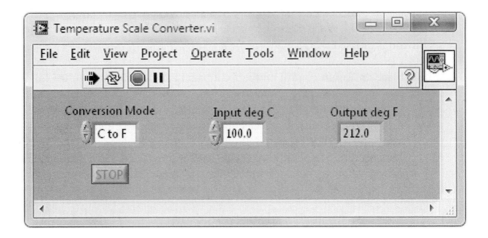

7. Write the program shown below called **Guess Number**. This VI randomly gener-
    ates an integer in the range from 1 to 10 and guides you in correctly guessing the
    integer. The front panel appears as shown next. Here, when the VI is running, you
    will type your guessed integer into the **Guess** control and then pass this value to
    the block diagram by clicking on the **OK Button** labeled **Enter Guess**. You will
    receive feedback about your guess (e.g., too low, too high, correct) in the **Message**
    string indicator.

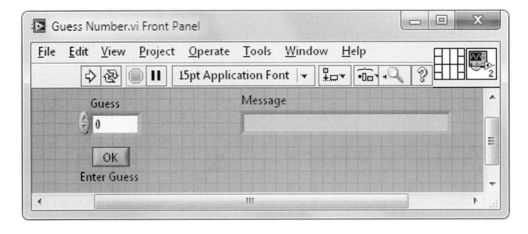

    Code the block diagram shown below, which is based on a nested While
Loop and Case Structure, a program architecture called a *state machine*. In a state
machine, the While Loop executes continuously until its ⬤ is set to TRUE, and

with each loop iteration, one of the Case Structure cases (termed a *state*) executes. The state that executes during a particular iteration performs some operation (e.g., checks the equality of two quantities) and, in addition, selects which state will be executed during the next iteration.

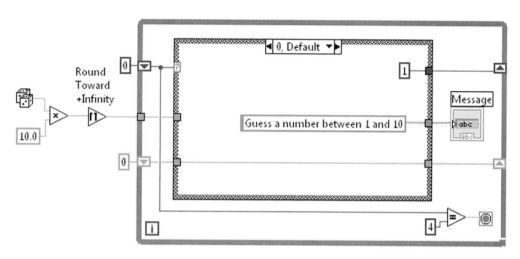

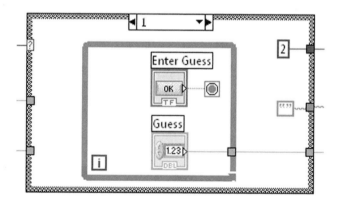

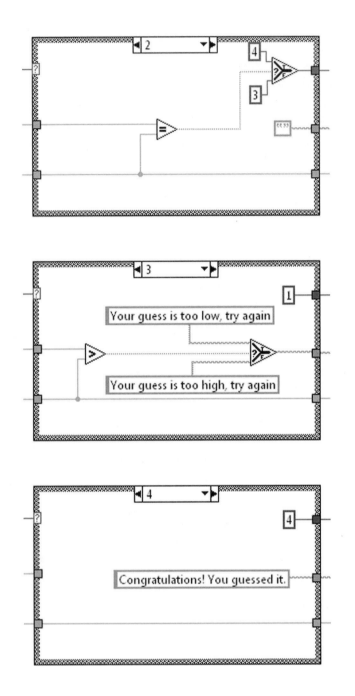

When completed, run **Guess Number** and verify that it performs as expected. For this state machine, what quantity is stored in the upper shift register? How about the lower shift register? In each of the five states (*0, 1, 2, 3,* and *4*), what operation is carried out and what state is selected as the next state?

8. The response of a low-pass filter, that is, its gain G as a function of frequency f, is given by

$$G = \frac{1}{\sqrt{1+\left(f/f_{3dB}\right)^2}}$$

where f_{3dB} is the filter's 3-dB frequency. The following Mathscript-based subdiagram executes a linear plot of G vs. f over the range from $f = 1$ Hz to $f = 10$ kHz with $f_{3dB} = 2000$ Hz. Remember that $.^\wedge$ is the element-wise power operator. Manually choose the data type for the G output to be **DBL 1D**.

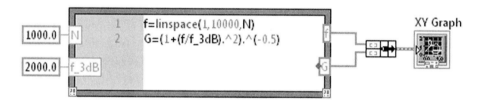

The gain in decibels is defined as dB $= 20\log_{10}(G)$. A *Bode magnitude plot* graphs *Gain (in dB)* vs. f, where the frequency axis is logarithmic. By modifying the subdiagram given above, write a program called **Low-Pass Filter Plot**, which allows a user to graph the low-pass filter response on an XY Graph either as a linear plot or as a Bode magnitude plot. Let the user choose the type of plot via a Boolean switch labeled **Bode?**, which selects between two cases in a block-diagram **Case Structure**. For the logarithmic frequency axis needed for the Bode magnitude plot, first create a Property Node by popping up on the XY Graph terminal, then selecting **Create>>Property Node>> X Scale>>Mapping Mode**. Then pop up on the Property Node, select **Change To Write**, and wire a constant *1* to it. To create a linear frequency axis, you will then need a similar Property Node with a *0* wired to it. For extra flair, program the *y*-axis label as *Gain* or *Gain (dB)* for the Linear or Bode plot, respectively, using the Property Node **Y Scale>>Name Label>>Text**.

CHAPTER 8

Data Dependency and the Sequence Structure

8.1 DATA DEPENDENCY AND SEQUENCE STRUCTURE BASICS

LabVIEW is a *dataflow* computer language. This approach to programming is contrary to the *control flow* mode in which most other programming languages operate. In control flow systems, the program elements execute one at a time in an order that is coded explicitly within the program. The series of sentence-like statements in, for example, C and BASIC programs, which describe the sequential execution of Procedure *A* followed by Procedure *B* followed by Procedure *C*, are manifestations of the inherent control flow fashion in which these text-based languages are structured. In contrast, LabVIEW program elements abide by the principle of dataflow execution. Obeying a condition termed *data dependency*, a given object on the block diagram (called a *node*) will begin execution at the moment *all* of its input data become available. After completing its internal operations, this node will then present processed results at its output terminals. The interesting advantage of dataflow programming is that several nodes (for example, *A*, *B*, and *C*) can, through LabVIEW's multitasking ability, effectively execute in parallel. Such parallel execution will occur whenever the receipt of node *A*'s inputs overlaps in time with the execution sequence of node *B* and/or *C*. This multitasking ability has the potential to enhance system throughput, especially in situations (not uncommon to data-acquisition operations) where a portion of a particular node's execution time involves waiting. In LabVIEW, useful parallel processes can be executed while one node waits for an event, whereas in a control flow system such waiting periods are simply dead time, putting the entire execution sequence on hold.

Despite the impression that you might glean from the previous paragraph, LabVIEW programs can, if desired, be written with a guaranteed one-step-at-a-time sequence of node execution. A proficient LabVIEW programmer may exploit the data dependency maxim in coding a block diagram where the output of node *A* is used as a source of input

338

data to node *B*. Such a diagram will mimic control-flow execution in that *A* must fully execute to enable the execution of *B*. The wiring configurations just described will many times provide an elegant solution to the need for ordered execution within a LabVIEW program. However, to create the correct configurations requires some skill on the programmer's part.

If an elegant data-dependent wiring configuration isn't forthcoming, LabVIEW provides the *Sequence Structure*, an explicit and foolproof way to obtain control flow within a program. The Sequence Structure comes in two styles—the *Flat Sequence Structure* and the *Stacked Sequence Structure*—both of which are found in **Functions>>Programming>>Structures**. In the developmental history of LabVIEW, the Flat form of the Sequence Structure was introduced later than the Stacked form, and most LabVIEW aficionados now agree that the Flat Sequence Structure produces the most readable code. Hence, we will only use the Flat Sequence Structure in the programs we develop in this chapter.

When it is initially placed on a block diagram, the **Flat Sequence Structure** looks like a single frame of movie film, as shown in the following illustration.

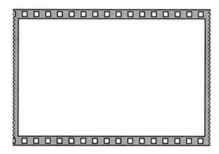

You can add a second frame, however, by popping up somewhere on the structure's border and selecting **Add Frame After**.

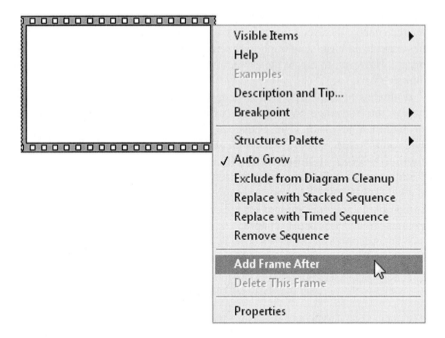

Then, the original and the newly created frame will appear side by side, as below. The area enclosed by each frame can be resized using the 🢖.

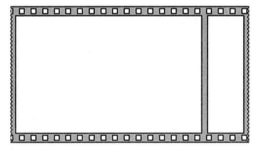

By repeating this procedure, you can add any number of new frames. For example, in the next illustration a Flat Sequence Structure with three frames has been created, with its frames labeled sequentially from left to right.

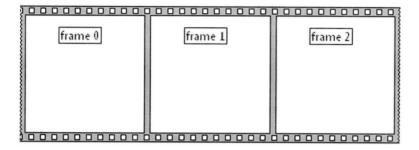

A Flat Sequence Structure executes sequentially from left to right. Hence, in a structure similar to that shown above, **frame 0** executes first, followed by **frame 1**, then **frame 2**, and so on until the last frame executes. Thus, by placing objects within different frames, you can use the Sequence Structure to control the order of execution among nodes that are not naturally coupled through data dependency. In addition, if a quantity produced in one frame is needed in a subsequent frame, that quantity can be passed between frames through a *tunnel*. For example, if one wanted to read the value of a front-panel Numeric Control, wait one-half second, and then display the value on a front-panel Numeric Indicator, that three-step sequential process can be accomplished by the following code.

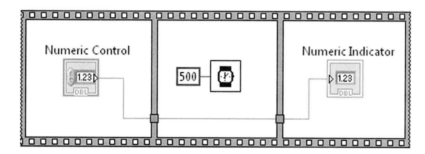

Finally, data can be input to any frame of the Flat Sequence Structure through a tunnel. Each frame can also emit output data through a tunnel, but, consistent with dataflow principles, these data will not be output from the structure until its last frame completes execution. Thus, the three-step sequential process described above can be alternately coded as shown next.

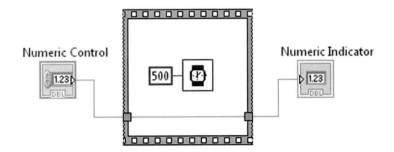

8.2   EVENT TIMER USING A SEQUENCE STRUCTURE

A timer, which measures the duration of a given event, provides an example of an instrument in which operations must proceed in a specific order—a START initiates the timing action, followed by a STOP that terminates it. In building a LabVIEW timer VI then, it is perfectly natural to employ a Sequence Structure on the block diagram. Let's build such a timer and use it to measure something interesting. We will use our timer to compare the time it takes each of LabVIEW's two repetitive-operation structures, the For Loop and the While Loop, to build a given array of data. Our findings may surprise you!

Build the following front panel. Use **File>>Save** to create a folder called **Chapter 8** within the **YourName** folder, and then save this VI in **YourName\Chapter 8** under the name **Loop Timer (Sequence)**. Change the data-type representation of the **Number of Elements** control and the **For Loop Array** and **While Loop Array** indicators to **I32** and then resize these objects so that they may accommodate large integer values. Format the **For Loop Time (ms)** and **While Loop Time (ms)** indicators as **U32**.

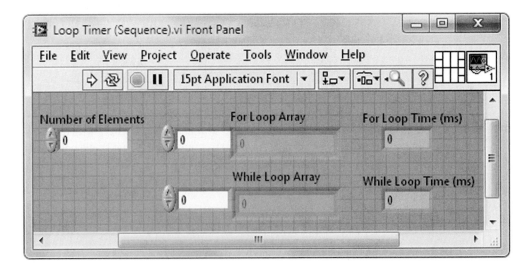

Switch to the block diagram. The basis of our timer will be the **Tick Count (ms)** icon. This VI is found in **Functions>>Programming>>Timing** and its Help Window is reproduced next. **Tick Count (ms)** outputs an unsigned 32-bit (**U32**) integer that denotes the number of milliseconds that have elapsed since your computer was powered on.

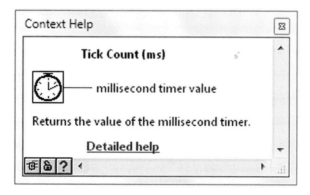

Our strategy for measuring elapsed time will be this: Call **Tick Count (ms)** at the START of an event and call it at the event's STOP. Then, by subtracting the two **Tick Count (ms)** output values, the event's elapsed time will be determined.

Place a **Flat Sequence Structure** on the block diagram. Inside the Sequence Structure's frame, put a **Tick Count (ms)** icon and wire its **millisecond timer value** output to a tunnel on the frame's border. This wire will contain your computer's internal clock value at the START of the For Loop, which will be contained in the next frame.

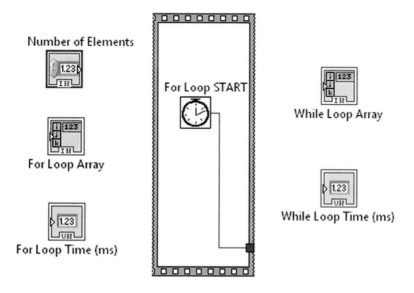

Create a second frame by selecting **Add Frame After** from the Sequence Structure's pop-up menu. Inside this frame, code a For Loop that creates an *N*-element array, where each array element has a numerical value equal to the value of its index. Output this array to the **For Loop Array** indicator on the front panel so that you may view its elements. *N* is determined by the **Number of Elements** control, which is wired from outside the Sequence Structure, through a tunnel, to the For Loop's count terminal.

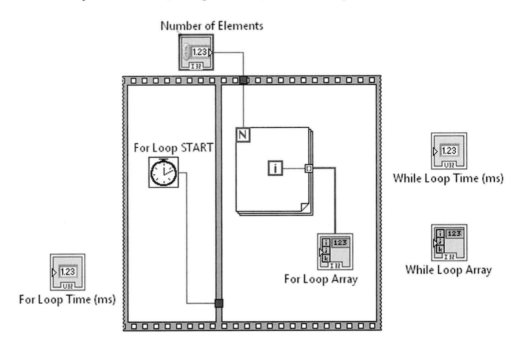

Create a third frame. Place a **Tick Count (ms)** icon there to ascertain your computer's clock value at the For Loop STOP. Then use **Subtract** to find the difference in milliseconds between the For Loop STOP time and START time (obtained in the "tunneled" wire from the initial frame), and output the result to the front panel's **For Loop Time (ms)** indicator. Finally, the **Tick Count (ms)** output in this frame will also be used as the While Loop START value. Wire the output of this icon to a tunnel, so that its value will be available for use in later frames.

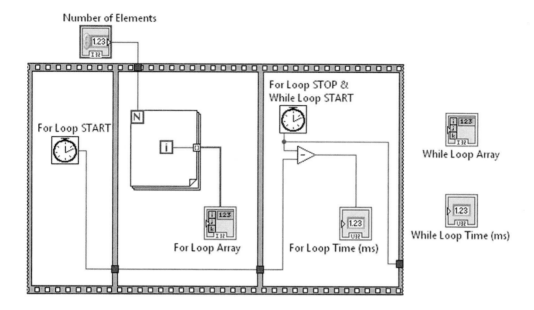

Create a fourth frame and enclose within it the While Loop–based code shown next. As we learned in Chapter 2, by wiring $N - 1$ to the lower terminal of the **Equal?** icon, the While Loop will create an N-element array. Remember to select **Enable Indexing** at the While Loop tunnel; the While Loop default is **Disable Indexing**.

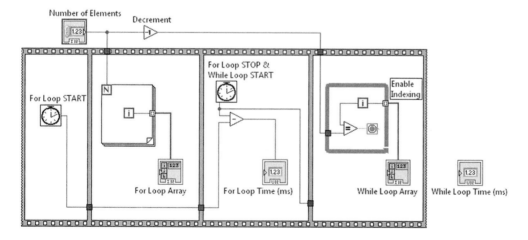

Finally, create a fifth frame. Here, determine the While Loop STOP through the use of a **Tick Count (ms)**, and then (using the "tunneled" wire from the third frame) subtract the While Loop START from it. Output this value for the elapsed While Loop execution time to the front panel's **While Loop Time (ms)** indicator.

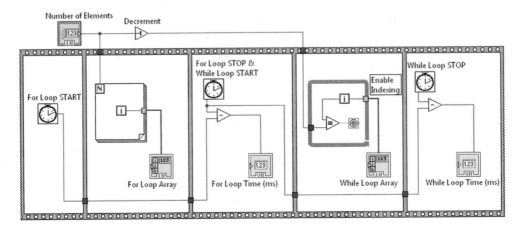

The VI is now complete, so return to the front panel and save your work. We will use **Loop Timer (Sequence)** to place the For Loop and While Loop in head-to-head competition to see if either provides superior performance in the task of assembling an *N*-element array.

Prior to running this program, you will need to input an appropriate value for *N* in the **Number of Elements** control. The proper choice for *N* will depend on the speed and memory capabilities of your computer. You should make *N* large enough so that it takes each of the two loops at least a sizable number of (at least 10) milliseconds to complete the array-building process. For times less than 10 milliseconds, the accuracy of **Tick Count (ms)** becomes an issue, as discussed below. Since the data type for **Number of Elements** is **I32**, the largest allowed value for *N* is $2^{31} - 1 = 2,147,483,647$. However, if you make *N* too large, you will find that **Loop Timer (Sequence)** produces a runtime error indicating that the program has overwhelmed your computer's free memory.

To get started, choose a value for *N* in the range of *1,000,000* to *10,000,000*. Then, through an iterative process, determine an appropriate *N* value for your system.

Once you've established a good value for *N*, get a fresh start by closing **Loop Timer (Sequence)** and then reopening it. Also, close as many other application programs as possible, so that your processor can devote the bulk of its resources to running LabVIEW. Input your *N* value, and then run the VI. Note the resultant values of **For Loop Time (ms)** and **While Loop Time (ms)**. You should find that the For Loop produces the array significantly faster than the While Loop does. You may wish to peruse the array indicators to verify that the exact same array is being produced by the For Loop and While Loop methods.

Without closing the VI, try rerunning it a second, third, fourth, and more times and noting **For Loop Time (ms)** and **While Loop Time (ms)** for each run. You should find that the second run produces output times noticeably shorter than the initial program execution (that is, the first run after opening the VI). Subsequent runs then reproduce nearly the same times, with only a small fluctuation on the order of milliseconds.

Why does the While Loop take significantly more time to produce an array than the For Loop? In comparing the array-producing code for each loop, note that the While Loop requires an **Equal?** icon to determine whether further iterations are required, but the For Loop has no such requirement. Perhaps then it is the extra execution times required for the **Equal?** operation during each iteration that account for the slower While Loop speed. To test this idea, make the loops equivalent (in regard to the **Equal?** icon) by adding an **Equal?** within the second frame's For Loop as shown below.

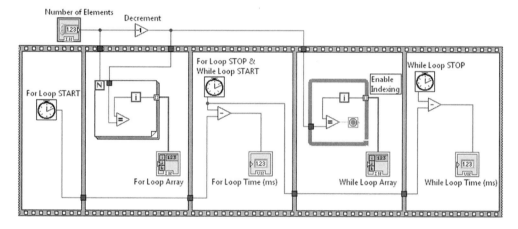

Run the VI. How do **For Loop Time (ms)** and **While Loop Time (ms)** compare once this addition is made to the VI? Does the extra time required for executing the **Equal?** icon significantly slow down the For Loop–based array-building code?

To understand the true reason why a While Loop takes more time to produce an array than a For Loop, you must first understand how LabVIEW uses memory. When LabVIEW launches, a single block of memory (either physical RAM or virtual memory) is allocated for all of the program's editing, compiling, and execution operations. During runtime, a memory manager allocates memory for tasks as needed, with the constraint that arrays and strings must be stored in contiguous blocks of memory. If the memory manager is unable to find a block of unused memory that is large enough for a particular string or array, a dialog box appears to indicate that LabVIEW was not able to allocate the required memory.

Now let's consider what happens when using the auto-indexing feature of a looping structure to build a data array. In the case of a For Loop, LabVIEW can predetermine the size of the array to be built based on the value wired to the Loop's count terminal. Thus, the memory manager is only called once as the loop initiates execution and at that time allocates the appropriate block of memory necessary to hold the array.

With a While Loop, however, the final array size cannot be known in advance. A new element is appended to the existing array with each loop iteration and this process continues until a TRUE Boolean value at the conditional terminal causes the loop to complete its execution. Because the array is constantly increasing in size as the loop iterates, the memory manager must be called continually to find an appropriately sized chunk of RAM to hold the ever-growing data array. These repeated calls to the manager take time, especially if memory starts to become scarce. In a memory-tight situation, the manager may take extra time as it tries to shuffle around other blocks until a suitable space opens up.

Thankfully, because of some built-in LabVIEW intelligence, a While Loop's array-building capability is not as handicapped by the memory manager as it might appear from the above description. To avoid calling the manager with each iteration, the While Loop auto-indexing feature instructs the manager to allot enough new memory to store not just a single new array element, but rather a large number of additional array elements each time it is called. Through this trick, the number of memory-manager calls necessary in building a given sized array becomes rather small. Of course, when the loop completes its execution, there most likely will be some unused memory associated with the array because of the overgenerous manner in which the manager delivered memory on its last call. Thus, when the loop terminates, LabVIEW simply directs the manager to dissociate this excess memory from the now-complete array. The result of this shrewd use of the memory manager is that the While and For Loops' array-building capabilities are not widely divergent (say, by powers of 10) in their performance, as you discovered through **Loop Timer (Sequence)**.

Above, we noted that the **Loop Timer (Sequence)** VI runs more slowly during its first run than it does during its subsequent executions. This observation can be traced to use of the memory manger. During the program's first run, the manager works out the memory allotments peculiar to the particular need of that VI. During subsequent runs, many of the first-run memory assignments are simply reused, diminishing the time-consuming use of the manager.

Finally, what causes the small fluctuations in execution times during the second, third, fourth, ... runs? Assuming that your processor is not being asked to multitask between many open applications, these fluctuations are caused by the intrinsic accuracy of LabVIEW's built-in timing functions. These icons mark time by counting interrupts to your system's CPU which occur every 1 ms on a Windows system. It is this one-millisecond resolution in the **Tick Count (ms)** icon that accounts for the observed timing fluctuations. The lesson to be learned here is that LabVIEW's timing VIs provide a simple and effective method of measuring the duration of an event with 1-ms resolution. In situations that demand submillisecond timing (such as is common in acquiring a sequence of analog-to-digital conversions), however, these icons are totally inadequate. For these high-accuracy timing applications, the submicrosecond-resolution clock on a National Instrument's data acquisition (DAQ) board can be used.

8.3 EVENT TIMER USING DATA DEPENDENCY

A skillful programmer can exploit LabVIEW's data-dependent mode of execution to force two or more nodes to execute sequentially. Thus, it is almost always possible to impose sequential execution on a block diagram without the use of a multiframe Sequence Structure.

In some cases, it is natural for node *A*'s output to be used as the input of node *B*. Then these "chained together" nodes will execute in the order *A-B*. LabVIEW's File I/O icons provide such an example. Let *A* and *B* be the two file-related icons **Open/Create/Replace File** and **Write To Text File**. Typical of all File I/O VIs, each of these icons has an input and output called **file path** (or **file**) and **refnum out**, respectively. If *A* is (somehow) provided an input **file path**, when it completes its execution, **refnum out** is output. By wiring this output to *B*'s **file** input, one guarantees that *A* executes fully before *B*. Look back at your work in Section 5.8. We implemented this method in the **Spreadsheet Storage (OpenWriteClose)**. Additionally, you will find that LabVIEW icons with an associated purpose (e.g., file I/O, DAQ) have **error in** input and **error out** output terminals that can be used to chain together several related nodes to ensure sequential execution.

Many times, however, the output of *A* does not constitute a natural input for *B*. In such situations, ordered execution can still be obtained by an alternate method called *artificial data dependency*. In this scheme, it is the mere arrival of data, with no regard to its actual value, that triggers the execution of a node. An example of artificial data dependency will be constructed on the following pages.

Let's try to write a timer diagram, whose ordered execution is controlled by artificial data dependency. First, open **Loop Timer (Sequence)** and use **Save As...** to create a new VI called **Loop Time (Data Dependency)** in **YourName\Chapter 8**. Keep the front panel unchanged as shown.

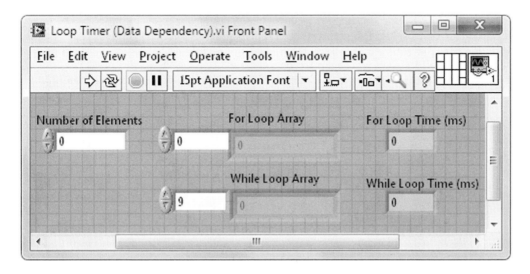

Switch to the block diagram. Eliminate the Flat Sequence Structure by popping up on its border and selecting **Remove Sequence**. Code the For Loop shown at the left in the next diagram, which creates an *N*-element array with the numerical value of each element equal to its index (the right portion of this diagram has not been changed yet). Wire the output of a **Tick Count (ms)** icon to the border of the For Loop. **Tick Count (ms)**'s output then is an input to the For Loop, meaning that the loop cannot initiate execution until the Tick Count (ms) icon produces a value. This clock value will be used as the For Loop START. The START value is never used within the For Loop, so it is simply the arrival of Tick Count's output that triggers the loop execution—an example of artificial data dependency.

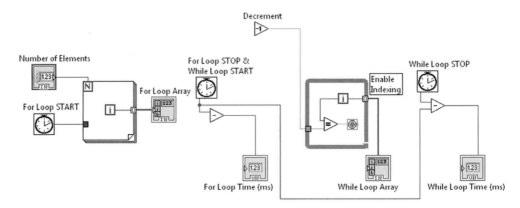

After the For Loop ceases execution, we want to obtain the For Loop STOP value using a **Tick Count (ms)** icon and then calculate the elapsed execution time in milliseconds. The trick for assuring this sequence of operations is to enclose the "STOP-value" code within a single-frame Flat Sequence Structure as shown next. The code within this Sequence Structure will execute upon receipt of a value at the structure's input tunnel. Wire the Start value, through the For Loop, to an output tunnel (disable the loop's auto-indexing feature). Then wire this For Loop output over to the Flat Sequence Structure input. The completion of the For Loop will then trigger the execution of the Sequence Structure's code, as desired.

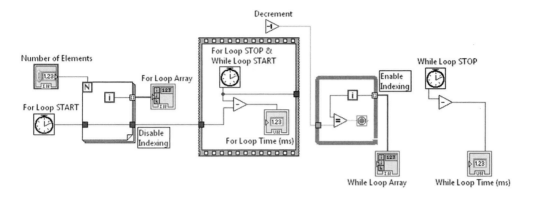

Also, when coding the Sequence Structure above, the enclosed **Tick Count (ms)**'s output is wired to an output tunnel on the structure's border. We will next turn our attention to a While Loop, which creates an *N*-element array. This tunnel outputs the While Loop START value and the arrival of this value at the While Loop will trigger the loop's execution—another example of artificial data dependency.

Code the array-building While Loop, whose execution is triggered by the arrival of the While Loop START value, as shown next. For array building, you will have to enable auto-indexing at the While Loop's border.

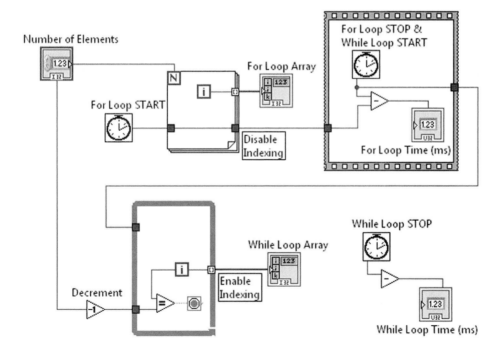

Complete the diagram as follows to calculate the time it takes for While Loop–based array building.

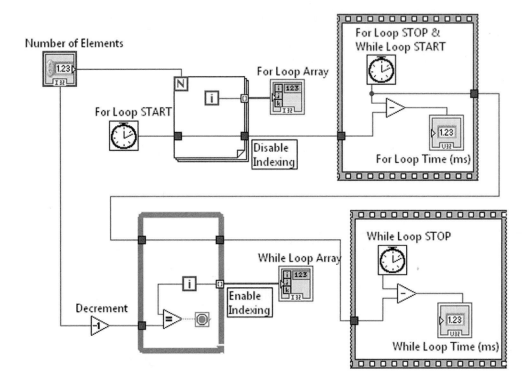

Return to the front panel and save your work. Run the VI with your optimized choice for **Number of Elements**. Check to see that the expected *N*-element array is built by both the For Loop and the While Loop. If your diagram is correct, the performance of **Loop Time (Data Dependency)** should be equivalent to that of **Loop Timer (Sequence)**.

8.4   HIGHLIGHT EXECUTION

To cement your understanding of the manner in which this VI executes, let's implement one of LabVIEW's handiest (and coolest!) debugging tools called *Highlight Execution*. Input something small, such as *10*, for **Number of Elements** on the front panel, and then switch to the block diagram. In the toolbar, enable **Highlight Execution** by clicking on the button containing the light bulb.

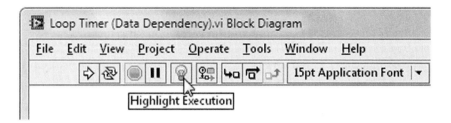

Then click on the **Run** button. The VI will execute in slow-motion animation, with the passage of data marked by bubbles moving along the wires and important data values given in automatic pop-up probes. Because the diagram executes in slow motion, the values for **For Loop Time (ms)** and **While Loop Time (ms)** obtained under Highlight Execution will be meaningless. However, this mode of execution provides a beautiful visual demonstration of the concept of artificial data dependency.

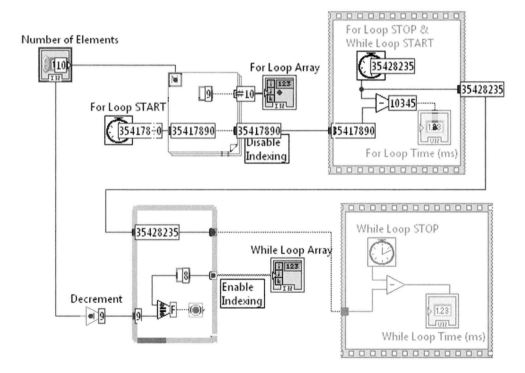

Use **Highlight Execution** liberally in your future work (but remember to turn it off once you are finished using it by clicking again on the light bulb button). This LabVIEW feature is an invaluable debugging tool for troubleshooting VIs that are causing you problems.

DO IT YOURSELF

Write a VI called **Reaction Time** that measures a user's reaction time, defined as the elapsed time from when **Square LED** is lit until the user presses the **Stop Button**. The front panel for this program includes a **Stop Button** and **Square LED** Boolean indicator suitably enlarged, as well as a **Numeric Indicator** labeled **Reaction Time (ms)**, as shown below.

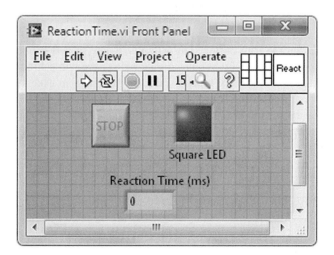

Reaction Time should execute in the following sequence of steps: (1) The user presses the **Run** button and then quickly moves the mouse cursor over to the **Stop Button**; when the **Run** button is pressed, **Square LED** and **Reaction Time (ms)** are initialized as *unlit* and *0*, respectively; (2) at a random time (within the range of *3* to *8* seconds) after the **Run** button has been pressed, **Square LED** becomes lit; (3) on the **Reaction Time (ms)** indicator, the value of the elapsed time in milliseconds since **Square LED** was lit is displayed continuously, and then the displayed value is halted at the moment that the user presses the **Stop Button**; and (4) **Square LED** becomes unlit.

A few helpful tips follow.

- A random number in the range from *0* up to *1* can be obtained using the **Random Number (0–1)** icon, which is found in **Functions>>Programming>>Numeric**.
- Think carefully about the appropriate choice for the **Mechanical Action** option in the Stop Button's pop-up window.
- Properties of a front-panel object (e.g., its color, visibility, position) can be controlled from the block diagram using a **Property Node**. For example, to create a **Property Node** that lights **Square LED**, pop up on its block-diagram icon terminal and select **Create>>Property Node>>Value**. The resultant **Property Node** can then be placed anywhere on the block diagram (i.e., removed from the **Square LED**'s icon terminal)

and will appear as shown at the left in the next illustration. The outward-directed arrow indicates that this icon is currently configured as an indicator, which reads **Square LED's** current **Value** (TRUE or FALSE). To change this icon into a control, pop up on it and select **Change To Write**. Its arrow now is directed inward, indicating the **Property Node** is a control, as shown next at the right. By wiring a TRUE Boolean constant to this icon, **Square LED's Value** will be TRUE, causing the front-panel indicator to be lit. Conversely, wiring a FALSE Boolean constant to the icon will make **Square LED** unlit. You are free to place an unlimited number of Property Nodes on the block diagram for any given front-panel object.

PROBLEMS

1. Write a VI called **Five Blinking Lights (Sequence Structure)**, which sequentially illuminates five LEDs on its front panel. As shown below, place five **Round LEDs** on the front panel and label them *0* through *4*. Using a Sequence Structure, develop a program that lights these LEDs one at a time in the order 0-1-2-3-4. Make each LED remain lit for 0.2 second and have the 0-1-2-3-4 lighting pattern repeat continuously until a front-panel **Stop Button** is pressed. To make multiple copies of an LED's block-diagram terminal, pop up on the terminal and select **Create>>Property Node>>Value** as described in the **Do It Yourself** project.

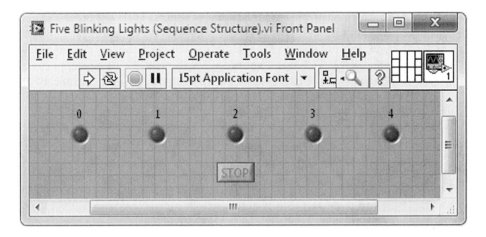

2. To demonstrate that a nested While Loop and Case Structure can accomplish the same function as a Sequence Structure, write a program called **Five Blinking Lights (State Machine)**. On the block diagram, use a nested While Loop and Case Structure to create code that repeatedly blinks five front-panel **Round LEDs** in the order 0–1–2–3-4 until a front-panel **Stop Button** is pressed, as outlined in Problem 1. You should find that this block-diagram architecture (which is called a *state machine*) provides a more elegant solution to the Five Blinking Lights problem than a Sequence Structure (e.g., no Property Nodes are required). You may find the following icon useful: **Quotient & Remainder** in **Functions>>Programming>>Numeric**.

3. Write a program called **Magic Stop Button**, whose front panel initially appears blank as shown in the next illustration.

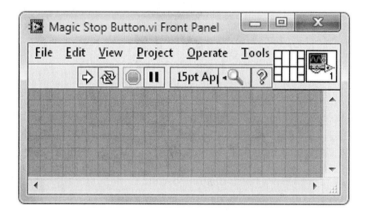

As shown next, when this VI is run, a blinking Stop Button appears on the front panel until the user presses it. The Stop Button then disappears and the program stops. To write this program, first place a Stop Button on the front panel, pop up on it and select **Visible>>Caption**, and then make the caption read *Press this button to stop the VI*. Then pop up on the Stop Button and select **Advanced>>Hide Control** and secure this choice by selecting **Edit>>Make Current Values Default** and saving the VI.

Code the block diagram so that when the program is run, first the Stop Button becomes visible, then it blinks until it is clicked, and finally the Stop Button is made not visible. This features of the Stop Button (e.g., its visibility, blinking) can be controlled by creating appropriate Property Nodes (see the discussion in the **Do It Yourself** project).

4. The subdiagram shown below generates and plots 100 data samples on a Waveform Chart.

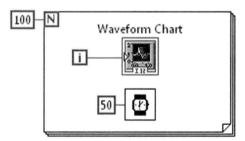

(a) Write a VI called **Two Charts (Simultaneous)**, which performs two such 100-sample plots simultaneously. That is, place two Waveform Charts on the front panel labeled **Chart 1** and **Chart 2**, and program the block diagram so that the 100-sample plots on **Chart 1** and **Chart 2** are performed over the same time interval.

(b) Write a VI called **Two Charts (Data Dependency)**, which first completes the 100-sample plot on **Chart 1** and then performs the 100-sample plot on **Chart 2**. Use artificial data dependency on the block diagram to accomplish this feat (i.e., do not use a Sequence Structure).

(c) Finally, add code to **Two Charts (Data Dependency)** so that the two Waveform Charts are cleared at the end of a run. To accomplish this task, place a single-frame Sequence Structure on the block diagram and use artificial data dependency to assure that this block-diagram object is the last item to execute before the VI completes a run. Within the Sequence Structure, place two plot-clearing Property Nodes, one associated with each Waveform Chart. The appropriate Property Node is made by popping up on the Waveform Chart's terminal and selecting **Create>>Property Node>>History Data**. After changing this Property Node to a control by selecting **Change to Write** in its pop-up menu, **Create>>Constant** will produce the needed input called an **Empty Array**, as shown below.

Waveform Chart

5. Write a program named **Parallel While Loops with Reset** in which two While Loops execute in parallel on the block diagram, simultaneously producing independent plots of random numbers (in the range 0 to 1) on two front-panel Waveform Charts, until stopped by the click of a single front-panel Stop Button. Label one of the Waveform Charts **Chart 1** and the other **Chart 2**.

(a) First, try coding **Parallel While Loops with Reset** using the following block diagram to accomplish the intended goal (i.e., halt simultaneously executing While Loops with a single Stop Button). Run this program and show that it does not produce two simultaneous plots. Describe briefly what this diagram does do instead and explain why.

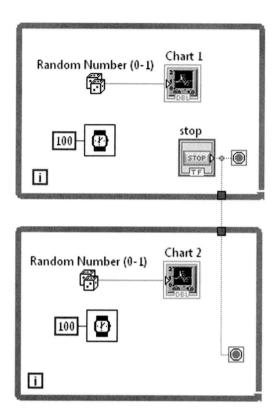

(b) For the correct diagram, one While Loop is stopped by wiring directly to the Stop Button's terminal and the other by wiring to a **Value** Property Node associated with this terminal. To produce the needed Property Node, pop up on the Stop Button's terminal and select **Create>>Property Node>>Value**. When wiring this Property Node, you will get a broken wire until the Stop Button's **Mechanical Action** is changed to something other than a "latch" mode (e.g., select **Mechanical Action>>Switch When Pressed**.) Complete the diagram for **Parallel While Loops with Rest** and verify that it runs as intended.

(c) When associated with a **Value** Property Node, the Stop Button must be in a "change value" (rather than a "latch") mode. Thus, when the Stop Button is pressed to stop the VI, the button will be left in its TRUE value as the VI completes execution, which is inconvenient for subsequent uses on this program. To remedy this problem, output a quantity from each While Loop and use artificial data dependency to reset the value of the Stop Button back to FALSE before the VI completes its execution. Run this final version of **Parallel While Loops with Rest** and verify that it executes as intended.

6. Explore the performance of the following alternate array-building diagrams. These diagrams purposely avoid the use of a loop's auto-indexing feature in an effort to make explicit what this feature does automatically. Use Help Windows to understand the functioning of the unfamiliar array-related icons (found in **Functions>>Programming>>Array**).

(a) Code the For Loop–based block diagram given below, which initializes an appropriate-sized array (with each element defined as zero) prior to the loop structure and then avoids further calls to the memory manager through the use of **Replace Array Subset** within the loop itself. Save this VI under the name **Loop Timer (Alternate#1)**. Run **Loop Timer (Alternate#1)** and then **Loop Timer (Sequence)** to compare the time it takes each program's For Loop to create the same N-element array of integers. Are the times comparable or is one program significantly faster?

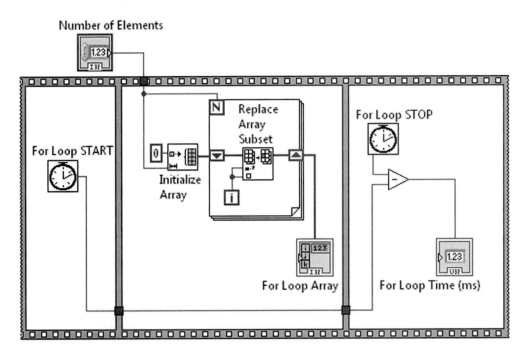

(b) Next, program the following While Loop–based diagram, which initializes the shift register with an empty array and then appends elements to it with each loop iteration. Save this VI under the name **Loop Timer (Alternate#2)**. Run **Loop Timer (Alternate#2)** and then **Loop Timer (Sequence)** to compare the time it takes each program's While Loop to create the same N-element array of integers. Why is **Loop Timer (Alternate#2)** such a poor performer? Although not time-efficient, in nondemanding applications (e.g., involving small-sized array), this diagram is commonly used to perform "real-time" graphing (with a calibrated *x*-axis) inside a While Loop by feeding the **Build Array** output to the terminal of a **Waveform Graph** (see Problem 7).

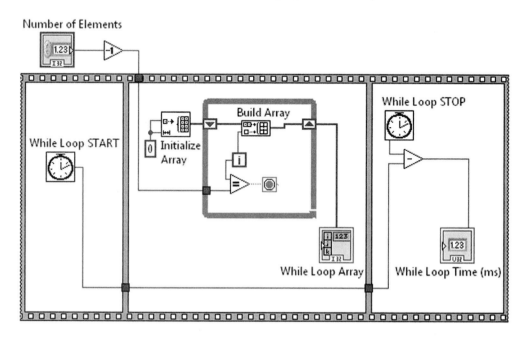

(c) This final While Loop–based diagram initializes a larger-than-needed array (a 15,000,000-element array is shown; your choice of *N* may require something different), sequentially places data values in the first *N* elements of this array through the use of **Replace Array Subset**, and then lops off the excess remainder of the array using **Array Subset**. Program this VI as shown in the next illustration and save it under the name **Loop Timer (Alternate#3)**. Run **Loop Timer (Alternate#3)** and then **Loop Timer (Sequence)** to compare the time it takes each program's While Loop to create the same *N*-element array of integers. Why does **Loop Timer (Alternate#3)** execute quickly?

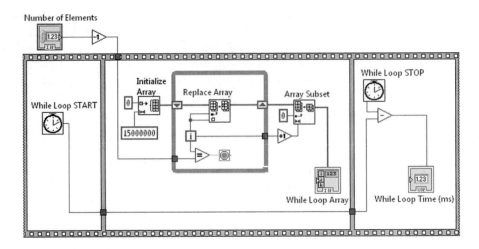

7. Code **Real-Time Waveform Graph**, whose block diagram is shown below. This diagram builds an array of data within a While Loop, and this ever-growing array is plotted on a Waveform Graph during each iteration.

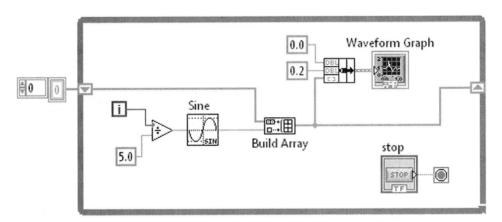

On the front panel, turn off autoscaling on the Waveform Graph's *x*-axis and manually scale this axis to run from *0* to something large (e.g., *20000*). Run **Real-Time Waveform Graph** and, based on your observations, identify the circumstances under which (e.g., after how many iterations) the design of this VI leads to poor performance. Explain why this diagram is a poor performer when run for a large number of iterations.

8. Write a VI called **For Loop Time (Icon vs. Mathscript)**, which compares the required array-building time *T* of a LabVIEW For Loop icon and a Mathscript Node. To build the arrays, use the diagrams shown next.

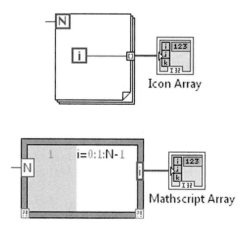

Icon Array

Mathscript Array

For various values of N, what is T for each diagram? Which diagram is the best performer?

9. The **error in** and **error out** terminals present on many LabVIEW icons can be used to sequence the execution of a collection of these icons via data degeneracy. As an example of this technique, use two **Elapsed Time** and one **Prompt User for Input** Express VIs (found in the subpalettes of **Functions>>Express**) to construct a program named Time To Press OK, which times how long it takes a user to click on a dialog-box **OK** button. Configure the Express VIs as shown next, and then wire them together so that data dependency resulting from the error terminals causes them to execute sequentially from left to right. The **Present (s)** terminal of Elapsed Time outputs the (universal) time in seconds at which this icon executes, so the difference of this value from the two Elapsed Time icons gives the time elapsed from when the VI was started until the user presses the dialog-box OK button. Add code that calculates and displays this elapsed time (in seconds) on a front-panel numeric indicator named Time To Press OK (s).

CHAPTER 9

Analysis VIs: Curve Fitting

In the coming lab exercises, you will be using a *thermistor* as a temperature sensor. A thermistor has the useful characteristic that its resistance vs. temperature $(R - T)$ curve can be fit to a well-known analytic function, facilitating its use as a calibrated thermometer. In Table 9.1, the $R - T$ data for a particular thermistor (Epcos Model B57863S0103F040) is given. If you will be using a different thermistor in the **Do It Yourself** project at the end of this chapter, please use the $R - T$ data for the thermistor that will be employed in your particular experimental setup.

9.1 THERMISTOR RESISTANCE-TEMPERATURE DATA FILE

In this chapter, you will learn how to calibrate a thermistor by fitting its $R - T$ data to a formula known as the *Steinhart–Hart Equation*. This accomplishment is contingent on the ability to read the calibration data into your fitting program. One method we will explore for inputting data into a VI is through reading a disk file that contains the relevant information.

It would be helpful if, prior to coming to lab, you create and save a spreadsheet-formatted file containing a thermistor's $R - T$ data (using either the provided table or the data for your own device). In Chapter 5 you learned that in the spreadsheet format, tabs separate columns and end of line (EOL) characters separate rows.

You can use either a spreadsheet data analysis program or a word processor to create the desired file and store it under the name **Thermistor R-T Data.txt**. For example, I used Microsoft Word to construct the ASCII spreadsheet file shown in the following illustration of the thermistor resistance (in kilohms) vs. temperature (in degrees Celsius) data. When saving the file, it is important to use the **Save** or **Save As...** command and then select **Plain Text (*.txt)**, as opposed to **Word Document (*.docx)**, in the **Format:** selection. This procedure creates an ASCII text file devoid of any Word formatting characters. Take care not to inadvertently add an extra EOL character at the end of the file.

TABLE 9.1   *R–T* Data for Epcos Model B57863S0103F040 Thermistor

Temperature (°C)	Resistance (kΩ)
−50.00	670.1
−45.00	471.7
−40.00	336.5
−35.00	242.6
−30.00	177.0
−25.00	130.4
−20.00	97.07
−15.00	72.93
−10.00	55.33
−5.00	42.32
0.00	32.65
5.00	25.39
10.00	19.90
15.00	15.71
20.00	12.49
25.00	10.00
30.00	8.057
35.00	6.531
40.00	5.327
45.00	4.369
50.00	3.603
55.00	2.986
60.00	2.488
65.00	2.083
70.00	1.752
75.00	1.481
80.00	1.258
85.00	1.072
90.00	0.9177
95.00	0.7885
100.00	0.6800
105.00	0.5886
110.00	0.5112

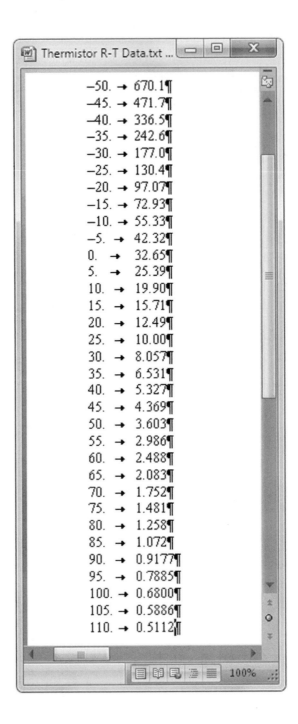

Alternately, if you use a spreadsheet-based program to create the file, you will want to save it as a **Text (Tab delimited) (.txt)** file. This option may be listed under a **Save**, **Save As...**, or **Export** menu item, depending on the program you use. Make sure that you save this file in the correct manner to avoid including a bunch of program-dependent formatting characters in the file. Perhaps the best way to check that you've produced the desired file is to open it using a word processor to see whether it appears as on the previous page.

Bring your spreadsheet file of $R - T$ data with you to lab. You'll need it during your work in this chapter.

9.2 TEMPERATURE MEASUREMENT USING THERMISTORS

Among the many electronic temperature sensors available—including platinum resistance thermometers, thermocouples, and semiconductor devices such as the Analog Devices AD590—thermistors are used in applications that require high sensitivity, small size, ruggedness, fast response time, and low cost. When utilized with the proper support circuitry, a thermistor can measure temperatures to an accuracy of 0.01 °C over a fairly wide range (about 100 °C). The only price to be paid for the thermistor's high sensitivity is that its resistance is a nonlinear function of temperature. In the past, the nonlinearity of thermistors was a distinct disadvantage, causing users to employ resistor-based linearizing networks or less-than-accurate resistance-to-temperature conversion methods such as the Beta formula (see below). Fortunately, the widespread availability of microprocessors has enabled the use of accurate (and somewhat complex) temperature-calibration models, greatly simplifying the use of thermistors in accurate temperature measurement.

The active layer of most thermistors is composed of a semiconducting metal-oxide alloy. As is characteristic of a semiconductor, when the thermistor's temperature is elevated, loosely bound (valence) electrons within its active layer are thermally released from their respective binding sites to become conduction electrons. By this mechanism the thermistor's resistance decreases with temperature and results in the utility of this device as a temperature sensor. Since the bound electronic state is of lower energy, an activation energy barrier E separates the conducting state from the insulating state. Thus, at a given temperature T (in Kelvin), we expect that the probability of such activation, and therefore the density n of conduction electrons, is proportional to a Boltzmann factor. That is,

$$n = n_0 \exp\left[-\frac{E}{kT}\right] \qquad [1]$$

where n_0 is the density of conduction electrons at very high temperatures, and k is Boltzmann's constant. It is easily shown that these conduction electrons imbue the material with a resistivity ρ given by

$$\rho = \frac{m}{ne^2 \tau} \qquad [2]$$

where e and m are the charge and mass of an electron, respectively. The scattering time τ is the average time traveled by a conduction electron until it is scattered by a lattice vibration, an impurity, or some other scattering mechanism peculiar to the material.

Thus, assuming that e, m, and τ are independent of temperature, we expect that the temperature dependence of ρ is entirely caused by the Boltzmann factor related to n. Putting [1] into [2], we predict that

$$\rho = \rho_0 \exp\left[\frac{E}{kT}\right] \qquad [3]$$

where ρ_0 is the temperature-independent constant given by

$$\rho_0 = \frac{m}{n_0 e^2 \tau} \qquad [4]$$

The macroscopic resistance R of a sample is determined by

$$R = \rho \frac{L}{A} \qquad [5]$$

where L and A are the sample's length and cross-sectional area, respectively.

Because a solid sample's geometric shape is little changed with temperature, we expect that the temperature dependence of R will simply mimic that of ρ. Thus,

$$R = R_0 \exp\left[\frac{E}{kT}\right] \qquad [6]$$

where R_0 is the (assumed) temperature-independent constant

$$R_0 = \frac{mL}{An_0 e^2 \tau} \qquad [7]$$

If all of our logic is correct, we predict that the temperature dependence of a thermistor's resistance should be described by Equation [6]. By taking the logarithm of both sides, this relation can be rewritten as the following linear relation

$$\frac{1}{T} = -\frac{k}{E}\ln R_0 + \frac{k}{E}\ln R \qquad [8]$$

or

$$\frac{1}{T} = A + B\ln R \qquad [9]$$

where A and B are constants defined to be $A \equiv -k \ln R_0/E$ and $B \equiv k/E$. Equation [9] is the so-called *Beta formula* (B is sometimes written as β) and it predicts that when a thermistor's temperature-dependent resistance data are plotted as $1/T$ (y-axis) vs. $\ln R$ (x-axis), a straight line will result with the y-intercept and slope equal to A and B, respectively.

Despite this expectation, it is found experimentally that a real thermistor's $R - T$ data exhibit some nonlinearity when plotted as $1/T$ vs. $\ln R$. Rather than being constant, the slope B decreases with decreasing temperature. Over a narrow temperature range (say, of $10\,°C$), B varies so slightly that the Beta formula can predict the thermistor's behavior with a temperature uncertainty of about $0.01\,°C$. However, for the much larger temperature span over which a thermistor is useful, the Beta formula inadequately models the data.

Something then must be missing in our theory. In reviewing the above derivation, one might suspect it is our assumption of a temperature-independent scattering time. It is easy to imagine that as temperature increases, the conduction electrons move faster and/or lattice vibrations become more pronounced and consequently shorten the scattering time τ. While the exponential Boltzmann factor activation of conduction electron density would still dominate the resistance's temperature dependence, the more mild temperature-dependent scattering time can possibly account for the observed slight nonlinearity of a $1/T$ vs. $\ln R$ plot.

How then can we more accurately model the $R - T$ data? Unfortunately, the theory for electrical conduction in metal oxide thermistors is still incomplete and cannot offer us a more refined relation to replace Equation [9]. (In fact, the semiconductor energy band model advocated above is not accepted by all researchers. Some investigators, instead, hold that electric conduction in these materials is caused by the "hopping" of charge carriers from one ionic site to the next.) In light of this theoretical vacuum, the most recent literature on thermistors accounts for the nonlinearity of the $1/T$ vs. $\ln R$ curve using the standard curve-fitting technique of considering $1/T$ to be a polynomial of $\ln R$. That is, the inverse temperature is written as the following nth-order polynomial:

$$\frac{1}{T} = A + B \ln R + C \left(\ln R\right)^2 + \ldots + Q \left(\ln R\right)^n \qquad [10]$$

From this point of view, Equation [9] is such a polynomial that has been truncated at the first-order term. One then is led to the conclusion that by retaining higher-order terms, the equation will remain valid over a wider range of temperatures.

Starting with this idea, researchers have found that the $100\,°C$ temperature span over which a typical thermistor is active can be accurately modeled by a third-order polynomial:

$$\frac{1}{T} = A + B \ln R + C \left(\ln R\right)^2 + D \left(\ln R\right)^3 \qquad [11]$$

This relation can be use to convert resistance to temperature with a precision equal to that of the original $R - T$ data (typically 0.005 to 0.01 °C).

In 1968, the two oceanographers Steinhart and Hart discovered that, when modeling up to 200 °C spans contained within the temperature range of –70 °C to 135 °C (they were especially interested in the oceanographic range of –2 °C to +30 °C), the second-order term in the above relation could be neglected without significant loss of accuracy. Then Equation [11] reduces to

$$\frac{1}{T} = A + B \ln R + D \left(\ln R \right)^3 \tag{12}$$

Equation [12] is called the *Steinhart–Hart Equation* and is widely used in thermistor calibration. For a typical thermistor, the constants A, B, and D have order of magnitude values of 10^{-3}, 10^{-4}, and 10^{-7}, respectively, when resistance is measured in ohms. The temperature must, of course, be on the Kelvin scale.

9.3 THE LINEAR LEAST-SQUARES METHOD

The process of scientific inquiry many times proceeds as follows: Researchers perform an experiment to investigate a particular physical phenomenon and to obtain a set of N data samples (x_j, y_j). After shutting off their instruments, the investigators apply appropriate curve-fitting techniques to determine the functional relationship $y = f(x)$ manifest in their data. Once armed with this $f(x)$, the scientific community can then judge the veracity of theoretical models posited to explain the phenomenon under investigation by testing each model's capacity for correctly predicting the experimentally observed $f(x)$. In this section, we will explore a curve-fitting method that allows one to extract an accurate $y = f(x)$ from a given data set (x_j, y_j), a skill that plays a crucial role in the above-described process. We will put this skill to immediate use in calibrating a temperature-sensing thermistor.

Quite often it is convenient to model a functional relationship $y = f(x)$ as the linear combination of a set of basis functions b_k. The general form of this kind of model is

$$y = f(x) = \sum_{k=0}^{M-1} a_k b_k (x) = a_0 b_0 + a_1 b_1 + \dots + a_{M-1} b_{M-1} \tag{13}$$

where the basis functions $b_k(x)$ are known functions of x and the a_k are the linear expansion coefficients. For example, if $b_k = x^k$, then $f(x) = a_0 + a_1 x + a_2 x^2 + \dots + a_{M-1} x^{M-1}$ is a polynomial of order $M - 1$. Alternately, if $b_k = \cos(kx)$ then $f(x)$ is a function described by a Fourier series. Note that the functions $b_k(x)$ can be nonlinear functions of x. However, the coefficients a_k appear in a linear fashion in Equation [13].

Once the decision is made to model $f(x)$ as a linear combination of a set of basis functions, one needs some method to determine the appropriate choice of the a_k in Equation [13], so that a given data set is accurately described. The theory of data analysis provides just such a method. Because all data-taking processes are subject to random errors, the repeated acquisition of a particular experimental quantity produces an array of values that is distributed as a Gaussian ("bell curve") distribution. By carefully considering the ramification of this fact, it can be shown that Equation [13] will best describe a given N data samples (x_j, y_j) when the a_k are chosen such that they minimize an error function e defined as

$$e \equiv \sum_{j=0}^{N-1} \left[y_j - f\left(x_j\right) \right]^2 = \sum_{j=0}^{N-1} \left[y_j - \sum_{k=0}^{M-1} a_k\, b_k\left(x_j\right) \right]^2 \qquad [14]$$

Note that e is a measure of the deviation between the experimentally determined y_j and the y obtained from the fitting function $y = f(x)$, which depends on the values chosen for the coefficients a_k. Various algorithms have been proposed to identify the a_k that minimizes e for a given data set, all of which are classified under the rubric of *linear least-squares* methods. In LabVIEW, the available methods are the Singular Value Decomposition (SVD), Givens, Householder, Lower Triangular–Upper Triangular (LU) Decomposition, and Cholesky algorithms. The interested reader may refer to a data analysis text such as *Numerical Recipes* (Cambridge University Press) for a description of the details of these methods.

LabVIEW provides built-in VIs that will implement a linear least-squares fit of your data to the functional form of Equation [13]. For applications that require high performance (e.g., fastest possible execution time, smallest possible files) or detailed control over the fitting process (e.g., choice of least-squares algorithm), the low-level curve-fitting icons found in **Functions>>Mathematics>>Fitting** should be employed. In less demanding situations, however, the high-level **Curve Fitting** *Express VI* will provide the same results as code written with low-level icons, with much more programming ease.

In the following pages, you will write a program that fits a thermistor's $R - T$ data to the Steinhart–Hart Equation. For this task, you will employ the **Curve Fitting** Express VI to carry out the linear least-squares fitting algorithm.

9.4 INPUTTING DATA TO A VI USING A FRONT-PANEL CONTROL

To write your curve-fitting VI, you must first understand how to input the given array of $R - T$ data into a LabVIEW program. The most straightforward method for inputting an array of values is from a front-panel Array Control. Let's write a VI that works this way. Construct the front panel shown here. Place two Array Controls (select **Array** shell first and then place **Numeric Control** inside) and label one T (Celsius) and the other R (kilohm). Next, add an **XY Graph** with its x- and y-axes labeled Temperature (Celsius) and Resistance (kilohm), respectively, so that the thermistor data may be viewed graphically. Using **File>>Save**, create a folder named Chapter 9 within the YourName folder, and then store this VI as R-T Data Input (Array Control) in YourName\Chapter 9.

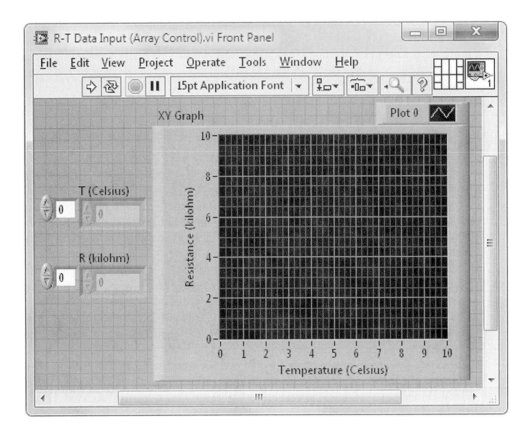

Switch to the block diagram. Write the following code that receives the input arrays, bundles them into a cluster, and then passes this cluster to the **XY Graph** for plotting.

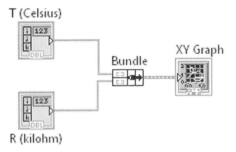

The **Bundle** icon in this diagram creates a cluster consisting of two elements. The input attached to Bundle's top and bottom terminals becomes the index-0 and index-1

elements of the cluster, respectively. When this cluster is passed to the **XY Graph** icon terminal, the index-0 and index-1 elements are plotted as the *x*- and *y*-axes variables, respectively. Thus, the above diagram yields a resistance (*y*-axis) vs. temperature (*x*-axis) plot.

Return to the front panel. OK, here would be the hard part where we manually type all of the temperature values from the thermistor calibration data table into the T (Celsius) control and then type all of the resistance values from the data table into the R (kilohm) control. Demonstrate to yourself how this procedure works by just entering the first five data values from the given thermistor calibration table.

Temperature:	−50.00	−45.00	−40.00	−35.00	−30.00
Resistance:	670.1	471.7	336.5	242.6	177.0

Remember that the small left-hand box portion of the Array Control (called the *index display*) indicates the index of the element being displayed, and the larger right-hand box (the *element display*) is the actual numerical value of that element. You increment or decrement the index display using the 🖑. Also use this tool to highlight the interior of the element display, and then type in the numerical value that you desire.

Now we want to store the information in these two arrays permanently. That is, we wish for the arrays to be initialized with these values each time the VI is opened. To accomplish this feat, select **Edit>>Make Current Values Default**. Save your work.

Run the VI and you should see a graph of the thermistor's resistance as a function of temperature over the range from −50 °C to −30 °C. Try closing and then opening the VI. Repeat this process a few times. Hopefully, you'll find the two front-panel arrays are initialized with the thermistor data each time the program is loaded.

For future reference, you can also input data into a program by placing an array directly on the block diagram using the **Array Constant** icon found in **Functions>> Programming>>Array**. The **Array Constant**, just like the front-panel **Array** shell, comes equipped with an index display as well as an element display receptacle that you stuff with a constant of desired data type (numeric, Boolean, string, or cluster). In this approach, you first position an **Array Constant** on your block diagram and then place a **Numeric Constant** from **Functions>>Programming>>Numeric** (with Representation **DBL**) within its element display receptacle. You then program the Array Constant with the relevant sequence of data values. The resulting block diagram would appear as below.

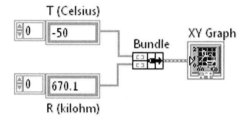

It is considered good LabVIEW style to group related programming objects in a cluster. This approach leads to well-organized front panels, reduces the quantity of block-diagram wires (termed *wire clutter*), and, coupled with the use of **Unbundle By Name**, produces a self-documented VI. Let's reform **R-T Data Input (Array Control)** in this manner. Place a **Cluster** shell (found in **Controls>>Modern>>Array, Matrix & Cluster**)

on a blank region of your VI's front panel and label it **R-T Data**. Using the ®, first drag **T (Celsius)** and then **R (kilohm)** into the **Cluster** shell. The cluster can then be automatically resized by popping up on its border and selecting **AutoSizing>>Size to Fit**.

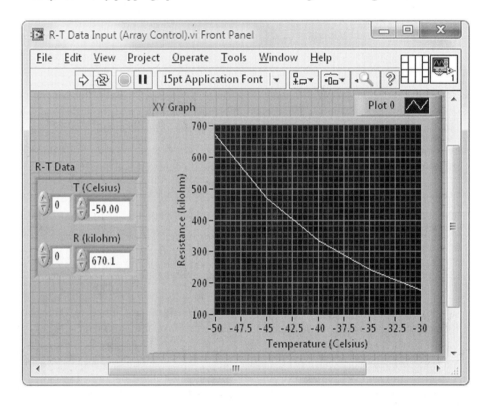

Switch to the block diagram. Remove the broken wires as well as the **Bundle** icon. Then, using the ®, connect the **R-T Data** and **XY Graph** icon terminals with a cluster wire.

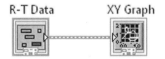

We're finished. That's our program. Return to the front panel and save your work. Run the VI to verify that it is working properly.

There is one subtle point that is important to understand so that you obtain a valid resistance (*y*-axis) vs. temperature (*x*-axis) plot. Namely, the elements in a cluster control are ordered. This ordering is not based on the position of the elements within the cluster shell, but initially is determined by the order in which the elements were placed in the Cluster shell during programming. The first element placed in the Cluster shell is indexed zero, the second element placed there is indexed one, and so on. Given a two-element cluster as input, an XY Graph associates the index-0 and index-1 elements with the *x*- and *y*-axes, respectively. Thus, in our present program, we want the T (Celsius) and R (kilohm) arrays within the Cluster shell to be indexed 0 and 1, respectively. To verify this ordering, pop up on the border of the Cluster shell and select **Reorder Controls In Cluster...** The front panel will change its appearance as shown next.

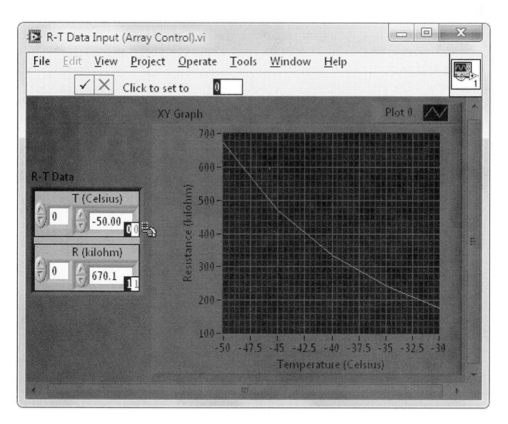

At the bottom right corner of each cluster element, its current cluster index is given in the white box. The mouse cursor has morphed into the cluster order cursor ⌗➘, which can be used to change the indexing of the cluster elements by clicking on their associated black boxes. Once the new index assignments have been made, they can be secured by clicking on the **Confirm** button ☑ or reverted to the original settings with the **Cancel** button ☒.

Using **Reorder Controls In Cluster...** in the Cluster shell's pop-up menu, first order the cluster so that your VI will plot temperature (y-axis) vs. resistance (x-axis). Click on the **Confirm** button, and then run the VI to view the $T - R$ plot. Next, by changing the cluster ordering, create a $R - T$ plot (i.e., graph resistance (y-axis) vs. temperature (x-axis)). Save your work as you close this VI.

9.5   INPUTTING DATA TO A VI BY READING FROM A DISK FILE

An alternate method for inputting data into a VI is by reading it in from a previously created disk file. Let's write a LabVIEW program that can receive and plot the thermistor spreadsheet data that you generated prior to coming to lab. If you haven't created this file yet, now is the time to do it.

Construct this front panel, which simply contains an **XY Graph** with x- and y-axes labeled **Temperature (Celsius)** and **Resistance (kilohm)**, respectively. Design an icon for this VI, and then save it in **YourName\Chapter 9** under the name **R-T Data Input (Spreadsheet)**.

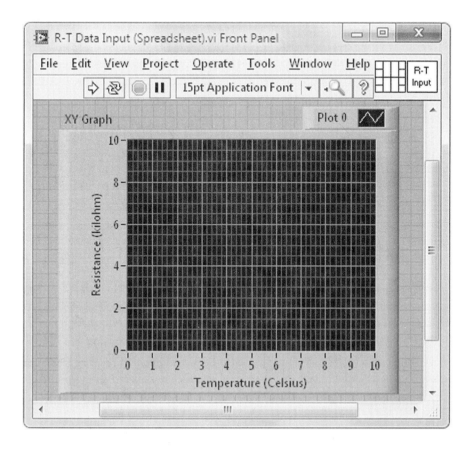

Switch to the block diagram. An ASCII spreadsheet file can be read into your VI using the **Read From Spreadsheet File.vi** icon found in **Functions>>Programming>>File I/O**. The Help Window for this icon (when configured for double-precision floating-point numbers as explained below) is shown in the following illustration. As with all of the more sophisticated LabVIEW functions, **Read From Spreadsheet File.vi** can be used for a lot of specialized purposes and therefore has many input and output connections. Default values for its inputs are shown in parentheses. On its Help Window, an icon's inputs are labeled in boldface, plain, or dimmed text to denote whether each input is *required*, *recommended*, or *optional*, respectively. Required inputs (**Read From Spreadsheet File.vi** has none of these) must be wired, or else the icon will not execute. Recommended and optional inputs control features available for your use, if desired. Optional inputs are less commonly used and so their labeling does not appear (as in the given Help Window) unless you click on the **Show Optional Terminals and Full Path** button ⊞ in the Help Window's bottom left-hand corner.

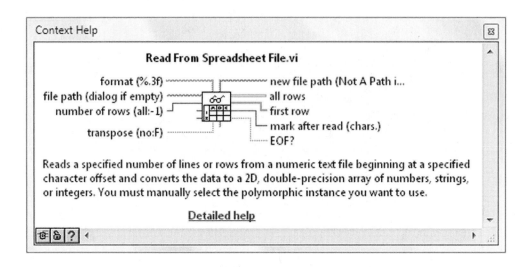

Place **Read From Spreadsheet File.vi** on your diagram. This icon is termed polymorphic, meaning that it can be configured to read spreadsheet files consisting of double-precision floating-point numbers, integers, or strings. This format choice is made using the icon's *Polymorphic VI Selector*. Operate the Polymorphic VI Selector either by clicking on it with the 🖑 as shown below at the left or by popping up on it and choosing **Select Type** as shown to the right. Since our data file consists of floating-point numbers, use one of these methods to program the icon for its **Double** mode.

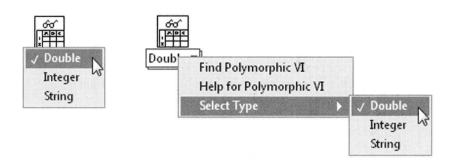

Use **Create>>Constant** to wire a **Path Constant** containing your spreadsheet-formatted $R - T$ data file's path to the **file path** input (to find a Windows file's exact path, right-click on the file's icon or name and select **Properties**). If you leave this input unwired, a dialog window will appear when the program is run, which will prompt you for the desired file.

Also, increase the precision of numerical values from the insufficient default value of 3 decimal places to 4 (or more, if needed) by wiring a **String Constant** containing *%.4f* to the **format** input. Finally, since we desire to read the entire file, which is the default value for the **number of rows** input, leave this input unwired.

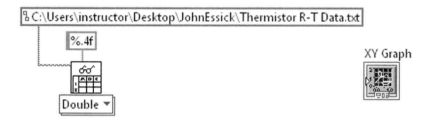

9.6 SLICING UP A MULTI-DIMENSIONAL ARRAY

Read From Spreadsheet File.vi will present your data at its **all rows** output in the form of a 2D array. This array's two columns then will have to be separated ("sliced off") into two 1D arrays to facilitate an XY plot. Separating out the columns is done using the **Index Array** icon (found in **Functions>>Programming>>Array**). Place this icon on your diagram. Note that it originates with just one **index** input.

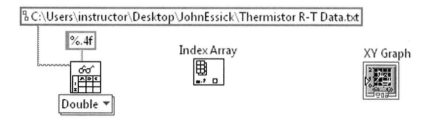

Next, wire Read From Spreadsheet File.vi's **all rows** output to the **n-dimension array** input of Index Array. The 2D array wire will appear and **Index Array** responds by automatically producing a second **index** input.

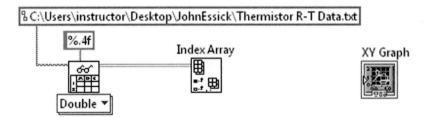

The top and bottom index inputs, which appear as small filled and hollowed-out black boxes, are the row and column indices, respectively. First, let's retrieve the temperature values from the 2D array. We know that the temperature values are all contained in the index-zero (first) column. So wire a **Numeric Constant** containing a *0* to the column (bottom) index of **Read From Spreadsheet File.vi**.

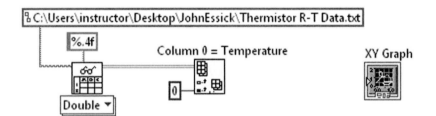

Note that once it is wired to the **Numeric Constant**, the column index input becomes a filled black box, which indicates the particular column selected. Now if we wanted to pick out, for example, the eighth temperature value in this column, which has the row index of seven, we would wire a **Numeric Constant** containing *7* to the top index input. **Index Array** would then output this single array element.

However, instead of a single temperature value, we want the entire column of temperature values. To instruct **Index Array** to "slice off" the entire column, we leave the row index input unwired. When unwired, row indexing is disabled (i.e., no particular row is selected), and the icon will output all rows in the selected column in the form of a 1D array. Note that the unwired row index input appears as a hollowed-out box, indicating that indexing at this input has been toggled off.

Repeat the above procedure to slice off the resistance data, which is in the index-one (second) column.

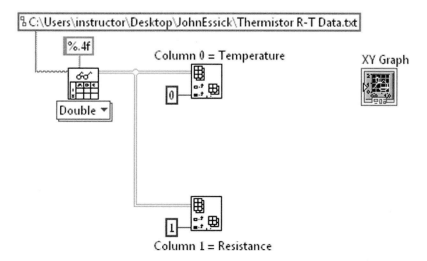

The **Index Array** icons below will each output a 1D array. Bundle these two 1D arrays together to form an *xy* cluster and wire this cluster to the XY Graph's terminal.

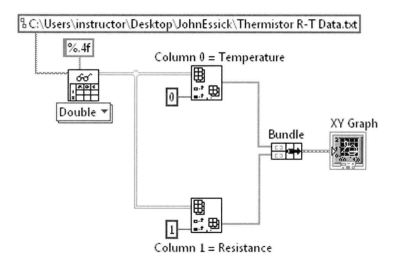

Return to the front panel, save your work, and then run your VI. You should be presented with a beautiful graph of your thermistor's resistance (*y*-axis) vs. temperature (*x*-axis). Note the exponential-looking decay of resistance with increasing temperature. This is a signature of the thermally activated process taking place within the material of which the thermistor is composed.

If you encounter an error when your VI tries to open the data file, carefully check that the file's path within the block diagram's **Path Constant** is perfectly correct (to find a Windows file's exact path, right-click on the file's icon or name and select **Properties**). Further check to make sure that your data file is correctly constructed—such an error may be easily detected by running your program with strategically placed Probes and/or using Highlight Execution. An extraneous EOL character at the end of the data file is a common problem.

In the following exercise, we will use **R-T Data Input (Spreadsheet)** as a subVI within a program that fits $R - T$ data to the Steinhart–Hart equation. We will need access to the given temperature and resistance data in two different guises: Celsius-kilohm as well as Kelvin-ohm.

To provide the data in the two required forms, add two indicator clusters to your VI as follows. First, pop up on the cluster wire emanating from the **Bundle** icon and create an indicator cluster called **Celsius-kilohm** using **Create>>Indicator**.

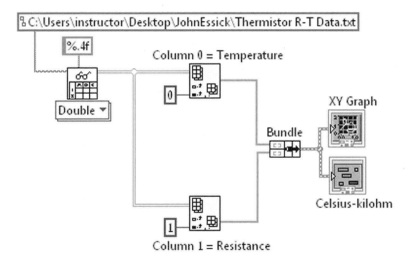

Switch to the front panel. Position the **Celsius-kilohm** cluster indicator appropriately. Make the owned label of each array within this cluster visible by popping up on its index display and selecting **Visible Items>>Label**. Label the top and bottom arrays

T (Celsius) and **R (kilohm)**, respectively. Do not carry out the labeling using the ![A icon] as this will create free, rather than owned, labels.

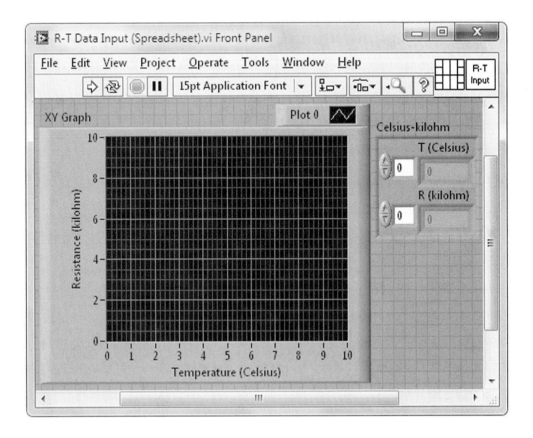

Switch back to the block diagram and complete it as shown next. This code converts the array of temperatures given in Celsius to the Kelvin scale (by adding the scalar constant *273.15* to each element of the Celsius temperature array). Also, the resistances given in kilohms are converted to ohms (by multiplying each element of the kilohm resistance array by the scalar constant *1000*). If you use **Create>>Constant** to produce the two required **Numeric Constant**s (programmed with *273.15* and *1000*), create these constants first (i.e., before connecting the array wires to the **Add** and **Multiply** icons), or else an Array Constant will be created, rather than a scalar Numeric Constant. After you connect the two inputs to **Bundle**, the **Kelvin-ohm** indicator cluster can be produced using **Create>>Indicator**.

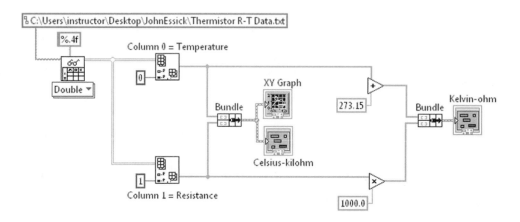

Switch to the front panel. Position the **Kelvin-ohm** indicator cluster appropriately, and then name the top and bottom arrays **T (Kelvin)** and **R (ohm)**, respectively, by selecting **Visible Items>>Label** in each array's pop-up menu. Finally, assign the connector pane's terminals consistent with the Help Window shown.

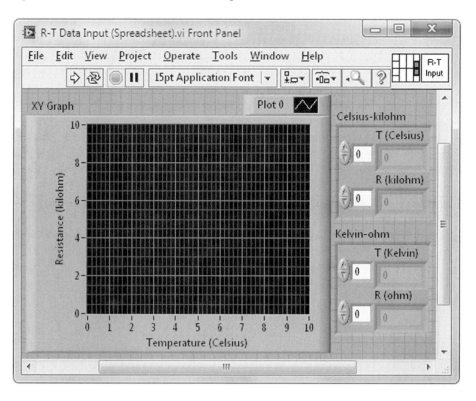

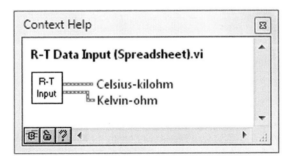

Run the VI and make sure that the correct temperature and resistance values appear in each cluster. For example, make sure that there is not an extra zero as the last element in an array (if an extra zero is present, you need to correct your **Thermistor R-T Data.txt** file). Then, save your work as you close this VI.

9.7 CURVE FITTING USING THE LINEAR LEAST-SQUARES METHOD

Now that you know how to input your thermistor data into a program, let's write a VI that uses the linear least-squares method to fit these data to the Steinhart–Hart Equation (Equation [12]). At first glance, the Steinhart–Hart Equation does not appear to follow the form assumed by the linear least-squares theory (Equation [13]). However, if we define $y \equiv 1/T$ and $x \equiv \ln(R)$, the Steinhart–Hart Equation becomes

$$y = A + Bx + Dx^3 \qquad [15]$$

which is of the form given in Equation [13] with the basis functions $b_0 = 1$, $b_1 = x$, and $b_2 = x^3$ and the coefficients $a_0 = A$, $a_1 = B$, and $a_2 = D$. Equation [15] is called the *linearized Steinhart–Hart Equation*. The strategy for our curve-fitting VI is this: Input the thermistor's $R - T$ data, recast these data as $x = \ln R$ and $y = 1/T$, and then use the **Curve Fitting** Express VI to perform a linear least-squares fit to obtain the values of A, B, and D.

The **Curve Fitting** Express VI is found in **Functions>>Express>>Signal Analysis**. The Help Window for this icon (when configured for linear least-squares method) is shown in the next illustration.

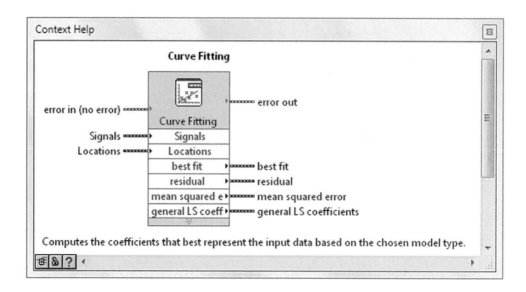

Here, it is assumed an experimentalist has acquired N data samples (x_j, y_j). These data are input to the Express VI in the form of an x and y N-element array at its **Locations** and **Signals** input, respectively. After programming it with M basis functions $b_k(x)$, the VI will find the particular linear combination of the given $b_k(x)$ that best describes the data. That is, the data input to this icon are fit to Equation [13] and the optimized values of the coefficients a_k are output at the **general LS coefficients** terminal. Beside the optimized values for a_k, this icon outputs the array of "best fit" y-values where

$$y_j^{\text{best fit}} \equiv f\left(x_j\right) = \sum_{k=0}^{M-1} a_k\, b_k\left(x_j\right) \qquad\qquad [16]$$

Also output are the N-element residual array R, where this array's jth element is defined as

$$R_j \equiv y_j - y_j^{\text{best fit}} \qquad\qquad [17]$$

and the *mean square error* (mse), which is defined as

$$\text{mse} \equiv \frac{1}{N}\sum_{j=0}^{N-1} R_j^{\,2} \qquad\qquad [18]$$

The residual array as well as the mse can be taken as a measure of the error in your fitted curve.

Build the front panel given below. To compare the resulting best fit *y*-values with the original data, include an **XY Graph** with two differentiated plots called **Data** and **Best Fit** (for example, select different colors or point styles and disable interpolation for one curve in the **Plot Legend**; for a review of how to place multiple plots on an XY Graph, see Section 7.9). Also, display the determined values of a_k in an Array Indicator labeled **Best-Fit Coefficients**. Pop up on this indicator and format it to display numerics in scientific notation (using **Display Format...**). Save this VI as **R-T Fit and Plot** in YourName\Chapter 9.

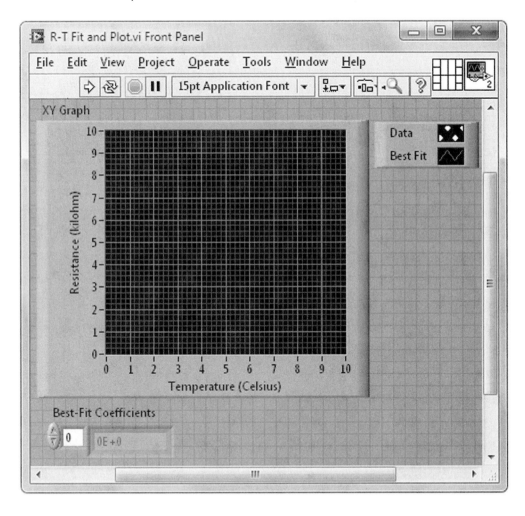

Switch to the block diagram and place the **Curve Fitting** Express VI on it. When first placed on the block diagram, the Express VI will open the dialog window shown next, where you will configure its operation. From among the several possible fitting options, select **General least squares linear**. The box labeled **Models** will become activated, and it is here that you define the basis functions to be used in the linear least-squares fitting operation. We will be fitting our data to the linearized Steinhart–Hart Equation, whose basis functions are $b_0 = 1$, $b_1 = x$, and $b_2 = x^3$, so enter 1, x, and x^3 on the first, second, and third lines of the **Models** box, respectively.

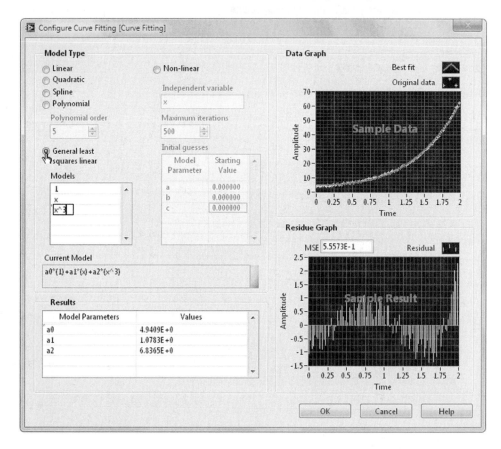

Click on the **OK** button. The dialog window will close and the **Curve Fitting** icon will appear on your block diagram with its input and output terminals expanded (except for **error in** and **error out**).

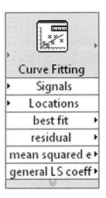

Place **R-T Data Input (Spreadsheet)** on the block diagram and then complete the code shown in the next illustration, which delivers the $y = 1/T$ and $x = \ln(R)$ arrays at Curve Fitting's **Signals** and **Locations** inputs, respectively. **Natural Logarithm** is found in **Functions>>Mathematics>>Elementary & Special Functions>>Exponential Functions**.

Express VIs accept and return information in a format called the *Dynamic Data Type (DDT)*, which is rendered as a dark-blue banded wire. The DDT bundles related information about the data it carries. For example, when carrying an array of numeric values, the DDT also includes the data's label name as well as the date and time the data were acquired. When you wire **Reciprocal**'s output to Curve Fitting's **Signals** input, the **Convert to Dynamic Data** icon will be inserted into the wire automatically. This icon converts the 1D floating-point numeric array produced by **Reciprocal** to the dynamic data type used by Express VIs such as Curve Fitting. Similarly, the **Convert to Dynamic Data** icon will be inserted automatically when wiring **Natural Logarithm**'s output to Curve Fitting's **Locations** input. If you ever need to place **Convert to Dynamic Data** on a block diagram manually, it is found in **Functions>>Express>>Signal Manipulation**.

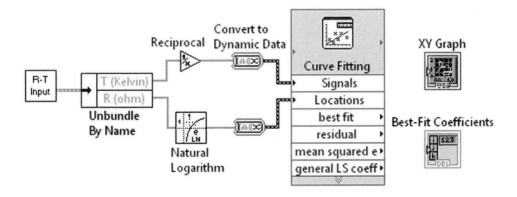

Finally, add the following code that plots the original data and the best-fit results for comparison. Here, you will have to place the **Convert from Dynamic Data** icons (found in **Functions>>Express>>Signal Manipulation**) on the diagram manually. When **Convert from Dynamic Data** is first placed on the block diagram, its dialog window will open; select **Conversion>>1D array of scalars-automatic** and **Scalar data type>>Floating point numbers (double)**, and then press the **OK** button. Curve Fitting's **best fit** output contains the array of $y_j^{\text{best fit}}$ values. Since $y = 1/T$, where T is in Kelvin, this array must be converted to Celsius temperatures prior to the comparison plot.

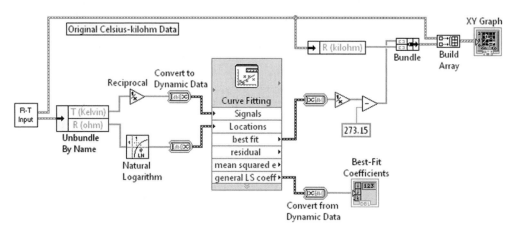

Return to the front panel and run your VI. Upon completion, the **Best-Fit Coefficients** array indicator will display the Steinhart–Hart's A, B, and D constants at its *0*, *1*, and *2* indices, respectively.

For convenient reading, you can make all three array elements visible at once, if you like, by the following procedure. First, place the ⬆ in the corner of the **Best-Fit Coefficients** indicator on the front panel so that it morphs into a *grid cursor* ⊞.

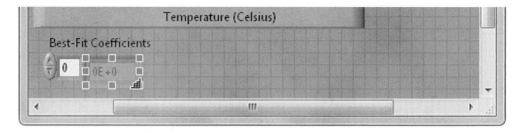

Then drag the grid cursor to the right, until it has created two additional indicators.

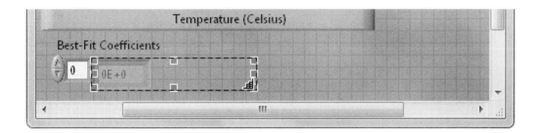

When you release the mouse button, indicators for array elements *0*, *1*, and *2* will then be visible. The index display provides the index for the element displayed in the leftmost indicator.

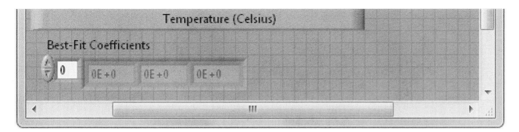

Record the resulting values of *A*, *B*, and *D* for your thermistor for future use. Do your *A*, *B*, and *D* have the order of magnitude values of 10^{-3}, 10^{-4}, and 10^{-7}, respectively, which are typical of most thermistors? Does the **XY Graph** indicate that the fitted Steinhart–Hart Equation accurately describes the temperature-dependent resistance of your thermistor?

9.8 RESIDUAL PLOT

Is there a quantitative method for judging how well the Steinhart–Hart Equation describes the thermistor's $R - T$ data? Yes, there is, and the Curve Fitting Express VI provides us with such a method. After it is has determined the array of best-fit *y*-values, Curve Fitting calculates the residual defined as $R_j \equiv y_j - y_j^{\text{best fit}}$ for each index *j*. We see that the residual is simply the deviation of the best-fit value from the actual *y*-value data input at **Signals** and so provides a quantitative measure of the "goodness" of the fitting procedure. All of the residual values R_j from $j = 0$ to $j = N - 1$ are output as an *N*-element array (contained within a dynamic data type wire) at the Curve Fitting's **residual** output.

In general, for a quantitative display of the deviation of Curve Fitting's best-fit values from the input experimental data, one can simply plot the **residual** array output vs. some appropriate quantity (such as the *x*-array input at **Positions**). However, in our present problem, we linearized the Steinhart–Hart Equation via $y = 1/T$, and so the residual calculated by Curve Fitting is the deviation of the inverse Kelvin temperature. That is, the **residual** output *R* is given by

$$R \equiv y - y^{\text{best fit}} = \frac{1}{T} - \frac{1}{T^{\text{best fit}}} \qquad [19]$$

where T is the experimental Kelvin temperature and $T^{\text{best fit}}$ is Curve Fitting's best-fit Kelvin temperature. Defining $\Delta y \equiv y - y^{\text{best fit}}$ and $\Delta(1/T) \equiv 1/T - 1/T^{\text{best fit}}$, we see that

$$R \equiv \Delta y = \Delta\left(\frac{1}{T}\right) \qquad [20]$$

To obtain a more intuitive measure, let's define the residual of Kelvin temperature as the "temperature residual" $R^T \equiv T - T^{\text{best fit}} = \Delta T$. Then, if $y = 1/T$, we know that for small ΔT (in comparison to T itself),

$$\Delta y \approx \frac{dy}{dT}\Delta T = \frac{d}{dt}\left(\frac{1}{T}\right)\Delta T = \left(-\frac{1}{T^2}\right)\Delta T \qquad [21]$$

Since $R = \Delta y$ and $R^T = \Delta T$, Equation [21] shows that the temperature residual can be determined from Curve Fitting's residual output using the relation

$$R^T \approx -T^2 R \qquad [22]$$

Since R^T measures a difference of Kelvin temperatures, its units can be in either Kelvin or Celsius degrees because the degree interval is of the same size on both of these temperature scales.

Place a second **XY Graph** on the front panel of **R-T Fit and Plot** and label its x- and y-axes **Temperature (Celsius)** and **Temperature Residual (Celsius)**, respectively.

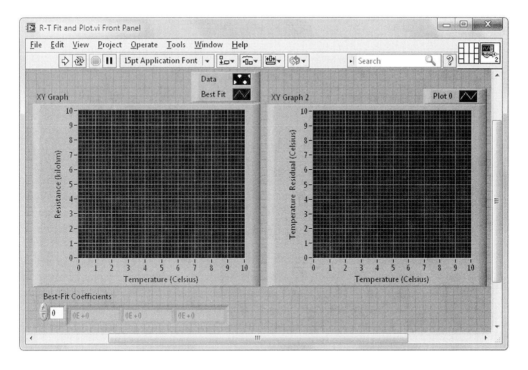

Switch to the block diagram and add the code shown below, which calculates the temperature residual and sends it to the front panel for display.

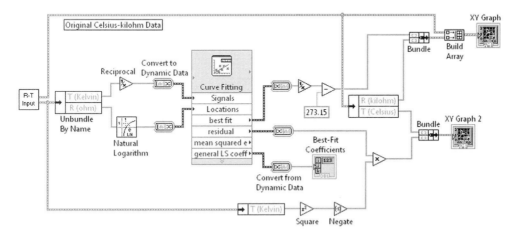

Return to the front panel and save your work. Then, run your VI. Is there a temperature range within which the Steinhart–Hart consistently overestimates the temperature value ($R^T < 0$) ? Is there a range within which the temperature is consistently underestimated ($R^T > 0$)? If you need your thermistor to be calibrated with an accuracy of ± 0.05 °C (i.e., the Steinhart–Hart value falls within ± 0.05 °C of the actual temperature), within what temperature range can you operate your thermistor?

When curve fitting, one expects that a better fit can be achieved if the number of basis functions used in the fitting algorithm is increased. Thus, we would expect that a third-order polynomial (Equation [11]) should fit our thermistor data better than the Steinhart–Hart Equation (Equation [12]), which deletes the second-order term.

Let's see if this conjecture is true. Switch to the block diagram and double-click on the **Curve Fitting** Express VI. When the dialog window opens, select **General least squares linear**, and then, in the **Models** box, program the VI to perform a fit using the four basis functions $b_0 = 1$, $b_1 = x$, $b_2 = x^2$, and $b_3 = x^3$. Then click on the **OK** button.

With this modification in place, return to the front panel. Expand **Best-Fit Coefficients** to include four indicators, and then run the VI. How has the accuracy (as determined from the Temperature Residual plot) of your fit improved? Is your result consistent with Steinhart–Hart's claim that the second-order term can be deleted from the fitting equation without significant loss of accuracy? Remember that this claim comes to us from an era when each additional term would have to have been calculated manually or with additional analog circuitry, and so any labor-saving tip would have been welcome.

Note that, by adding the $(\ln R)^2$ term, the new coefficients for the 1, $\ln R$, and $(\ln R)^3$ terms become different from those derived in the Steinhart–Hart case. This phenomenon results from the fact that the functions x^k that we are using as a basis are not *orthogonal*. Take a linear algebra course for more details.

Finally, try fitting the thermistor data to the Beta formula (Equation [9]). In this first-order polynomial fit, only two basis functions are required: $b_0 = 1$, and $b_1 = x$. Run **R-T Fit and Plot**, and then comment on the accuracy achieved by the Beta formula in comparison to the Steinhart–Hart Equation.

DO IT YOURSELF

Build a computer-based digital thermometer that uses the thermistor that you have calibrated in this chapter as its temperature sensor.

Construct this instrument in the following steps:

1. Build the circuit shown in Figure 9.1, which produces a constant current of 0.1 mA through the thermistor. You may wish to include the high input-impedance unity-gain op-amp buffer to isolate the thermistor circuit from the voltage-sensing circuitry. If available, use an ammeter to determine the current through your thermistor precisely. It probably will differ slightly from 0.1 mA because of resistor

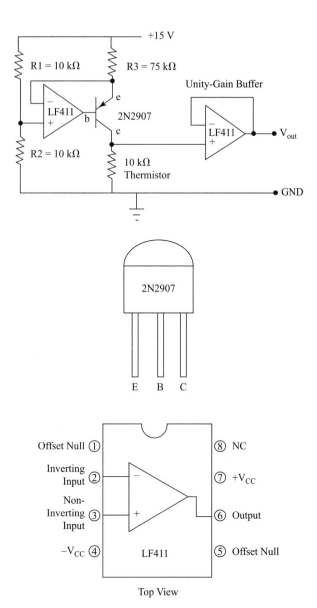

FIG. 9.1 Hardware circuit for Do It Yourself project. Constant current of 0.1 mA flows through the thermistor.

tolerances. With a voltmeter, check that the voltage difference between V_{out} and GND is within the expected range. For example, the voltage difference across a thermistor whose resistance is about 10 kΩ near room temperature should be $V_{out} = IR \approx (0.1\text{mA})(10\ \text{k}\Omega) = 1$ V.

2. Connect the positive and negative pins of a DAQ device's differential analog input channel to V_{out} and GND, respectively.

3. Write a VI called **Digital Thermometer**, whose front panel has a **Numeric Indicator** labeled **Temperature (deg C)** and a **Stop Button** as shown next. Save this VI in the **YourName\Chapter 9** folder.

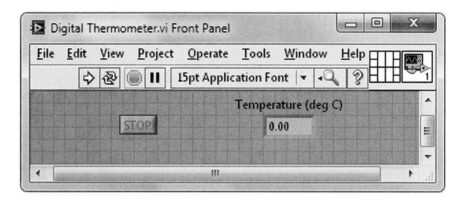

On the block diagram, place a **DAQ Assistant** icon. When the **Create New Express Task...** dialog window appears, configure it to perform an Analog Input operation that outputs temperature from a current-excited thermistor (**Acquire Signals>>Analog Input>> Temperature>>Iex Thermistor**) on the particular AI channel that you have connected to the circuit (e.g., **ai0**). When the **DAQ Assistant** dialog window then appears, configure it with the Steinhart–Hart values appropriate for your thermistor. The three Steinhart–Hart coefficients, which we called A, B, and D in this chapter, are respectively labeled A, B, and C in the DAQ Assistant dialog window. Also choose **Selected Units>>deg C**, **Iex Source>>External**, **Iex Value>>100 u** (which means 100 μA = 0.1 mA; use your measured value for the current, if available), **Configuration>>4-Wire**, and **Acquisition Mode>>1 Sample (On Demand).** Then close the dialog window by clicking the **OK** button. When the DAQ Assistant icon appears on your block diagram, enclose it within a **While Loop** and complete the required extra code, so that temperature readings are produced and displayed on a front-panel **Temperature (deg C)** indicator once every, say, 0.5 or 1 second, and this process repeats continually until a front-panel **Stop Button** is pressed.

For myDAQ users:

A myDAQ does not support the Temperature option of DAQ Assistant. Thus, configure DAQ Assistant to acquire the voltage across the thermistor, and then output this value to a MathScript Node where Ohm's Law and the Steinhart–Hart Equation are used to determine the thermistor's resistance and temperature, respectively.

Once completed, run **Digital Thermometer** and use it to measure the room's temperature and the temperature of your skin.

PROBLEMS

1. According to the theoretical considerations discussed at the beginning of this chapter, the value of B determined from fitting a thermistor's $R - T$ data to the Steinhart–Hart Equation is related to the activation energy barrier E for electronic conduction in the thermistor. Use the value you obtained for B to obtain E for your thermistor in the units of electronvolts (eV). The value for Boltzmann's constant is $k = 8.6 \times 10^{-5}$ eV/K.

2. A type T thermocouple is constructed by joining the ends of a copper and a constantan wire. This junction between the two dissimilar metals produces a temperature-dependent voltage, which can be used as a temperature sensor. The type T thermocouple's voltage (in microVolts) as a function of temperature (in degrees Celsius) is given in the following table.

Temperature $T(°C)$	Voltage $V(\mu V)$
0	0.000
50	2036
100	4279
150	6704
200	9288
250	12013
300	14862
350	17819
400	20872

The calibration equation for this temperature sensor is commonly obtained by fitting these *Temperature* vs. *Voltage* data to the following equation

$$T = a_1 V + a_2 V^2 + a_3 V^3 + a_4 V^4 + a_5 V^5 + a_6 V^6 \qquad [23]$$

Write a program called **Thermocouple Fit** that uses the **Curve Fitting** Express VI to determine the coefficients a_1 through a_6 in Equation [23]. Consider entering the given $T - V$ data into a spreadsheet and then using **Spreadsheet Read** (the **Do It Yourself** project in Chapter 5, or a slightly modified version of it) to input the data to your program. Alternately, you can use Array Controls to input the data. (You might be interested in comparing your values for a_1 through a_6 with those used in the **Volts to Temperature.vi** subVI within **Convert Thermocouple Reading.vi**, which is found in **Programming>>Numeric>>Scaling**. These values of a_1 through a_6 are derived from a more complete $T - V$ data set than that used in this problem.)

3. The following subdiagram implements **Wait Until Next ms Multiple**, which is found in **Functions>>Programming>>Timing**, to produce a y-array of millisecond timer values and a corresponding x-array of integers that index the y-values.

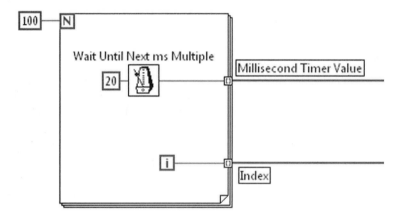

(a) Refer to **Wait Until Next ms Multiple**'s Help Window and then explain why the (x, y) data this subdiagram produces should obey the equation of a straight line $y = mx + b$, where m and b are the line's slope and y-intercept, respectively. What is the expected value for m of this line?

(b) Place the above subdiagram on the block diagram of a VI called **Wait Until Next ms Fit**. On this same block diagram, place a **Curve Fitting** Express VI and program it to perform a linear fit on the given (x, y) data and output the resulting value for the slope to the front panel. When you run your completed VI, do you obtain the expected value for m?

4. A resonant circuit composed of an inductor, a resistor, and a capacitor is powered by an ac input voltage whose amplitude V_{in} is constant, but whose frequency f can be varied. An experimentalist measures the circuit's output voltage amplitude V_{out} at several frequencies f and then calculates the circuit's gain at each frequency,

where gain is defined as the ratio of the output to input voltage amplitudes (i.e., $G \equiv V_{out}/V_{in}$). The results from this experiment are shown in the following data table.

Frequency f(Hz)	Gain G
1500	0.070
1600	0.084
1700	0.102
1800	0.128
1900	0.168
2000	0.235
2100	0.374
2200	0.740
2300	0.861
2400	0.442
2500	0.281
2600	0.206
2700	0.164
2800	0.136
2900	0.117
3000	0.103

A theoretical analysis of this resonant circuit predicts that the frequency-dependent gain $G\,(f)$ is given by

$$G = \frac{1}{\sqrt{1+\left[\dfrac{1}{\Delta f}\left(f-\dfrac{f_o^2}{f}\right)\right]^2}} \qquad [24]$$

where the constants f_0 and Δf are the resonant frequency and full-width at half-power maximum, respectively (there is one frequency below f_0 and one frequency above f_0 where $G = 1/\sqrt{2}$; the full-width is defined as the difference between these frequencies).

Write a VI called **Resonant Circuit** that fits the given experimental $G - f$ data to the theoretically predicted relation for $G\,(f)$ (Equation [24]). On the VI's front panel, compare the data and the best-fit curves by plotting both on the same

XY Graph and also display the best-fit values for constants f_0 and Δf in an Array Indicator.

A few suggestions for your work:

- Consider entering the given $G - f$ data into a spreadsheet and then using **Spreadsheet Read** (the **Do It Yourself** project in Chapter 5, or a slightly modified version of it) to input the data to your program. Alternately, you can use Array Controls to input the data.
- Since the parameters f_0 and Δf appear in Equation [24] in a nonlinear fashion, use the **Model Type>>Non-linear** option of the Curve Fitting Express VI to perform the fitting procedure. For the **Model Parameters**, let $f_0 = a$ and $\Delta f = b$. You will have to supply **Starting Values** for a and b. For the nonlinear fitting algorithm implemented by Curve Fitting to function properly, these initial guesses for a and b need to be somewhat close to the final best-fit values.
- After you run your program once with the dialog window for **Curve Fitting** closed, the **Signals** and **Locations** arrays will be loaded into this Express VI. You can then open **Curve Fitting** (by double-clicking on its icon) and run it interactively. Thus, you can iteratively execute **Curve Fitting** with a succession of choices for **Starting Values**, until you determine a choice that yields a successful fit.

5. The Steinhart–Hart Equation can be written as $T = \dfrac{1}{A + Bx + Cx^3}$, where $x = \ln R$. In this form, the constants A, B, and D appear in a nonlinear manner. Starting with **R-T Fit and Plot**, use **File>>Save As...** to create a new VI named **R-T Fit and Plot (Nonlinear)**. Modify the block diagram so that **Curve Fitting** fits the input thermistor data to the above form of the Steinhart–Hart Equation (you will have to program **Curve Fitting** to perform a nonlinear fit; see suggestions given in Problem 4). Verify that you obtain (nearly) the same values for A, B, and D as well as the temperature residual $R^T \equiv T - T^{\text{best fit}}$ as you obtained from the linear least-squares method implemented in this chapter.

6. Write a program called **Noisy Sine Fit**, which fits a sine function to a noisy sine wave input.

   (a) First, place a **Simulate Signal** Express VI, found in **Functions>> Express>> Input**, on the block diagram. When its dialog window opens, program this Express VI to produce 100 samples of a sine wave with **Uniform White Noise** added at a sampling frequency of 10000 Hz. Close the window and configure the **Simulate Signal** icon so that you have front-panel control of the sine wave frequency and amplitude as well as the noise amplitude shown next.

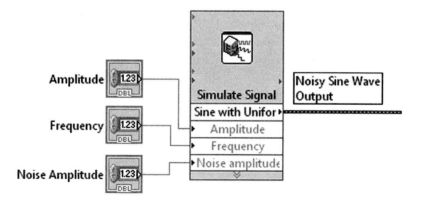

Place a Waveform Graph on the front panel and plot the noisy sine-wave output from **Simulate Signal**. Run the VI with **Amplitude**, **Frequency**, and **Noise Amplitude** set to *1*, *250*, and *0.6*, respectively, to verify that your program functions correctly.

(b) Put an Array Indicator on the front panel and label it **Best-Fit Amplitude and Frequency**. Then on the block diagram, place a **Curve Fitting** icon and program it to perform a nonlinear fit to the function $a*\sin(2*3.14*b*x)$. You will also need to supply initial guesses for the parameters a and b. After closing the dialog window, connect the output of Simulate Signal to the **Signals** input of Curve Fitting. The resulting dynamic data type wire will automatically create the needed independent variable x for Curve Fitting (i.e., there is no need to explicitly wire anything to the **Locations** input). Wire the nonlinear coefficients output to the front-panel array indicator. Finally, if interested, simply join a wire from Curve Fitting's **best fit** output to the noisy sine-wave wire emanating from **Simulate Signal**. A **Merge Signals** Express VI will automatically appear, causing both curves to be plotted on the Waveform Graph.

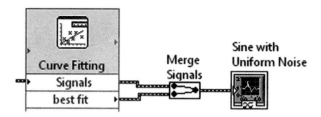

(c) Run your completed VI with various choices for **Amplitude**, **Frequency**, and **Noise Amplitude**. To achieve a good fit of the sine function to the noisy sine wave, you will most likely have to experiment with the choice of initial guesses to the fitting parameters. For which parameter must the initial guess be most carefully chosen?

7. Beginning at time $t = 0$, a capacitor of unknown value C is discharged through a resistor $R = 47$ kΩ. To determine C, as the capacitor is discharging, the voltage difference V between its plates is measured at 5 millisecond intervals, resulting in the following data set.

t (s)	0.005	0.010	0.015	0.020	0.025	0.030
V (V)	8.7	6.3	4.5	3.3	2.4	1.7

Theoretically, the relation between V and t is given by

$$V = V_0 e^{-t/RC} \qquad\qquad [25]$$

where V_0 is the voltage across the capacitor at $t = 0$.

(a) By taking the logarithm of both sides, rewrite Equation [25] in a linearized form, that is, in the form $y = mx + b$. Then, write a program called **Discharging Capacitor (Linear)**, which determines C (in μF) and also V_0 from the given V vs. t data. Build this VI so that the data set is input to its block diagram, where the data are used to construct appropriate x and y arrays, which are then input to a **Curve Fitting** icon. Program **Curve Fitting** to fit the input arrays to the equation of a straight line, use the resulting best-fit parameters m and b to calculate C and V_0, and then display these values on the front panel.

(b) Write another program called **Discharging Capacitor (Nonlinear)**, which also determines C as well as V_0 from the given V vs. t data. Build this VI so that the data set is input to its block diagram, where a **Curve Fitting** icon is programmed to perform a nonlinear fit (see suggestions in Problem 4) of the input arrays to the equation $y = ae^{-x/b}$. Use the resulting best-fit parameters a and b to calculate C (in μF) and V_0, and then display these values on the front panel.

(c) Run **Discharging Capacitor (Linear)** and **Discharging Capacitor (Nonlinear)**. What values do you obtain for C and V_0 from each VI? Do the two approaches yield the same result?

8. Write a Mathscript-based program called **Discharging Capacitor (Mathscript)** that analyzes the discharging capacitor experiment described in Problem 7. In particular, use the Mathscript command *[p, s] = polyfit(x, y, n)* to perform a fit of the *Voltage* vs. *Time* data to a linearized form to Equation [25]. A description of

this command can be found by typing *help polyfit* in the Mathscript Interactive Window. Construct this VI in the following steps:

(a) Enter the given V vs. t data as row vectors in a Mathscript Node. The V row vector is created with the command $V=[8.7\ 6.3\ 4.5\ 3.3\ 2.4\ 1.7]$. Then plot V vs. t on a front-panel XY Graph with the data points as solid dots without interpolation.

(b) By taking the logarithm of both its sides and defining $y = \ln(V)$ and $x = t$, Equation [25] can be linearized in the form of the first-order polynomial $y = a_0 + a_1 x$, where a_0 and a_1 are constants. Add the required code within your Mathscript Node to form the y array (use *help basic* to find the Mathscript command for the natural logarithm) and then use *[p, s] = polyfit(x, y, 1)* to determine the constants a_1 and a_0. These constants will be given as the first and second elements of the p row vector, that is, $p(1) = a_1$ and $p(2) = a_0$ (unlike LabVIEW arrays, Mathscript-array indexing begins with 1, rather than 0). Using these best-fit a_1 and a_0, calculate C (in μF) and V_0 and display these values in front-panel indicators.

(c) Finally, generate an array of time values called *tfit* from $t = 0$ to 0.030 s in 0.001 s increments. Then create another array called *Vfit*, which is produced by evaluating the best-fit form of Equation [25], that is, $Vfit = V_0 \exp(-tfit/RC)$, at all of the *tfit* values. Finally, plot *Vfit* vs. *tfit* as an interpolated line on the same XY Graph used to plot the V vs. t experimental data.

(d) Run **Discharging Capacitor (Mathscript)**. What values are obtained for C and V_0? Does the best-fit curve accurately describe the V vs. t data?

9. Curve fitting of data must be guided by a theory describing the phenomenon under investigation. Consider the following hypothetical situation: An experimenter, investigating the relation between two quantities y and x, uses an experimental system to acquire seven samples of y as function of x, with the results given in the following table. These data, of course, reflect the true mathematical function $y(x)$ that relates the two quantities, but this relation is somewhat obscured by noise in the experimental system. The experimenter knows that a theory of the phenomenon being studied suggests that the measured quantities should be related by a second-order polynomial $y(x) = a_0 + a_1 x + a_2 x^2$, where $a_0 = 6.0$, $a_1 = 4.0$, and $a_2 = 2.0$ in appropriate units.

x	1.0	2.0	3.0	4.0	5.0	6.0	7.0
y	10	24	34	54	75	101	133

To compare the experimentalist's results to the theoretically predicted relation, analyze the data in the following ways:

(a) Write a program called **Polynomial Fit** that plots the given data on an XY Graph and also inputs these data to a **Curve Fitting** Express VI. Program the Express VI to fit the data to a second-order polynomial and display the resulting **polynomial coefficients** a_0, a_1, and a_2 in an indicator on the front panel. Also, on the block diagram, include the code shown next to generate a well-resolved representation of the best-fit polynomial y_{fit} and plot this curve, along with the given data, on the front-panel XY Graph (the icon **Reverse 1D Array** is needed because the function *polyval* requires the polynomial coefficients to be in reverse order of that provided by **Curve Fitting**). How do your fitted values of a_0, a_1, and a_2 compare with the theoretically expected values? Why do you suppose that the experimentally determined values for these coefficients don't perfectly match the theoretical values? Characterize qualitatively (in a sentence or two) how well the best-fit polynomial describes the measured data points.

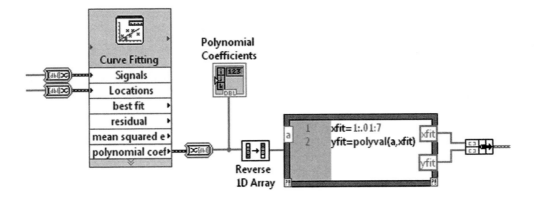

(b) A better fit to the acquired data can be obtained using a higher-order polynomial. To demonstrate this fact, reprogram **Curve Fitting** so that it fits the given data to a fourth-order polynomial. From the XY Graph, does y_{fit} from this fourth-order fit agree with the experimental data more favorably than the second-order fit carried out in part (a)? What values do you obtain for a_0, a_1, a_2, a_3, and a_4? How do the first three of these values compare with the values expected from theory? Explain why, even though the fourth-order polynomial provides a closer match to the experimental data, the second-order fit carried out in part (a) provides the best comparison between the experimental findings and the given theory.

(c) Interestingly, a fitted polynomial of order $N-1$ will perfectly pass through N given data samples. Demonstrate this fact using your program. Here, we are magnifying the effect studied in part (b). By providing too much flexibility in the fitting function, we are fitting experimental noise, at the expense of uncovering the sought-after relation $y(x)$.

CHAPTER 10

Analysis VIs: Fast Fourier Transform

10.1 THE FOURIER TRANSFORM

Imagine you are investigating a physical phenomenon by measuring the analog quantity x that varies continuously with time t, that is, $x = x(t)$. Based on our discussion at the beginning of the previous chapter, we expect that $x(t)$ can be modeled as the linear combination of a set of M basis functions $b(t)$ such that

$$x(t) = \sum_{k=0}^{M-1} a_k\, b_k(t) \tag{1}$$

The Fourier transform is an example of just such a model. The pioneering work of nineteenth-century mathematician Baron Jean Baptiste Joseph Fourier, with subsequent elaboration by others, showed that the infinite set of complex exponentials $\exp(i2\pi ft)$, where f is all the frequencies from $-\infty$ to $+\infty$ and $i = \sqrt{-1}$, is a complete basis set with which one can form linear combinations to model any arbitrary function. Since an infinite number of exponentials with infinitesimally close frequency spacing is required to model an arbitrary function, the summation in Equation [1] becomes an integral, yielding

$$x(t) = \int_{-\infty}^{+\infty} X(f)\, e^{i2\pi ft}\, df \tag{2}$$

where the "*Fourier component*" $X(f)$ is the amplitude (I'll be much more precise in my use of the word *amplitude* shortly) of the exponential with frequency f. Generally, $X(f)$ is a complex-valued function, carrying both *magnitude* and *phase* information, a point we will carefully consider in this chapter.

Conversely, the Fourier components $X(f)$ may be obtained by the following equation

$$X(f) = \int_{-\infty}^{+\infty} x(t) e^{-i2\pi ft} dt \qquad [3]$$

Equations [3] and [2] are called the *Fourier transform* and the *inverse Fourier transform*, respectively.

10.2 DISCRETE SAMPLING AND THE NYQUIST FREQUENCY

In a real experiment, one cannot sample the quantity x continuously, but must instead settle for discretely sampling it N times. If the samples are taken at equally spaced times, where the time difference between adjacent samples is Δt, then the sampling frequency f_s is given by

$$f_s = \frac{1}{\Delta t} \qquad [4]$$

and the N consecutive sampling times are

$$t_j = j\,\Delta t \qquad j = 0, 1, 2, ..., N-1 \qquad [5]$$

Given such a sampling scheme, there is a maximum-frequency sine wave that we can expect to detect. This limitation is simply due to the fact that to observe a sine signal we must at the very least, for example, first sample its positive peak, then its negative trough at the next sample, followed by its positive peak at the third sample, and so on. Thus, the minimal sampling of a sine wave is two sample points per cycle. In an experiment then, where your instruments are acquiring data every Δt seconds (at a sampling rate $f_s = 1/\Delta t$), the maximum frequency sinusoidal signal you can detect has a period $T = 2\,\Delta t$. This detection limit is called the *Nyquist frequency* $f_{nyquist}$ and is given by

$$f_{nyquist} = \frac{1}{2\Delta t} = \frac{f_s}{2} \qquad [6]$$

Additionally, by sampling a waveform $x(t)$ at N equally spaced times t_j, one can only expect to determine the Fourier components $X(f)$ of N equally spaced frequencies f_k in the Fourier transform. This statement follows from the tacit assumption that $x(t)$ is periodic such that $x(t) = x(t + N\,\Delta t)$, a criterion that forces the waveform $x(t)$ to be composed only of component frequencies that fit an integer number of cycles into the sequence of N samples. The constant (zero-frequency) function obviously meets this criterion. The next lowest acceptable frequency has a period of $N\,\Delta t$ and thus a frequency of $f_1 = 1/N\Delta t = f_s/N$. Within the acceptable range of $-f_{nyquist}$ to $+f_{nyquist}$, the rest of the frequencies f_k are the integer multiples of f_1. Thus, defining the spacing between adjacent frequencies to be $\Delta f = f_s/N$, the N equally spaced frequencies f_k determined by discretely sampling data are

$$f_k = k\left(\frac{f_s}{N}\right) = k\Delta f \qquad k = -\frac{N}{2}+1,....,0,....,+\frac{N}{2} \qquad [7]$$

Although Equation [7] appears deficient in its neglect of the frequency $-f_{nyquist}$ (given by $k = -N/2$), the two extreme frequencies $\pm f_{nyquist} = \pm f_s/2$ actually describe the same basis function due to the periodic nature of $x(t)$. Thus, the range of k-values in the above relation rightfully runs from $-N/2 + 1$ to $+N/2$.

10.3 THE DISCRETE FOURIER TRANSFORM

With the above-described constraints caused by discrete sampling, we then approximate the Fourier transform integral (Equation [3]) by the following sum:

$$X(f_k) = \int_{-\infty}^{+\infty} x(t)e^{-i2\pi f_k t}\,dt \approx \sum_{j=0}^{N-1} x(t_j)e^{-i2\pi f_k t_j}\,\Delta t \qquad [8]$$

Note, using Equations [4], [5], and [7], that

$$f_k\,t_j = (k\Delta f)(j\Delta t) = k\,j\frac{f_s}{N}\,\Delta t = \frac{k\,j}{N} \qquad [9]$$

Putting [9] into [8], and realizing that Δt is a constant, we find

$$X(f_k) \approx \Delta t \sum_{j=0}^{N-1} x(t_j)e^{-i2\pi jk/N} \equiv \Delta t\,X_k \qquad [10]$$

where we have defined the *discrete Fourier transform* to be

$$X_k \equiv \sum_{j=0}^{N-1} x(t_j)e^{-i2\pi jk/N} \qquad k = -\frac{N}{2}+1,...,0,...,+\frac{N}{2} \qquad [11]$$

Finally, we approximate the inverse Fourier transform integral (Equation [2]) by the following sum

$$x(t_j) = \int_{-\infty}^{+\infty} X(f)e^{i2\pi f t_j}\,df \approx \sum_{k=0}^{N-1} X(f_k)e^{i2\pi f_k t_j}\,\Delta f \qquad [12]$$

Putting [10] into [12],

$$x_j \approx \sum_{k=0}^{N-1} X_k\,\Delta t\,e^{i2\pi f_k t_j}\,\Delta f \qquad [13]$$

Then using Equation [9] and also noting that $(\Delta t)(\Delta f) = (1/f_s)(f_s/N) = 1/N$, Equation [13] becomes the *discrete inverse Fourier transform*

$$x_j \approx \sum_{k=0}^{N-1} \frac{X_k}{N} e^{i2\pi f_k t_k} = \sum_{k=0}^{N-1} \frac{X_k}{N} e^{i2\pi k j/N} \equiv \sum_{k=0}^{N-1} A_k e^{i2\pi k j/N} \qquad [14]$$

Equations [11] and [14] provide us with a method of performing spectral analysis on an experimental signal x. Given that x has been sampled discretely at N evenly spaced times t_j, we first calculate the N values of X_k using the definition of the discrete Fourier transform (Equation [11]). Then, in the spirit of Fourier, viewing the signal x as really the composite of oscillations at N different frequencies f_k, Equation [14] tells us that the amplitude of oscillation A_k at each f_k is given by

$$A_k = \frac{X_k}{N} \qquad [15]$$

Since A_k is derived from the complex-valued X_k, in general, A_k is a complex-valued function. Let's then name A_k the *complex-amplitude*. The plot of complex-amplitude A_k vs. frequency f_k is called the *frequency spectrum* of the sampled signal x_j.

Up to now we have taken k to be all of the integers within the range of $-N/2 + 1$ to $+N/2$. However, it is easy to show that Equation [11] is periodic in k, with a period of N:

$$X_{k+N} = \sum_{j=0}^{N-1} x(t_j) e^{-i2\pi j(k+N)/N} = \sum_{j=0}^{N-1} x(t_j) e^{-i2\pi j k/N} e^{-i2\pi j} = X_k \qquad [16]$$

where we have used the fact that $\exp(-i2\pi j) = 1$ because j is an integer. Based on Equation [16], the negative k-values in the range of $-N/2 + 1$ to -1 are equivalent to the k-value range of $+N/2 + 1$ to $+N - 1$. Because the array indices in computer programs generally are positive integers, algorithms to evaluate Equation [11] typically let k vary from 0 to $N - 1$ (which covers one complete period). This is the convention that LabVIEW's Fourier transform–based VIs follow. Thus, in the N-element array of discrete Fourier transform values X_k that such VIs output, the elements with indices from 0 to $N/2$ contain the sequence of X_0 to $X_{N/2}$ (affiliated with positive frequencies in the range of $0 \leq f \leq +f_{nyquist}$) and elements with indices from $N/2 + 1$ to $N - 1$ contain the sequence of $X_{-N/2+1}$ to X_{-1} (affiliated with negative frequencies in the range of $-f_{nyquist} < f < 0$).

10.4 THE FAST FOURIER TRANSFORM

Since the mid-1960s, a clever algorithm to calculate the discrete Fourier transform (Equation [11]) in an efficient manner has been widely used. This algorithm, known as the *fast Fourier transform (FFT)*, exploits inherent symmetries in the calculation of

Equation [11] that arise because of the periodic way in which the original data were taken (N evenly spaced samples taken with time spacing Δt). Where the "brute-force" evaluation of Equation [11] would require on the order of N^2 complex multiplication operations, the FFT manages this task with only $N \log_2 N$ such operations. The savings in time are immense, especially when N is large. As an example, consider a digital oscilloscope configured to produce a data set of 1024 points (a typical mode of operation). To analyze the frequency spectrum of this set, the FFT algorithm will be 100 times faster than the straightforward evaluation of Equation [11]. For a very large data set of 10^6 points, the FFT is 50,000 times faster!

In addition to the assumption of evenly spaced data points, there is one extra restriction to the use of FFTs. In deriving a FFT algorithm, one must assume that the size of the data set is a power of two (that is, $N = 2^m$, where $m = 1,2,3,\ldots$) to obtain maximum calculational efficiency. If N is not equal to a power of two, algorithms (called *discrete Fourier transforms*, or *DFTs*) exist to evaluate the discrete Fourier transform, but with a speed much slower than that of the FFT.

10.5 FREQUENCY CALCULATOR VI

The LabVIEW FFT icon that you will implement observes the following convention: The discrete Fourier transform values X_k are output as an N-element array, whose elements are indexed 0 to $N-1$, which are associated with N evenly spaced frequencies (where $\Delta f = f_s/N$) in the range of $-f_{nyquist} < f \leq +f_{nyquist}$. The array elements with indices from 0 to $N/2$ contain the sequence of X_0 to $X_{N/2}$ associated with the positive frequencies from 0 to $+f_{nyquist}$, and the elements with indices from $N/2+1$ to $N-1$ contain the sequence of $X_{-N/2+1}$ to X_{-1} associated with the negative frequencies from $-f_{nyquist} + \Delta f$ to $-\Delta f$.

Let's first write a program, to be used later as a subVI, that calculates the frequency associated with each array element of the FFT output. Start by creating the following front panel. Use **Select a Control...** in the Controls Palette to obtain the **Digitizing Parameters** control cluster in **YourName\Controls** (if you haven't created this cluster yet, see Section 3.12, "**Control and Indicator Clusters**"). Assign the connector pane's terminals consistent with the Help Window shown and design an icon. Use **File>>Save** to create a folder called **Chapter 10** in the **YourName** folder, and then save this VI under the name **Frequency Calculator** in **YourName\Chapter 10**.

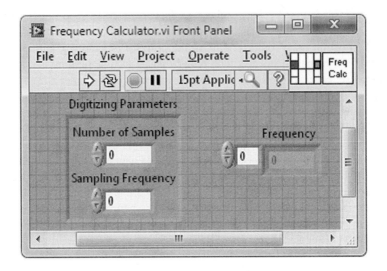

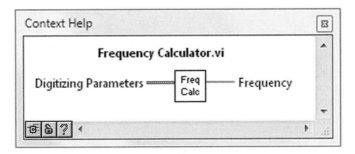

Now write a block diagram that creates an N-element array of frequencies f_k, indexed from 0 to $N-1$. We desire the frequency at index k to be the f_k associated with the X_k at that same index in the FFT icon's output array. The separation between adjacent frequencies f_k is $\Delta f = f_s/N$, where f_s is the sampling frequency. From the above paragraphs' discussion, we see that the f_k can be calculated as follows:

$$
f_k = \begin{cases} k\,\Delta f & k = 0, 1, 2, \ \ldots , \dfrac{N}{2} \\[2mm] (k-N)\Delta f & k = \dfrac{N}{2} + 1, \ \ldots , N-1 \end{cases}
$$
[17]

Program this formula using a **Case Structure** as shown in the following illustration.

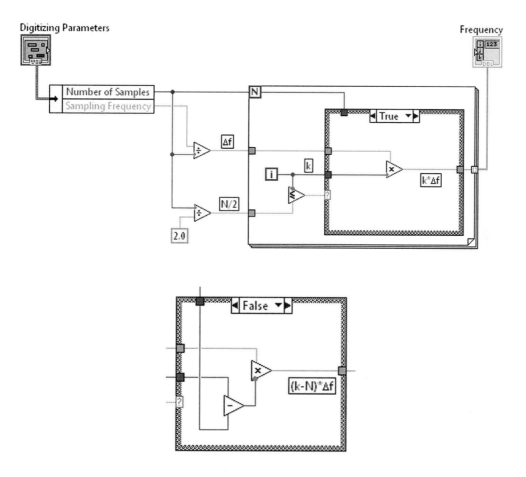

Alternately, Equation [17] can be coded using a **Mathscript Node** as shown next. Note that, unlike LabVIEW arrays, Mathscript-array indexing begins with 1, rather than 0. Thus, within the Mathscript Node, the N-element f array runs from index 1 to index N. However, at the f output of the Mathscript Node, this array becomes an N-element LabVIEW-formatted array, that is, its indexing runs from 0 to $N - 1$. You can check that this is so using the **Frequency** array indicator.

Digitizing Parameters

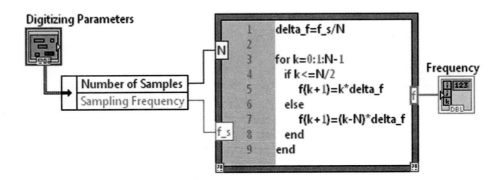

```
     delta_f=f_s/N

     for k=0:1:N-1
       if k<=N/2
         f(k+1)=k*delta_f
       else
         f(k+1)=(k-N)*delta_f
       end
     end
```

Frequency

Return to the front panel. Input the values *1024* and *2000* to the **Number of Samples** and **Sampling Frequency** controls, respectively. Run the VI. If it functions properly, the positive frequencies of 0 to +1000, in increments of $2000/1024 \cong 1.95$, will appear in the **Frequency** array at the indices 0 through 512. The negative frequencies of (approximately) −998.05 to −1.95 will reside at the array elements with indices from 513 through 1023.

10.6 FFT OF SINUSOIDS

To get a handle on how FFT signal processing works, let's synthesize some data consisting of sinusoidal waves with known complex-amplitudes A_k and frequencies f_k, feed these data to the FFT algorithm, divide the resultant X_k values by N to produce the spectrum of A_k (see Equations [14] and [15]), and determine whether this spectrum matches the known input. We will use our old friend **Waveform Simulator**, which you created in Chapter 3 and saved in **YourName\Chapter 3**, to synthesize an N-element array of data and our new friend **Frequency Calculator** to produce the array of f_k values.

Construct the front panel shown next, saving it under the name **FFT of Sinusoids** in **YourName\Chapter 10**. You can obtain the control clusters in **YourName\Controls** using **Controls>>Select a Control…** or else through the **Create>>Control** pop-menu option once you place **Waveform Simulator** on the block diagram. We will display the complex-amplitudes A_k on the **XY Graph**, so label its x- and y-axes **Frequency** and **Complex-Amplitude**, respectively. Since the A_k are complex numbers, we will display the real and imaginary parts of the A_k via two plots. Using the **Plot Legend**, set up these two plots, and label them **Re(A)** and **Im(A)**. Distinguish these plots from each other by color (for a review of how to place multiple plots on an XY Graph, see Section 7.9).

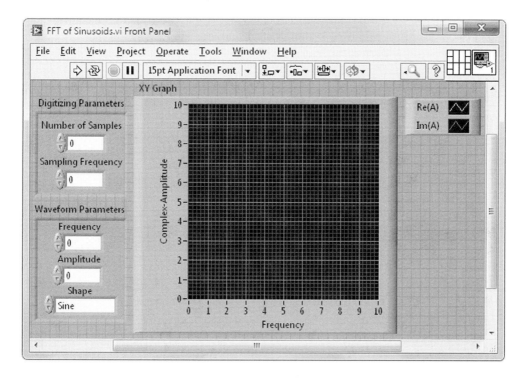

The discrete Fourier transform values X_k output by the FFT algorithm are complex numbers. LabVIEW includes a complex number data representation, so X_k is output from the **FFT.vi** icon as this data type. The Help Window for **FFT.vi** (found in **Functions>>Digital Signal Processing>>Transforms**) is shown next. In our program, the data array we will input at **x** is purely real. Since the FFT icon is polymorphic, it will adapt itself automatically to receive this purely real input. Alternately, if we input a complex-valued array at **x**, the icon would again automatically adapt. The complex-valued X_k are output at **FFT {x}**.

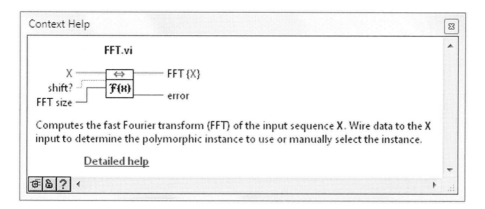

Now write the block diagram as follows. The **Complex To Re/Im** icon is found in **Functions>>Programming>>Numeric>>Complex.**

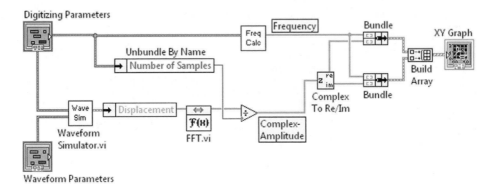

Return to the front panel of **FFT of Sinusoids** and save your work.

10.7 APPLYING THE FFT TO VARIOUS SINUSOIDAL INPUTS

Using the **Waveform Parameters** control, program your VI to produce a *Cosine* function with amplitude of *4.0* and frequency of *250* Hz. Note: I will use the word *amplitude* to mean the peak-height (relative to the zero level) of a sinusoidal function. For the **Digitizing Parameters**, choose *1024* and *2000* for **Number of Samples** and **Sampling Frequency**, respectively. Remember that the FFT algorithm requires the number of samples in your input data to be a power of two, so we have chosen $N = 2^{10} = 1024$. Run your program. The VI will find the real and imaginary parts of the complex-amplitude A_k for positive and negative frequencies f_k up to the Nyquist frequency $\pm f_{nyquist} = \pm f_s/2$. For our choice of $f_s = 2000$ Hz, the Nyquist frequency is 1000 Hz, so the 250 Hz cosine oscillation frequency that you have input will fall within the FFT's range of detection.

If all works well, the VI should output the Fourier transform of your cosine input in the following form: The real part of the complex-amplitude A is equal to zero at all frequencies f, except $A = +2$ at $f = \pm 250$ Hz. The imaginary part of A is zero at all frequencies f.

Why is this output the correct representation for the cosine input? Shouldn't the amplitude A be 4, not 2? To understand your result, plug the output back into Equation [14]. Then

$$x_j \approx \sum_{k=0}^{N-1} A_k e^{i2\pi f_k t_j} \quad = (+2)e^{i2\pi(250)t_j} + (+2)e^{i2\pi(-250)t_j}$$

$$= 4\left[\frac{e^{i2\pi(250)t_j} + e^{i2\pi(-250)t_j}}{2}\right]$$

$$= 4\cos\left[2\pi(250)t_j\right]$$

where we have used the identity $\cos x = (e^{ix} + e^{-ix})/2$ for the last step. So we see that the output complex-amplitudes are indeed the correct representation for an input cosine function of amplitude 4.0.

Try a sine-wave input and see what you get. Using **Waveform Parameters**, program **FFT of Sinusoids** to produce a *Sine* function with amplitude of *4.0* and frequency of *250* Hz, while keeping *1024* and *2000* for **Number of Samples** and **Sampling Frequency**, respectively. Now run the VI. You should find that the imaginary part of the complex-amplitude is $A = -2$ at $f = +250$ Hz, $A = +2$ at $f = -250$ Hz, and is zero elsewhere. The real part of A is zero at all frequencies. To see that this is the correct representation of your input, we again start with Equation [14] and note the following:

$$x_j \approx \sum_{k=0}^{N-1} A_k e^{i2\pi f_k t_j} \quad = (-2i) e^{i2\pi(250)\,t_j} + (+2i) e^{i2\pi(-250)\,t_j}$$

$$= 4 \left[\frac{e^{i2\pi(250)\,t_j} - e^{i2\pi(-250)\,t_j}}{2i} \right]$$

$$= 4 \sin\left[2\pi(250) t_j \right]$$

where we have used the facts that $i = -1/i$ and $\sin x = (e^{ix} - e^{-ix})/2i$.

There are two frequencies that behave differently from the rest, zero-frequency and the Nyquist frequency. First, program **FFT of Sinusoids** with the constant *DC Level* (also called *zero-frequency*) function

$$x_j = 4.0$$

and then run the VI with **Number of Samples** and **Sampling Frequency** equal to *1024* and *2000*, respectively. Note that, rather than finding half of the amplitude at a positive and negative frequency as in the previously studied nonzero-frequency cases, the full amplitude (called the *DC component*) appears at the single frequency $f = 0$.

Now, program **FFT of Sinusoids** with the following function that oscillates at the Nyquist frequency

$$x_j = 4.0 \cos [2\pi(1000)t_j]$$

and then run the VI again with **Number of Samples** and **Sampling Frequency** equal to *1024* and *2000*, respectively. If a mysterious diagonal line appears in the resulting FFT plot, remove it by popping up on the **Plot Legend** and deselecting the "connect-the-dots" mode of plotting in the **Interpolation** option. Similar to the zero-frequency case, you'll discover that the full oscillatory amplitude appears at the single frequency $f_{nyquist}$.

Here is a synopsis of our last observation: The cosine function, which is used to generate the data set, takes on its peak value when the 2000 Hz sampling process begins at time $t = 0$. The 1000 Hz Nyquist-frequency waveform then is sampled twice each cycle, producing a data set that has the following sequence–peak value, trough value, peak value, trough value....and so on. We find that by applying the FFT algorithm of this data set, the correct value for the amplitude of oscillation (i.e., 4.0) is determined at $f_{nyquist}$.

Unfortunately, by exploring the Nyquist-frequency situation a little further, a troublesome problem emerges. Try programming **FFT of Sinusoids** to produce a Nyquist-frequency oscillation, but this time one that follows the sine function

$$x_j = 4.0 \sin [2\pi(1000)t_j]$$

and then run the VI with **Number of Samples** and **Sampling Frequency** equal to *1024* and *2000*, respectively. You'll find that the FFT detects no amplitude of oscillation at $f_{nyquist}$ in this case. Can you explain why the 1000 Hz sinusoidal oscillation is invisible in this data set?

If interested, you can try including a phase constant δ (in radians) in the sine function's argument to produce a data set that takes on neither its peak value nor its zero-value (like the cosine and sine function, respectively) at time $t = 0$. To create the desired function, open **Waveform Simulator** by double-clicking on its icon on the block diagram of **FFT of Sinusoids**. Then program Waveform Simulator's **User-Defined** function to be

$$x_j = 4.0 \sin [\ 2\pi(1000)t_j + \delta\ ]$$

Run **FFT of Sinusoids** with **Shape** selected as *User-Defined* in the **Waveform Parameters** control cluster. Does the FFT yield the correct amplitude of oscillation (i.e., 4.0) for any nonzero value of δ?

To summarize our Nyquist-frequency findings, in taking the FFT of a discretely sampled data set, the resultant value for the amplitude at $f_{nyquist}$ is only accurate when the Nyquist-frequency oscillation is "in phase" with the sampling process. Since one cannot guarantee that this will be the case in an actual experimental situation, the complex-amplitude at $f_{nyquist}$ should always be viewed with suspicion. Thus, in our work below, we will ignore the Fourier component determined at the Nyquist frequency.

Next, program the **User-Defined** function on **Waveform Simulator**'s block diagram to be a function that includes two simultaneous sinusoidal oscillations, along with a DC level, such as

$$x_j = 8.0 + 4.0 \sin [2\pi(250)t_j] + 6.0 \cos [2\pi(500)t_j]$$

and then run **FFT of Sinusoids**. Do you understand the output?

Finally, what happens when the input contains a sine and cosine oscillation, both at the same frequency? Run **FFT of Sinusoids** with the **User-Defined** function

$$x_j = 4.0 \sin [2\pi(250)t_j] + 6.0 \cos [2\pi(500)t_j]$$

You may have to change, for example, the **Point Style** of each plot to view the output accuracy. Do you understand the real and imaginary plots?

10.8 MAGNITUDE OF THE COMPLEX-AMPLITUDE

To this point, we have been representing a complex number as the sum of a real and an imaginary part, that is, $z = x + iy$. However, as you know, the complex number z may also be represented as a *magnitude r* times a *phase factor* $e^{i\phi}$, where $r = \sqrt{x^2 + y^2}$ and tan $\phi = y/x$. Let's say that the oscillation of a system at frequency f is, as in the above example, the composite of a sine and cosine function with (real) amplitudes B and C, respectively:

$$x = B\sin [2\pi ft] + C \cos [2\pi ft] \tag{18}$$

Then, defining $\theta = 2\pi f t$ and using the complex exponential representation of the sine and cosine functions,

$$x = B\frac{e^{i\theta} - e^{-i\theta}}{2i} + C\frac{e^{i\theta} + e^{-i\theta}}{2}$$

$$= \frac{1}{2}(C - iB)e^{i\theta} + \frac{1}{2}(C + iB)e^{-i\theta}$$

From Figure 10.1, we see that $(C + iB) = re^{+i\phi}$ and $(C - iB) = re^{-i\phi}$, where $r = \sqrt{B^2 + C^2}$ and $\phi = \tan^{-1} (B/C)$.

Thus,

$$x = \frac{1}{2}(re^{-i\phi})e^{i\theta} + \frac{1}{2}(re^{+i\phi})e^{-i\theta}$$

$$= r\frac{e^{i(\theta - \phi)} + e^{-i(\theta - \phi)}}{2}$$

$$= r\cos(\theta - \phi) \tag{19}$$

Then, equating [18] with [19] and remembering $\theta = 2\pi f t$, $r = \sqrt{B^2 + C^2}$, and $\phi = \tan^{-1} (B/C)$, we find that

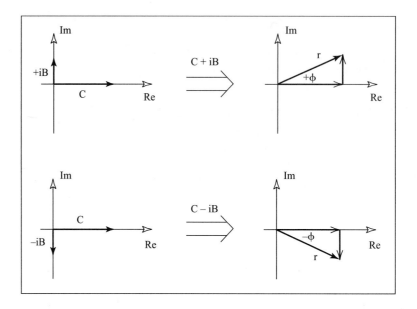

FIG. 10.1

$$x = B\sin(2\pi f\,t) + C\cos(2\pi f\,t) = \sqrt{B^2 + C^2}\;\cos(2\pi f\,t - \phi) \qquad [20]$$

Equation [20] tells us that a sine and cosine oscillation at the frequency f combine to produce a single phase-shifted cosine oscillation at frequency f with a net amplitude equal to the vectorial sum of the sine and cosine amplitudes.

Let's then write a VI called **FFT (Magnitude Only)** that, given an input data set x_j, finds the net amplitude of oscillation at the frequencies f_k, but ignores the phase information (that is, neglects whether this oscillation is in the form of a pure sine wave or a pure cosine wave or a composite of sine plus cosine). With **FFT of Sinusoids** open, select **File>>Save As...**, and create a new VI named **FFT (Magnitude Only)** in **YourName\Chapter 10**. Use the **Plot Legend** to request only a single plot—labeled **Mag(A)**—on the **XY Graph**. Relabel the XY Graph's y-axis as **Magnitude of Complex-Amplitude**.

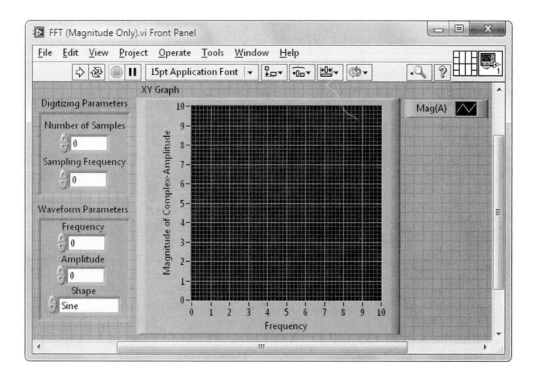

Now modify the block diagram so that, using the complex-amplitude's real and imaginary parts, its magnitude is calculated and displayed. Replace the **Complex To Re/Im** icon with **Complex To Polar** (found in **Functions>>Programming>>Numeric>>Complex**). The simple way to make this swap is by popping up on **Complex To Re/Im** and then selecting the **Replace** menu item. It will be obvious what to do from that point. Complete the modification necessary to produce the block diagram shown here.

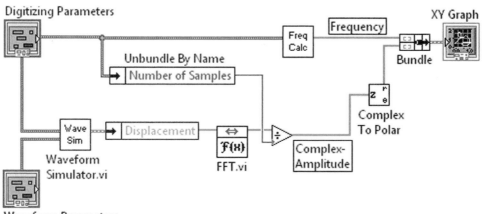

419

Program the **User Defined** function of **Waveform Simulator** to calculate the sum of a sine and cosine function, both having frequency $f = 250$ Hz, with amplitudes 3.0 and 4.0, respectively:

$$x = 3.0\sin\left[2\pi(250)t\right] + 4.0\cos\left[2\pi(250)t\right]$$

From Equation [20], the above waveform should be equivalent to

$$x = \sqrt{3^2 + 4^2}\,\cos\left[2\pi(250)t - \tan^{-1}(3/4)\right]$$

Run **FFT (Magnitude Only)** with **Number of Samples** and **Sampling Frequency** equal to *1024* and *2000*, respectively. Is the output representative of a single 250 Hz cosine curve with net amplitude $\sqrt{3^2 + 4^2} = 5$?

The Fourier transform method represents a function as the linear combinations of the complex exponential basis set. Because $\cos(2\pi f t) = (e^{i\,2\pi f t} + e^{-i\,2\pi f t})/2$, a cosine function's amplitude is equally divided between the complex-amplitudes of the positive and negative frequency basis functions $e^{\pm i2\pi f t}$. Two exceptions to this "equal division of amplitude" occur, of course, for the basis functions at zero frequency (i.e., $e^{i\,2\pi(0)t} = 1$) and at the Nyquist frequency $f_{nyquist}$. When displaying the spectrum of a given input data set, this symmetry of complex-amplitudes in frequency space is commonly exploited by simply doubling the magnitude of the positive frequency's complex-amplitude and only plotting the positive-frequency axis. Such a plot displays the net amplitude of sinusoidal oscillation at each frequency $|f_k|$ directly, while phase information is completely neglected.

Modify **FFT (Magnitude Only)** to plot the input data's spectrum in the manner described above. Let's ignore the Fourier component at $f_{nyquist}$ because, as we saw previously, its value is sometimes suspect. Since the y-axis values are now equivalent to the net amplitude of sinusoidal oscillation at each (positive) frequency f_k, relabel the XY Graph's y-axis and Plot Legend as **Amplitude**. The following block diagram accomplishes the desired features.

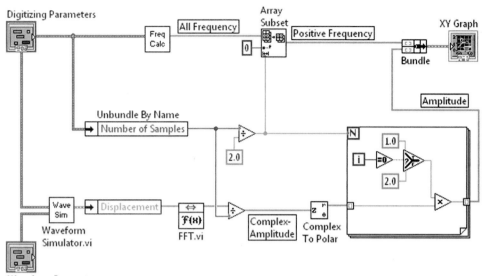

Run the VI and verify that it performs correctly. Once satisfied, save your work.

10.9 OBSERVING LEAKAGE

I have good news and I have bad news. First, the bad news. Reprogram **FFT (Magnitude Only)** with the function

$$x = 4.0 \cos [2\pi(250)t]$$

and then run the VI using the values *1024* and *2000* for **Number of Samples** and **Sampling Frequency**, respectively. Its spectrum will appear as shown.

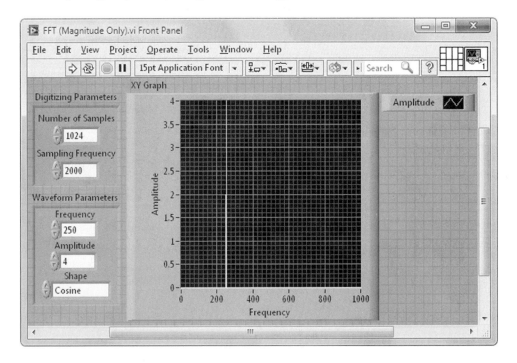

Using the **Plot Legend**, make the individual plotted points visible by changing the **Point Style**. Then magnify the spectral peak at $f = 250$ Hz by changing the x-axis endpoints. An easy way to zoom in on a peak is to use the XY Graph's *Zoom Tool*. To access the Zoom Tool, first pop up on the XY Graph's interior and select **Visible Items>>Graph Palette**. Using the 🖑, click on the Zoom Tool 🔍, which looks like a magnifying glass, and select the **X-Axis Zoom** option.

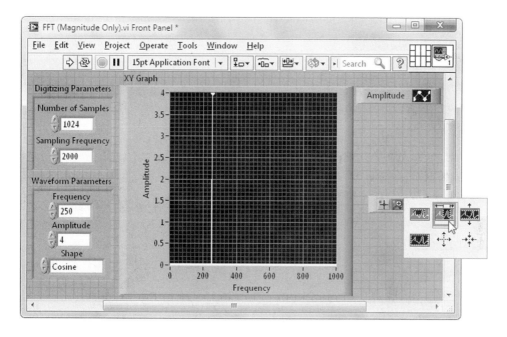

The mouse cursor will become a small magnifying glass when positioned over the plot. Click and drag it to select the region that you would like to zoom in on.

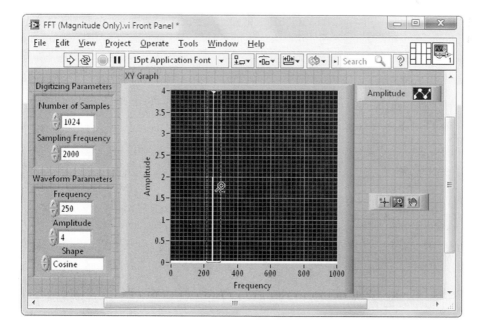

When you release the mouse button, the *x*-axis will be rescaled in the desired manner.

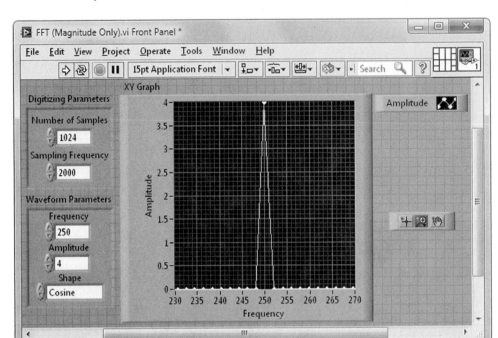

If you want to return to the original plot, select **Zoom to Fit** from the Zoom Tool menu, which will cause all of the data to be displayed again by autoscaling both axes.

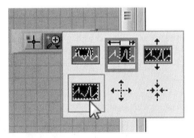

The spectrum looks perfect, doesn't it? The amplitude is equal to zero everywhere, except at the frequency of 250 Hz, where it equals +4.0, just as we expected.

So, what's the problem? Try repeating the above procedure after programming FFT (Magnitude Only) with the following function

$$x = 4.0 \cos \left[2\pi(249)t \right]$$

Surprisingly, the resulting spectrum will no longer appear as a "delta-function" spike with a height of 4.0, but instead will be a broadened peak of maximum height less than 4.0, as shown.

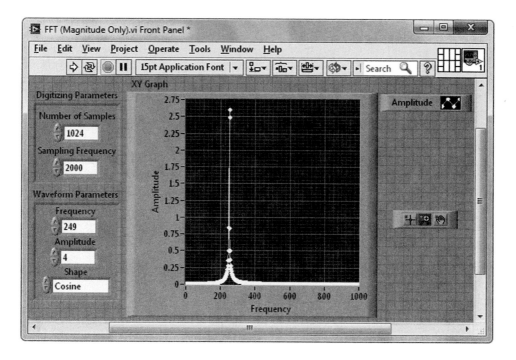

Zoom in on the peak to get a closer look.

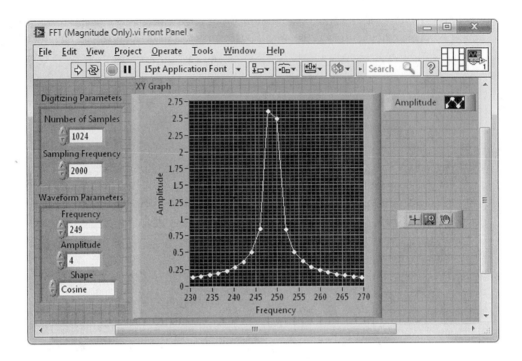

At face value, this spectrum seems to be telling us that the input data are oscillating at several different frequencies in the approximate range of 230 to 270 Hz. However, we know that this is simply untrue. We input a pure sine wave with the exact frequency of 249 Hz.

We are then faced with a mystery. Why does our FFT-based VI work perfectly for an input frequency of 250 Hz, but fails when that frequency is changed to 249 Hz? Well, here's the resolution to this paradox: Above, I've been sneaky and always, with the exception of $f = 249$ Hz, asked you to generate input data using sine and cosine functions that oscillate at one of the N discrete frequencies f_k. For example, via Equation [7], when acquiring 1024 equally spaced data samples at a sampling rate of 2000 Hz, the FFT algorithm produces complex-amplitudes A_k at the N frequencies f_k given by

$$f_k = k \left(\frac{2000 \, \text{Hz}}{1024} \right) \qquad k = -511, \dots, 0, \dots, +512$$

You can verify that 250 Hz and 500 Hz are both one of the f_k. In particular, they are f_{128} and f_{256}, respectively.

A way of summarizing our above observations is this: If you input a sinusoidal oscillation at exactly one of the frequencies f_k, for example, $f_{128} = 250.00$ Hz or $f_{256} = 500.00$ Hz, the FFT algorithm will perfectly produce the frequency spectrum of that data (that is, a delta-function spike of the correct height at the correct frequency). However, if instead one inputs a sinusoidal oscillation at a frequency f not equal to one of the f_k, such as 249 Hz, which falls between $f_{127} = 248.05$ Hz and $f_{128} = 250.00$ Hz, the resulting spectrum is imprecise. In fact, it is as if its spectral amplitude, which rightfully should be a spike at frequency f, has diffused from this central point and distributed itself among the neighboring f_k. This smearing of spectral information—termed *leakage*—is an artifact of the finite number N of data samples contained in our discretely sampled data set. You can prove this for yourself by rerunning **FFT (Magnitude Only)** programmed to calculate a 249 Hz cosine wave, first making **Number of Samples** equal to *512*, then *1024*, then *2048*, followed by *4096*, and so on. You will find that as N increases, the spectrum becomes much more delta-function-like (it will be helpful here to pop up on the XY Graph and deselect **X Scale>>AutoScale X**).

10.10 ANALYTIC DESCRIPTION OF LEAKAGE

It's not too hard to derive analytic expressions that explain these observations. Consider an oscillatory waveform with frequency f that is described by the complex exponential $x(t) = A \exp(i2\pi f t)$, where A is a constant. Once the features of this waveform's "single-spike" Fourier transform are understood, the properties of a sinusoid's "dual-spike" spectrum will be obvious. Applying the discrete Fourier transform (Equation [11]) to the complex-exponential waveform, the values of X_k are

$$X_k = \sum_{j=0}^{N-1} A e^{i2\pi f(j\Delta t)} e^{-2\pi j k/N}$$

so

$$X_k = A \sum_{j=0}^{N-1} \left[e^{i2\pi(f\Delta t - k/N)} \right]^j \tag{21}$$

This series is the well-known finite geometric series, whose form and summation value are given by

$$\sum_{j=0}^{N-1} x^j = \frac{1 - x^N}{1 - x} \tag{22}$$

After using [22] to evaluate the sum in [21], a few lines of algebra and trigonometric relations yield the following expression for the magnitude of the discrete Fourier transform values X_k

$$|X_k| = \left| A \frac{\sin\left[\pi N \left(f\Delta t - k/N \right) \right]}{\sin\left[\pi \left(f\Delta t - k/N \right) \right]} \right| \tag{23}$$

This equation then describes the quantity determined by FFT (Magnitude Only).

Let's investigate the meaning of Equation [23]. First, consider the case when the input frequency f exactly equals one of the frequencies f_k, say $f_{k'}$. Setting $f = f_{k'} = k'(f_s/N) = k'(1/N\Delta t)$, Equation [23] becomes

$$|X_k| = \left| A \frac{\sin[\pi(k'-k)]}{\sin[\pi(k'-k)/N]} \right|$$ [24]

For $k \neq k'$, the difference $(k' - k)$ will be an integer less than N, making the numerator and denominator of [24] zero and nonzero, respectively. Thus, $|X_k| = 0$ for $k \neq k'$. However, for $k = k'$, both numerator and denominator of [24] are zero and l'Hôpital's rule gives $|X_k| = AN$. So we expect the resulting frequency spectrum to be a delta function with a single spike at frequency $f_{k'}$ of height $|A_{k'}| = |X_{k'}|/N = A$. This prediction is consistent with what we observed when we input the oscillations at $f = f_{128} = 250$ Hz and $f = f_{256} = 500$ Hz inputs to the FFT algorithm.

Second, consider the case when the input frequency f is not equal to one of the frequencies f_k. We can simplify the appearance of Equation [23] by noting from Equations [4] and [7]

$$f\Delta t - \frac{k}{N} = \frac{f}{f_s} - \frac{f_k}{f_s} = \frac{f - f_k}{f_s}$$

Then Equation [23] becomes

$$|X_k| = \left| A \frac{\sin\left[\pi N\left(\dfrac{f - f_k}{f_s}\right)\right]}{\sin\left[\pi\left(\dfrac{f - f_k}{f_s}\right)\right]} \right|$$

and, since $\Delta f = f_s/N$,

$$|X_k| = \left| A \frac{\sin\left[\pi\left(\dfrac{f - f_k}{\Delta f}\right)\right]}{\sin\left[\pi\left(\dfrac{f - f_k}{f_s}\right)\right]} \right|$$ [25]

Inspecting Equation [25], we see that, under our assumption that f does not equal one of the f_k, none of the $|X_k|$ will be zero. However, the frequencies f_k that fall closest to f will make the denominator of Equation [25] the smallest. Thus, the $|X_k|$ associated with the f_k neighboring f will take on the largest values. To better visualize the meaning of Equation [25], I use it to plot the magnitude of the complex-amplitude $|A_x| = |X_k|/N$ vs. f_k with the following familiar choice of parameters: $f_s = 2000$ Hz, $N = 1024$, $A = 4.0$, and $f = 249$ Hz. Then $\Delta f = f_s/N = 2000/1024$ and $f_k = k \, \Delta f$, where k = $-511,\ldots, 0,\ldots, +512$. This plot will predict the output of by **FFT (Magnitude Only)** when a 249 Hz cosine function of amplitude 4 is input.

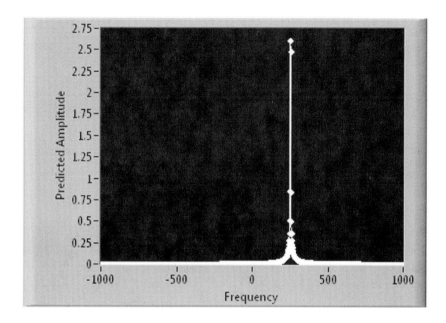

Zooming in on the peak to get a closer look, we see that Equation [25] provides a perfect theoretical prediction of the spectral leakage we observed previously with a 249 Hz input to **FFT (Magnitude Only)**.

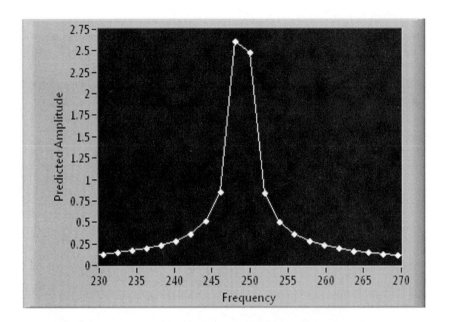

10.11 DESCRIPTION OF LEAKAGE USING THE CONVOLUTION THEOREM

The *Convolution Theorem*, a powerful result from higher mathematics, provides another vantage point from which to understand the problem of leakage. From this point of view, the situation appears as follows: When we acquire a finite number N of discretely sampled points for FFT spectral evaluation, we are in effect observing an infinite set of data d_j (where $j = -\infty, \ldots, -1, 0 + 1, \ldots, +\infty$) through a rectangular viewing window in time. Mathematically, we can define the rectangular window function $w(t_j)$ to be zero at all times $t_j = j\,\Delta t$, except during the "data-viewing" time interval from $j = 0$ to $j = N - 1$, when it is equal to one. Then our finite set of N sampled data points x_j is given by the product $x_j = d_j\,w_j$. This idea is illustrated in Figure 10.2.

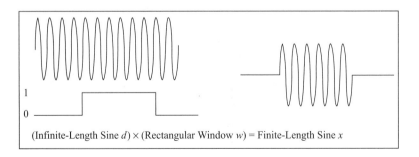

(Infinite-Length Sine d) × (Rectangular Window w) = Finite-Length Sine x

FIG. 10.2

Let's say that the Fourier transforms of d_j and w_j are D_k and W_k, respectively. The question relevant to our finite-length data set then becomes, "What happens when one takes the Fourier transform of the product $x_j = d_j w_j$?" Well, according to the famous Convolution Theorem, the Fourier transform of the product of two functions $d_j w_j$ is equal to the *convolution* of the two Fourier transforms D_k and W_k. The convolution for continuous functions, denoted by $D*W$, is defined by

$$D(f) * W(f) = \int_{-\infty}^{+\infty} D(\phi) W(f - \phi) d\phi \tag{26}$$

In the discrete case, this definition becomes

$$(D*W)_k = \sum_{m=-N/2+1}^{N/2} D_m W_{k-m} \tag{27}$$

The convolution of two functions, although complicated in general to determine, is simple to ascertain in the following important case. Let D_m be a unit-amplitude delta function located at the frequency f_n, that is, $D_m = 0$ for all m, except $D_{m=n} = 1$. Then, from [27], we find that $(D * W)_k = W_{k-n}$, meaning that the convolution is just the function W displaced so that it is now centered at f_n rather than $f = 0$. This idea is illustrated in Figure 10.3 using continuous functions.

This example provides new insight into the leakage phenomenon. Consider the case of a finite-length complex-exponential input of the form $x(t_j) = A exp (i2\pi f t_j)$, which can be described as the product of an infinite-length complex-exponential $d(t_j)$ and a rectangular window function $w(t_j)$. The Fourier transform D of the infinite complex-exponential

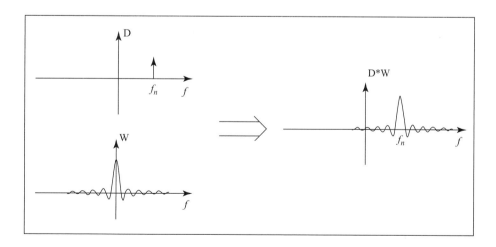

FIG. 10.3

is, of course, a delta function of height A located at frequency $+f$, while the discrete Fourier transform of the rectangular window (apart from a unity-amplitude phase factor) is easily shown to be

$$\left(W_{rectangle}\right)_k = \frac{\sin\left[\dfrac{\pi f_k}{\Delta f}\right]}{\sin\left[\dfrac{\pi f_k}{f_s}\right]} \qquad [28]$$

A plot of Equation [28] (treating frequency as a continuous variable) with $f_s = 2000$ Hz and $N = 1024$ is shown in the following diagram. Note the substantial amplitudes of $W_{rectangle}$ at high frequencies, which we know qualitatively result from the sharp turn-on and turn-off at the edges of the rectangular window (recall from the Fourier analysis of a square wave, its sharp edges are produced by the presence of high-frequency harmonics). The values of $W_{rectangle}$ at the frequencies f_k, that is, $(W_{rectangle})_k$, are given by the dots.

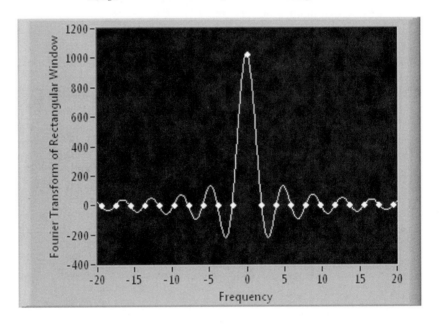

The magnitude of the finite-length complex-exponential's Fourier transform X is formed from the convolution of $W_{rectangle}$ with the height-A delta function at $+f$. This convolution is simply $W_{rectangle}$, shifted along the frequency axis so that it is centered at frequency $+f$, and then multiplied by A. Thus,

$$|X_k| = |(D * W)_k| = \left| A \frac{\sin\left[\dfrac{\pi(f_k - f)}{\Delta f}\right]}{\sin\left[\dfrac{\pi(f_k - f)}{f_s}\right]} \right| \qquad [29]$$

Note that Equation [29] is equivalent to the "leakage description" given by Equation [25].

Next, I plot Equation [29] for $f_s = 2000$ Hz and $N = 1024$ with f equal to one of the f_k, namely, $f = f_{128} = 250$ Hz. I take $A = 1$ for simplicity. Note that all of the values of $|X_k|$, which are given by the dots, fall on the zero-crossings of Equation [29], except $|X_{128}|$ at $f_{128} = 250$. Thus, the discrete Fourier spectrum will be a delta function in this case.

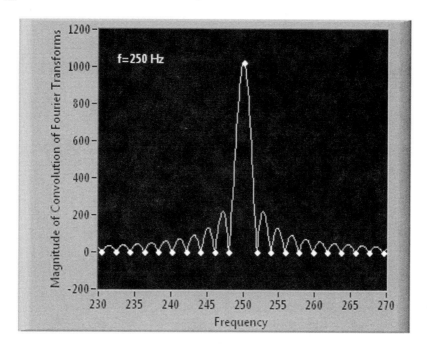

Changing f to the non-f_k value of 249 Hz, the following plot of Equation [29] is obtained. Note that the $|X_k|$, again indicated by the dots, now fall at nonzero locations on the curve defined by Equation [29], resulting in the spectral leakage phenomenon.

433

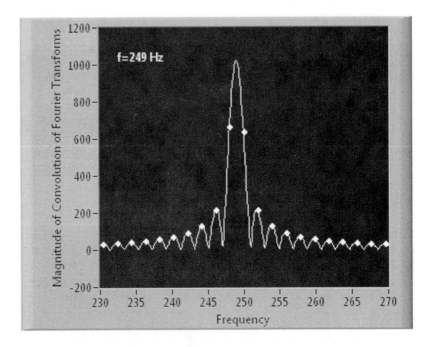

The Convolution Theorem then gives us the following insight: It is the wide-ranging frequency content of the rectangular window function (as given in Equation [28] and shown in the previous plot) that produces leakage of spectral amplitude to frequencies f_k far removed from the input frequency f.

10.12 WINDOWING

Now the good news. Thanks to the power of your high-speed computer, you have the option of windowing your finite-length data set with some function other than a rectangle. Once the rectangular-windowed data set $(x_{rectangle})_j$ has been collected into your computer, all you have to do is construct a more desirable window w_j in software and then form the product $x_j = w_j (x_{rectangle})_j$. Leakage can simply be suppressed by choosing a window whose Fourier transform has minimal high-frequency components. Qualitatively, we know that the leakage-producing high-frequency components result from the discontinuous jumps at the edges of the rectangular function, so the "desirable" software window functions should be constructed to have a much smoother turn-on and turn-off. The *Hanning* window, which is defined to be

$$\left(w_{hanning}\right)_j = 0.5\left[1 - \cos\left(\frac{2\pi j}{N}\right)\right] \qquad j = 0, 1, ..., N-1$$

is a popular choice for such a software window. The Hanning and Rectangle windows are plotted here for comparison.

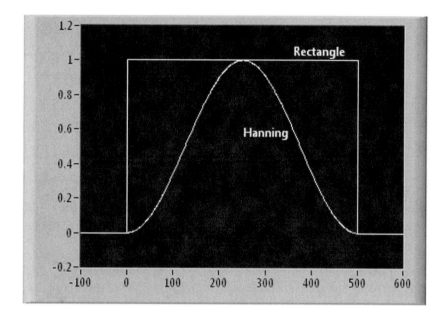

Let's try windowing our data set in **FFT (Magnitude Only)** and see how this improves things. To window the **Waveform Simulator**–produced data set, you will use **Scaled Time Domain Window.vi**, found in **Functions>>Signal Processing>>Windows**. Its Help Window is shown next.

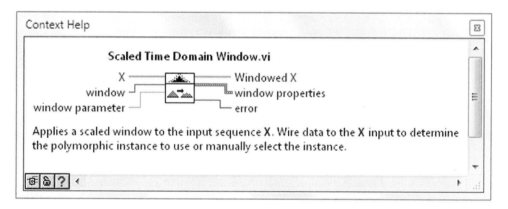

TABLE 10.1   Commonly Used Windows

Window	Optimal Application
Hanning	General-purpose applications
Hamming	Resolution of closely spaced sine waves
Flat Top	Accurate amplitude measurement of isolated frequency
Kaiser*	Resolution of two closely spaced frequencies with widely differing amplitudes
Rectangle	Resolution of two closely spaced frequencies with almost equal amplitudes

* This window's central-lobe width is controlled by value input at **window parameter**.

For this icon, you provide the "to-be-windowed" data array at the **x** input and select the desired window at **window** using an Enumerated Type ("Enum") control. The resultant array (equal to the product of the data and the window) is output at **Windowed x**. The **window properties** cluster is a two-element bundle of the selected window's *equivalent noise bandwidth (ENBW)* and *coherent gain*. These window parameters are important for use in some calculations, as will be shown in a few minutes.

When using **Scaled Time Domain Window.vi**, the user must choose one window, from among the approximately 20 available, to be applied to the input data array. Each window has its own properties such as central-lobe width and roll-off rate, which are designed to optimize its use in particular applications. Generally, the Hanning window is a satisfactory choice in most situations. This window suppresses spectral leakage, allowing frequencies within the input signal to be well resolved. Table 10.1 lists some of the commonly used windows, along with the applications for which they are most suitable.

Switch to the block diagram of FFT (Magnitude Only). Include **Scaled Time Domain Window.vi** as shown, so that the data set obtained from Waveform Simulator is windowed prior to being passed to the **FFT.vi** icon. Additionally, pop up on Scaled Time Domain Window.vi's **window** input and select **Create>>Control** to create the front-panel Enum control that facilitates window selection.

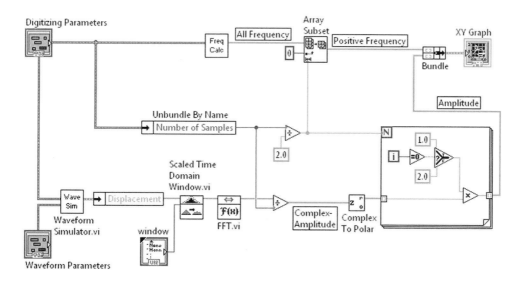

Switch to your front panel, find the Enum control labeled **window** that now appears there, position it aesthetically, and then save your work.

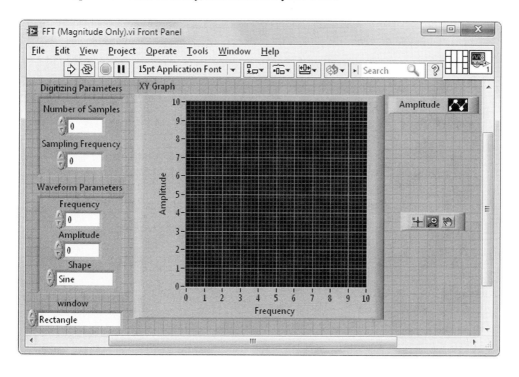

One of the nice benefits derived from having implemented **Create>>Control** to produce the front-panel **window** control is that, while creating it, LabVIEW automatically loads this Enum with the sequence of text messages that label the available options at Scaled Time Domain Window.vi's **window** input. Using the 🖑, review the sequence of selections offered by the **window** control.

Let's explore the positive effects of windowing. Let **Number of Samples** and **Sampling Frequency** be *1024* and *2000*, respectively, and program your VI to produce the waveform 4.0 sin [2π (249 Hz)t]. Zoom in on the expected peak by scaling the axes appropriately, say, *230* to *270* for the *x*-axis and *0* to *4* for the *y*-axis, and then pop up on the XY Graph and use its pop-up menu to turn off the *x*- and *y*-axes autoscaling options. Run the VI without the benefit of windowing by selecting *Rectangle* in the **window** control.

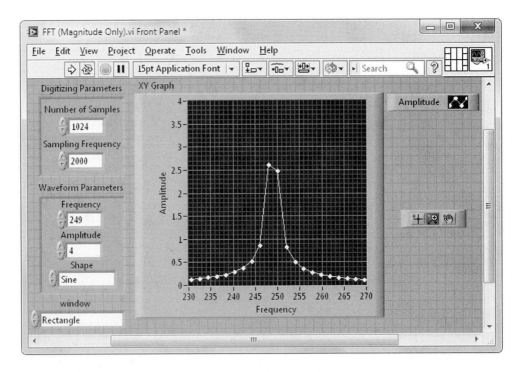

Now select the *Hanning* window and rerun the VI.

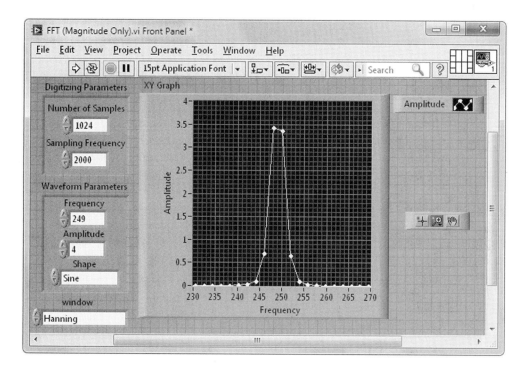

Note the dramatic decrease in leakage through the use of a Hanning window. See, I told you I had good news!

To demonstrate that certain windows perform better in a particular application, consider the challenge of resolving two sine waves of nearly the same frequency with widely differing amplitudes, a problem that might be encountered in, say, speech recognition research. Use the **User-Defined** option of **Waveform Simulator** to input the following waveform to **FFT (Magnitude Only)**:

$$x = 4.0\sin\left[2\pi(249)t\right] + 0.1\sin\left[2\pi(253)t\right]$$

With this input waveform (you will need to adjust the XY Graph's *y*-axis scaling appropriately), run **FFT (Magnitude Only)** several times, each time with a different choice for **window**. For the Kaiser window, you must supply a value (try *3*) at the **window parameter** input of Scaled Time Domain Window.vi. Which windows perform best for this particular application? For which windows does the smaller 253 Hz peak get obscured by the 249-Hz peak (because of large spectral leakage)?

Change the input waveform to two sine waves of nearly the same frequency with equal amplitudes:

$$x = 4.0\sin\left[2\pi(249)t\right] + 4.0\sin\left[2\pi(253)t\right]$$

Which window best resolves these peaks now (e.g., shows the deepest dip between the peaks)?

Finally, the *Flat Top* window is optimized to produce a peak with an amplitude that closely matches that of the input waveform. Input $4.0\sin[2\pi(249\text{ Hz})t]$ to **FFT (Magnitude Only)** with various choices for **window**. Is the amplitude of the spectral peak produced from this input closest to the value of 4.0 when using the Flat Top window (in comparison to when other windows are used)?

10.13 ESTIMATING FREQUENCY AND AMPLITUDE

I have even more good news. Define the *peak power* at frequency f_k to be $P_k \equiv A_k^2$. Then the sinusoidal input frequency f that generated the finite-width spectral peaks observed above can be estimated by a weighted sum, where each observed f_k is weighted by the peak power at that frequency. That is,

$$f \approx \frac{\sum_k f_k P_k}{\sum_k P_k} = \frac{\sum_k f_k A_k^2}{\sum_k A_k^2} \tag{30}$$

where the sum is over the k-values that span the peak.

Also, the peak power $P \equiv A^2$ of the sinusoid with amplitude A can be estimated by

$$P \approx \frac{\sum_k P_k}{\text{ENBW}} = \frac{\sum_k A_k^2}{\text{ENBW}} \tag{31}$$

where *ENBW* is the effective noise bandwidth of the window used in the analysis process that produced the peak. Once P is determined, then the sinusoidal amplitude is $A = \sqrt{P}$.

Write a VI called **Estimated Frequency and Amplitude** that implements Equations [30] and [31]. Store this program in **YourName\Chapter 10**. The front panel should look as shown next. The **Positive Frequency** and **Amplitude** arrays need to be the index-0 and index-1 elements, respectively, of the **Input Arrays** control cluster to be consistent with how these arrays are bundled together on the **FTT (Magnitude Only)** block diagram. Pop up on this cluster and use **Reorder Controls In Cluster...** to make sure the two arrays are indexed properly. Assign the connector pane's terminals consistent with the Help Window shown.

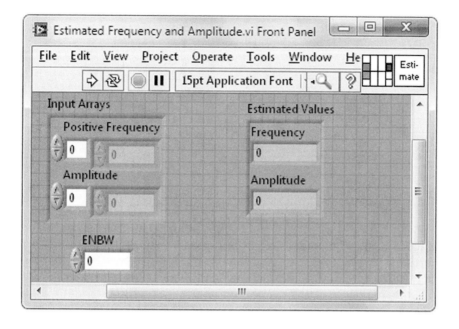

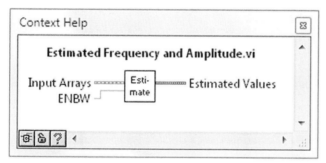

Now code the following block diagram. Because it is easy to do, we will perform the sums in Equations [30] and [31] over all k, but it is only necessary to include k-values for which the A_k are significantly nonzero, that is, k in the neighborhood of the peak.

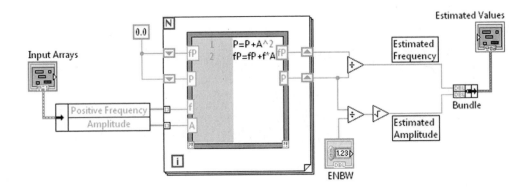

Save this VI as you close it.

Return to the block diagram of **FTT (Magnitude Only)** and incorporate **Estimated Frequency and Amplitude** as shown below. Use **Unbundle By Name** to obtain the selected window's ENBW (labeled **eq noise BW**) from the **window properties** cluster that is output from Scaled Time Domain Window.vi. Use **Create>>Indicator** to create the **Estimated Values** cluster.

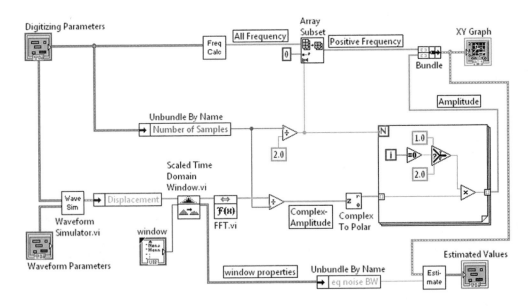

Switch to the front panel and position **Estimated Values** nicely. Within this cluster, pop up on both the **Frequency** and the **Amplitude** indicators and, using the **Display Format...** option, deselect **Hide trailing zeros**. Save your work.

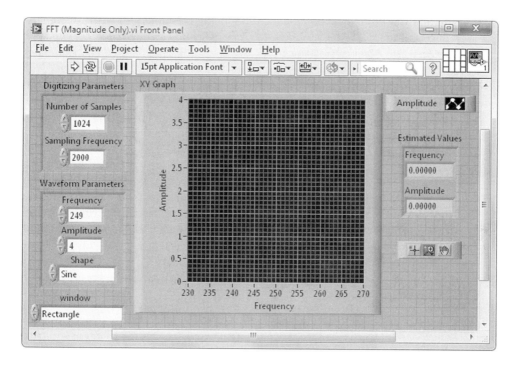

With the values for **Digitizing Parameters** and **Waveform Parameters** shown above, run the VI using various windows. I think you'll be very pleased with the estimated values of **Frequency** and **Amplitude**!

10.14 ALIASING

Finally, let's use **FFT (Magnitude Only)** to demonstrate aliasing, the phenomenon in which an incoming signal is digitally sampled too infrequently, resulting in a digitized waveform whose frequency is much lower than that of the actual signal. In Chapter 4, we found that when sampling a signal of frequency f at sampling frequency f_s, the digitized waveform will also have the frequency f only if $f \leq f_{nyquist}$, where $f_{nyquist} = f_s/2$ is the Nyquist frequency. If, instead, the frequency of the incoming signal is $f > f_{nyquist}$, then the digitized waveform will have frequency f_{alias} ($\neq f$) given by

$$f_{alias} = |f - nf_s| \qquad n = 1, 2, 3, \ldots \qquad [32]$$

where f_{alias} is in the range $0 \leq f_{alias} \leq f_{nyquist}$.

Program **FFT (Magnitude Only)** so that the input waveform is a sine wave of amplitude 4.0. Choose your favorite window, say, the general-purpose *Hanning*. In **Digitizing**

Parameters, set **Number of Samples** and **Sampling Frequency** to be *1024* and *2000*, respectively. Then, we expect aliasing to occur when the frequency of our input sine wave exceeds the Nyquist frequency $f_{nyquist} = f_s/2 = 2000$ Hz/2 = 1000 Hz. Run the VI several times with the following sequence of input sine-wave frequencies: *1200 Hz, 1500 Hz, 1800 Hz, 2000 Hz, 2800 Hz, 3300 Hz, 7700 Hz*. Does Equation [32] correctly predict the aliased frequency you observe for each input?

In this exercise, we have the benefit of knowing the input signal's true frequency and so can use Equation [32] to figure out why the digitizing process is producing an incorrect lower-frequency output. In a real experiment, however, where we'd have no prior knowledge of the input signal's true frequency, we would have no other recourse than to assume that the digitized waveform has the same frequency as the incoming signal. That is, we'd assume the signal is not being aliased in the digitizing process. To assure that this "no-aliasing" assumption is valid, it is imperative that an experimentalist first determine the digitizing system's Nyquist frequency (based on the known sampling rate) and then assure (e.g., though the use of a low-pass filter) that no signals with frequencies greater than the Nyquist frequency are input to the digitizer.

DO IT YOURSELF

Build a computer-based spectrum analyzer that digitizes an incoming voltage waveform, takes the Fourier transform of these acquired data, and displays the resulting spectrum.
Complete the following steps.

1. Write the **Spectrum Analyzer** VI:
    With **FFT (Magnitude Only)** open, use **File>>Save As...** to create a new VI called **Spectrum Analyzer** and save it in the **YourName\Chapter 10** folder.
    Switch to the block diagram of **Spectrum Analyzer**. Delete the **Waveform Simulator** and **Waveform Parameters** icons. Place **DAQ Assistant** on your diagram and configure it to perform an analog input voltage operation on a particular differential AI channel of your DAQ device (e.g., **ai0**) with **Acquisition mode>> N Samples** selected. Include the option of digital triggering, if you wish. On your block diagram, expand **DAQ Assistant**'s terminals to include **data**, **number of samples**, and **rate**. Complete the necessary modifications so that **DAQ Assistant** receives the values of **Number of Samples** and **Sampling Rate** from the **Digitizing Parameters** cluster and then, after obtaining *N* data samples at the instructed sampling rate, passes the acquired data array to **Scaled Time Domain Window.vi**. Finally, enclose all of the block diagram code within a **While Loop**, so that spectra are produced and plotted repeatedly, until a front-panel **Stop Button** is pressed. Save your VI.
2. Operate the computer-based instrument:
    Using a function generator, input a sine-wave voltage with a frequency of about 100 Hz input into the analog input channel of your DAQ device that was programmed into DAQ Assistant (e.g., **ai0**). If you configured DAQ Assistant for

digital triggering, also input your function generator's sync output to the DAQ device's appropriate pins. With **Number of Samples** and **Sampling Frequency** set to *1024* and *2000*, respectively, and your choice for **window**, run **Spectrum Analyzer** to see the spectral makeup of your voltage signal. How do the values of **Frequency** and **Amplitude** in **Estimated Values** compare with the known value (from the function generator's control panel) of your input signal? Try changing the input sine-wave frequency f to values within the range $0 \leq f \leq 1000$ Hz. Does the instrument perform well within this frequency range?

3. Explore interesting inputs:

Increase the input sine wave's frequency beyond the Nyquist frequency to see the aliasing effect. Explain your observations using Equation [32].

Try inputting a square wave with a frequency of about 180 Hz and amplitude of 1 V (i.e., oscillates between -1 V and $+1$ V). Fourier analysis of a square wave of frequency f and unity amplitude predicts that this waveform is equivalent to the following sum of sine waves:

$$\frac{4}{\pi} \sum_{n \text{ odd}} \frac{1}{n} \sin[2\pi nft] = \frac{4}{\pi} \left\{ \sin[2\pi ft] + \frac{1}{3}\sin[2\pi(3f)t] + \frac{1}{5}\sin[2\pi(5f)t] + ... \right\}$$

Note that this sum of sine waves only includes the odd harmonic frequencies f, $3f$, $5f$, and so on. With a square-wave input, is the output of **Spectrum Analyzer** onsistent with this Fourier sum? Do you note any aliasing of the square wave's higher harmonics?

PROBLEMS

1. Use **FFT (Magnitude Only)** to investigate aliasing. Program the **User-Defined** option of **Waveform Simulator** to produce the digitized signal x of four incoming sine waves of frequency $f_1 = 25$ Hz, $f_2 = 70$ Hz, $f_3 = 160$ Hz, and $f_4 = 510$ Hz, each with amplitudes of 1 (i.e., $x = \sin[2\pi(25)t] + \sin[2\pi(70)t] + \sin[2\pi(160)t] + \text{w}$ $\sin[2\pi(510)t]$). Then, on the front panel of **FFT (Magnitude Only)**, set **Number of Samples**, **Sampling Frequency**, **Shape**, and **window** equal to *4096*, *100*, *User-Defined*, and *Rectangle*, respectively. Run **FFT (Magnitude Only)** and record the frequencies of the four peaks that you observe in the resulting spectrum. Identify each peak, that is, tell whether the peak is associated with a true frequency in the signal x or with an aliased frequency produced by the digitizing process. For each aliased frequency, use Equation [32] to identify the true frequency ($f_1, f_2, f_3,$ or f_4) that was aliased.

2. Use **FFT (Magnitude Only)** to explore the improved frequency resolution that results from long-time sampling of an input signal. Program the **User-Defined** option of **Waveform Simulator** under the assumption that the input signal x from

an experiment consists of two closely spaced sine waves of frequency 399 Hz and 401 Hz, each with amplitudes of 4, that is, $x = 4\sin\left[2\pi(399)t\right] + 4\sin\left[2\pi(401)t\right]$. Then, on the front panel of FFT (Magnitude Only), set Sampling Frequency, Shape, and window equal to *2000*, *User-Defined*, and *Rectangle*, respectively.

(a) Simulate sampling the input signal for longer and longer times by running FFT (Magnitude Only) with the following succession of values for Number of Samples—128, 256, 512, 1024, 2048, 4096—each time noting whether the 399-Hz and 401-Hz peaks are resolved in the resulting spectrum.

(b) At what threshold value for Number of Samples do the two peaks begin to be resolved? From your knowledge of FFT theory, explain how this threshold value is determined by the various parameters values used in this investigation.

3. Observe the Fourier transform of a square wave. Program FFT (Magnitude Only) with the following choices: Number of Samples and Sampling Frequency equal to *4096* and *5000*; and Frequency, Amplitude, and Shape equal to *180*, *1*, and *Square*, respectively, to simulate a square wave with a frequency $f = 180$ Hz, which oscillates between 0 V and +1 V. Fourier analysis predicts that this waveform is equivalent to a DC component (of magnitude 0.5) plus a sum of sine waves as follows:

$$0.5 + \frac{2}{\pi}\sum_{n\ \text{odd}}\frac{1}{n}\sin\left[2\pi n f t\right] = 0.5 + \frac{2}{\pi}\left\{\sin\left[2\pi f t\right] + \frac{1}{3}\sin\left[2\pi(3f)t\right] + \frac{1}{5}\sin\left[2\pi(5f)t\right] + ...\right\}$$

Note that the sum of sine waves only includes the odd harmonic frequencies f, $3f$, $5f$,..., with amplitudes of $2/\pi$, $2/3\pi$, $2/5\pi$,.... Finally, set window to *Flat Top* so that the Fourier peaks will have accurate amplitudes.

Run FFT (Magnitude Only) and record the frequency and amplitude of every peak that you observe in the resulting Fourier spectrum. Then, identify the harmonic to which each peak corresponds, including those peaks that are present as a result of aliasing. Finally, comment on whether each observed peak has the amplitude expected from Fourier analysis.

4. Use FFT (Magnitude Only) to explore the *frequency doubling* (and *tripling*) phenomenon that occurs when a pure sine wave signal is input to a nonlinear detector/amplifier (e.g., radio wave on diode detector, monochromatic light on nonlinear crystal). To investigate this effect, use the User-Defined option of Waveform Simulator to simulate the following process: a pure sinusoid $x(t) = A\sin(2\pi f t)$ is passed through a slightly nonlinear amplifier producing an output signal $a(x)$ given by

$$a(x) = \alpha x + \beta x^2$$

where α and β are constants with α significantly larger than β. Remember that .^ is Mathscript's element-wise power operator.

Run **FFT (Magnitude Only)** with some choice of f, α, and β ($f = 250$ Hz, $\alpha = 10$, and $\beta = 1$ might be a good place to start) to find the spectrum of $a(x)$.

(a) In terms of f, what frequencies do you observe in the spectrum?

(b) The x^2 term in $a(x)$ produces a product of sine waves that can be rewritten using the following trigonometric identity:

$$\sin\theta\sin\phi = \frac{1}{2}\left[\cos\left(\theta-\phi\right) - \cos\left(\theta+\phi\right)\right]$$

With the help of this relation, predict the frequencies that should appear in the spectrum of $a(x)$. Are these the frequencies that you observed?

(c) Explore what happens when the amplifier is even more nonlinear so that $a(x) = \alpha x + \beta x^2 + \gamma x^3$. In terms of f, what frequencies do you observe now in the spectrum?

[In experimental optics, laser light of frequency f is commonly input on a nonlinear crystal to produce output light of frequency $2f$ and $3f$.]

5. Observe the *sidebands* that exist in the frequency spectrum of an *amplitude modulated* (AM) wave as follows.

(a) On the block diagram of **FFT (Magnitude Only)**, replace the subVI **Waveform Simulator** (and its input control clusters) with **AM Wave** (written as Chapter 3's **Do It Yourself** project). Then, with *1024, 2000, 250,* and *50* for **Number of Samples, Sampling Frequency, Signal Frequency**, and **Modulation Frequency**, respectively, run **FFT (Magnitude Only)**. The spectrum that you observe consists of a central peak and "sidebands." In terms of f_{sig} and f_{mod}, what is the central peak's frequency and the spacing of the sidebands from the central peak?

(b) In an AM radio receiver, the AM wave received from an antenna is "demodulated" by passing it through a nonlinear diode detector. By making an appropriate modification of the block diagram of **AM Wave**, simulate this demodulation method by applying a nonlinear amplification $a(x) = x^2$ to your AM wave $x(t)$. Run **FFT (Magnitude Only)**, and identify all of the frequencies in the resultant spectrum in terms of f_{sig} and f_{mod}. [Note that one of these frequencies is the modulation frequency f_{mod} and that by proper (low-pass) filtering it could be selected out for distribution to the radio's speaker. In optics, this technique is implemented to produce two optical laser frequencies separated by a small "radio-frequency" difference.]

6. Given an input consisting of the sum of many AC voltages with various frequencies, a *lock-in amplifier* can selectively measure the oscillatory amplitude of a single one of those frequencies. The theory of operation of this instrument can be simulated as follows. First, assume that the "input signal" y_{sig} to the lock-in consists of the sum of three sine waves with frequencies 100, 200, and 300 Hz and amplitudes 4, 6, and 8, respectively, that is, $y_{sig} = 4 \sin [2\pi (100)t] + 6 \sin [2\pi (200)t] + 8 \sin [2\pi (300)t]$. Then, if a user programs the lock-in to measure the amplitude of the 200 Hz sine wave, the instrument internally produces a "reference signal" $y_{ref} = 2\sin [2\pi (200)t]$ and then the product $x = y_{sig} y_{ref}$. Finally, the instrument filters x to determine the magnitude of the DC component present in x and outputs this DC value.

   (a) Use **FFT (Magnitude Only)** to gain insight into the above-described theory. First, program the **User-Defined** option of **Waveform Simulator** to produce the given y_{sig} and y_{ref} and then the product $x = y_{sig} y_{ref}$. Remember that .* is Mathscript's element-wise multiplication operator. Next, run **FFT (Magnitude Only)** with **Number of Samples** and **Sampling Frequency** equal to *1024* and *2000*, respectively, and **Shape** set to *User-Defined*. You will find that several frequencies, including $f = 0$, are present in x. How does the DC component compare with the amplitude of the 200 Hz sine wave present in y_{sig}?

   (b) Change the reference signal to $y_{ref} = 2\sin [2\pi (300)t]$ and rerun **FFT (Magnitude Only)**. What is the resulting DC component equal to now?

   (c) From the trigonometric identity $\sin \theta \sin \phi = \dfrac{1}{2}\left[\cos (\theta - \phi) - \cos (\theta + \phi)\right]$, can you explain the frequencies you observe in x?

7. Write a program called **FFT (Express)**, which uses the **Spectral Measurements** Express VI (found in **Functions>>Express>>Signal Analysis**) to perform a fast Fourier transform on a given input from your **Waveform Simulator** VI. Place two Waveform Graphs on the front panel. Label one **FFT (Peak Magnitude)** with x- and y-axes as **Frequency** and **Amplitude**; label the other **FFT (Phase)** with x- and y-axes as **Frequency** and **Phase (Degrees)**. Complete the block diagram as shown next. Here the output of **Waveform Simulator** is packaged together as a *waveform* data type, a data type that can be directly input to an Express VI. The **Build Waveform** icon, found in **Functions>>Programming >>Waveform**, combines an initial time **t0** (called a *time stamp*), the sampling time increment **dt**, and the waveform array. The time stamp is created using **Get Date/Time In Seconds** (found in **Functions>>Programming>> Timing**). When the dialog window for **Spectral Measurements** opens, choose **Selected Measurement>>Magnitude (Peak)**, **Window>>Hanning**, and **Phase>>Convert to degree**.

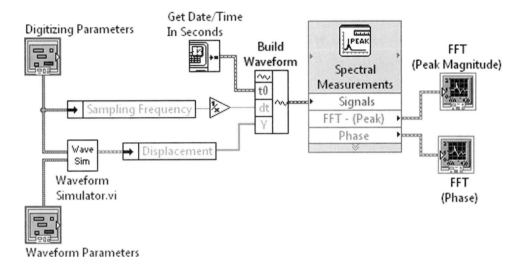

Digitizing Parameters

Get Date/Time In Seconds

Build Waveform

Spectral Measurements
Signals
FFT - (Peak)
Phase

FFT (Peak Magnitude)

FFT (Phase)

Sampling Frequency

Wave Sim

Displacement

Waveform Simulator.vi

Waveform Parameters

On the front panel of **FFT (Express)**, choose *1024* and *2000* for **Number of Samples** and **Sampling Frequency**, respectively.

(a) Program your VI to produce a *Cosine* function with amplitude of *4.0* and frequency of *250* Hz, and then run it. Use the two plots to determine the resulting values of amplitude and phase at $f = 250$ Hz (it may be helpful to turn off autoscaling on the x-axis and manually scale it to display a restricted range of frequencies). Then program the VI to produce a 250 Hz sine function of amplitude 4 and run the program. Use the two plots to determine the values of amplitude and phase at $f = 250$ Hz that result from this input.

(b) In this chapter, you found that the complex-amplitude of a 250 Hz cosine and sine function at $f = 250$ Hz is $A = +2$ and $A = -2i$, respectively. Rewrite these two complex numbers in the form $A = |A| e^{i\theta}$ and thereby predict the phase θ expected for the FFT of a cosine and a sine function. How do these predicted values compare with the values you observed in part (a)?

(c) Program the **User-Defined** option of **Waveform Simulator** to be $x = 4\cos[2\pi(250)t + \pi/6]$, and then run **FFT (Express)**. Are the values for amplitude and phase at $f = 250$ Hz on the two front-panel plots in agreement with the values you expect? Repeat this procedure for the input $x = 4\sin[2\pi(250)t + \pi/6]$.

8. Using a function generator, input a square-wave voltage with a frequency of about 180 Hz input into the low-pass op-amp filter shown in Figure 10.4 with $R = 15$ kΩ and $C = 0.015$ μF, whose 3-dB frequency is given by $f_{3dB} = 1/2\pi RC$. Attach the output of this filter to the analog input channel of your DAQ device that was

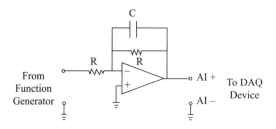

FIG. 10.4   Low-pass filter for Problem 8.

programmed into DAQ Assistant (e.g., **ai0**) on the **Spectrum Analyzer** block diagram. With **Number of Samples** and **Sampling Frequency** set to *1024* and *2000*, respectively, run **Spectrum Analyzer** to see the spectral makeup of your filtered voltage signal.

(a) Calculate f_{3dB}. Given this frequency, what harmonics of the square wave do you expect to observe and what should be the ratio of their amplitudes compared with the first harmonic (see discussion in the **Do It Yourself** project)? Which harmonics do you actually observe and what are their amplitudes (in comparison with the amplitude of the first harmonic)? Comment on any deviation between your expectations and your observations.

(b) Are any of the observed peaks caused by aliasing of harmonics with frequencies higher than $f_{nyquist}$? If so, should you increase or decrease C in your filter to eliminate these peaks?

Data Acquisition and Generation Using DAQmx VIs

> Fundamental concepts of data acquisition are presented in Chapter 4. After reviewing that material, use the **Measurement & Automation Explorer (MAX)** to identify the DAQ device (e.g., model PCI-6251 with name *dev1*) connected to your computer. From the pinout diagram for your device, determine the pins associated with the differential analog input channel ai0, digital triggering channel PFI0 (if available), analog output channel ao0, and GROUND. Have appropriate cabling on hand to connect to these pins as well as a function generator, DC voltage source, and oscilloscope.

Since your work in Chapter 4, you have been able to control the operation of a multi-function data acquisition (DAQ) device using programs based on LabVIEW's **DAQ Assistant**. As you know, DAQ Assistant is a sophisticated, high-level Express VI that, through the use of the dialog windows, can be configured to perform all manner of data acquisition and generation tasks (e.g., analog input, digital output). In this chapter we will show that after DAQ Assistant has been configured for a particular task, below its glossy, easy-to-operate user interface, LabVIEW automatically generates customized code that accomplishes the requested task. This customized code is based on a collection of low-level icons called the *DAQmx VIs*. Here, you will learn how to write your own programs using the DAQmx VIs and explore the added flexibility and performance this low-level approach to programming offers in comparison to the use of DAQ Assistant.

11.1 DAQmx VIs

A large assortment of data acquisition (also called *read* or *input*) operations as well as data generation (aka *write* or *output*) operations are within the capability of any multifunction DAQ device manufactured by National Instruments. When a programmer chooses to perform one such operation, that choice (e.g., analog input voltage on channel ai0), along with its particular timing (e.g., sampling rate) and triggering (e.g., analog trigger) selections, is termed a *task*. To program a DAQmx task, one follows a template of five sequential steps: *create and configure the task*, *start the task*, *read or write data*, *stop the task*, and *clear the task*. Each of the DAQmx icons found in **Functions>>Measurement I/O>>NI-DAQmx** is designed to execute one of these required steps and so, by properly configuring a sequence of these icons, a complete DAQmx task can be coded. Thus, most DAQmx-based data acquisition and generation programs are organized on the block diagram as shown in the next illustration.

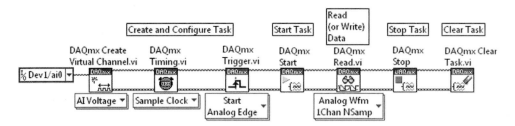

In this diagram, the task is explicitly started by **DAQmx Start Task.vi** and stopped by **DAQmx Stop Task.vi**. However, if **DAQmx Read.vi** is not preceded by **DAQmx Start Task.vi** and followed by **DAQmx Stop Task.vi**, **DAQmx Read.vi** will automatically start and, after it has completed its work, stop the task. An analogous statement applies to **DAQmx Write.vi**. Hence, a DAQmx-based diagram can often be written in the following simplified form.

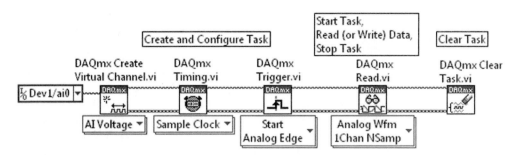

Many of the DAQmx icons are *polymorphic VIs*. When placed on a block diagram, a polymorphic VI has an associated *polymorphic VI selector* that consists of a drop-down menu of possible operational modes for the VI. For example, **DAQmx Read.vi** can be programmed to execute an analog input voltage or a digital input operation by simply making the appropriate selection on its polymorphic VI selector. To make the selection, either click on the selector with the ⟨hand icon⟩ or pop up on the selector and use the **Select Type** option. Through this process, **DAQmx Read.vi** can be configured to operate in nearly 50 different modes. Below, we show how the ⟨hand icon⟩ is used to select **Analog>>Single Channel>>Multiple Samples>>1D DBL**, which instructs **DAQmx Read.vi** to perform an *N*-sample Analog Input operation on one channel and output the data in the form of a 1D double-precision floating-point array.

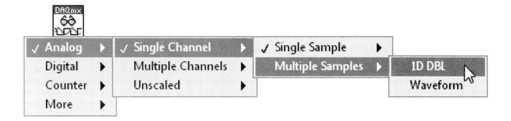

11.2 SIMPLE ANALOG INPUT OPERATION ON A DC VOLTAGE

As a step toward understanding how to program a DAQmx-based diagram, let's build a VI that turns your computer into a voltmeter. For this VI, we will use three DAQmx icons, so we begin by briefly describing the function of each of these block-diagram objects.

First, consider the polymorphic **DAQmx Create Virtual Channel.vi**. When placed on the block diagram, this VI can be used to define ("create") close to 70 different types of tasks (e.g., determine the frequency of the waveform at the counter input). The next illustration shows the selection on the polymorphic VI selector required to create a "voltmeter" task, that is, a task that digitizes voltage at a particular AI channel.

DAQmx Create
Virtual Channel.vi

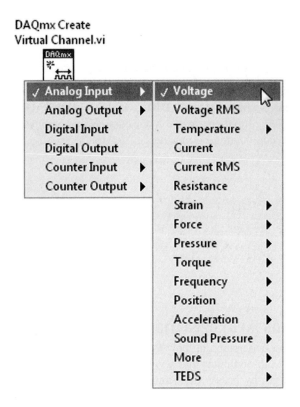

The Help Window of this polymorphic VI adapts to the selection on its polymorphic VI selector. After being configured as **Analog Input>>Voltage**, **DAQmx Create Virtual Channel.vi**'s Help Window appears as shown next. Note that the only required input (shown in boldface text) is **physical channels**; all other inputs (labeled in plain text) are recommended inputs, that is, available for your discretionary use, but may be left unwired (so that default values are used). Given the analog input channel (or channels) specified at **physical channels**, the job of this icon is to create a *virtual channel*, which consists of configurational information including the AI channel name(s) as well as the allowed input voltage range and input terminal configuration (e.g., differential mode). This virtual channel becomes part of the information that is packaged together to define a task. If a task already exists and is provided at the **task in** input, the newly created virtual channel is added to that task. However, if **task in** is left unwired, a new task is created based on the newly created virtual channel. The (modified or newly created) task is then passed out of the **task out** output for use by other DAQmx icons.

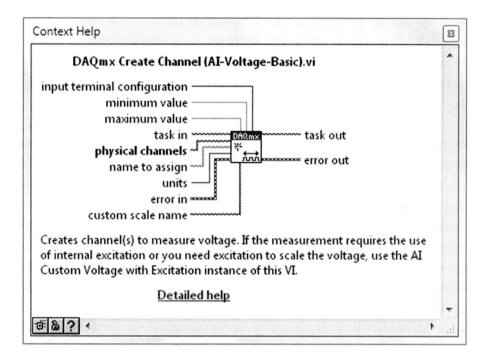

The polymorphic **DAQmx Read.vi** performs the actual data acquisition. To configure this icon to perform a single-sample analog input voltage operation on one channel and output the data in the form of a double-precision floating-point number, select **Analog>>Single Channel>>Single Sample>>DBL** on its polymorphic VI selector. After this selection has been made, the Help Window for this icon appears as shown next. Once presented with the task configurational information at its **task/channels in** input, this icon outputs the digitized voltage value as a double-precision numeric at its **data** output. Additionally, the task definition is available at the **task out** output terminal. As mentioned previously, this icon will start and stop the task if **DAQmx Start Task.vi** and **DAQmx Stop Task.vi** do not appear explicitly on the block diagram.

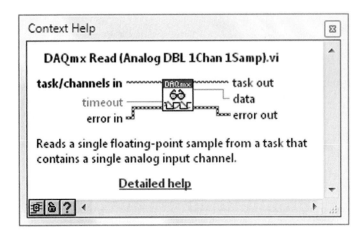

Finally, the Help Window for **DAQmx Clear Task.vi** is given below. This icon discharges the task specified at its **task in** input, releasing any computing resources such as an allotment of RAM that the task has reserved. Once a task is cleared, it cannot be run again; a new similarly defined task must be created and then run.

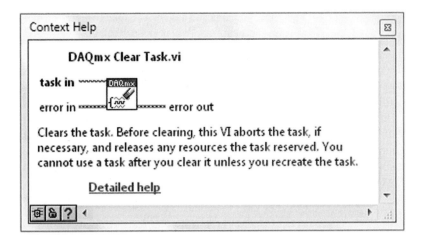

Note that all three of the above icons include error reporting via the error cluster, which appears at the **error in** and **error out** terminals.

A single voltmeter reading then can be accomplished by wiring these three icons together as follows.

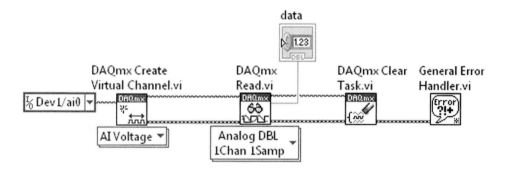

This wiring scheme takes advantage of the principle of LabVIEW programming called *data dependency*. Simply stated, data dependency means that an icon cannot execute until data are available at *all* of its inputs. In this diagram, all of the inputs to **DAQmx Create Virtual Channel.vi** are wired (**physical channel**, the only required input, is explicitly wired; all of the unwired recommended inputs are considered by LabVIEW to be implicitly wired to their default values). So, when this diagram is run, **DAQmx Create Virtual Channel.vi** executes immediately. Upon completion, **DAQmx Create Virtual Channel.vi** outputs a task at its **task out** terminal, which is passed through the wiring to DAQmx Read.vi's **task in** input. Because of data dependency, **DAQmx Read.vi** cannot execute until it receives the task from **DAQmx Create Virtual Channel.vi.** Then, after **DAQmx Read.vi** completes its execution, it passes the task from its **task out** terminal to DAQmx Clear Task.vi's **task in** input. Only then can **DAQmx Clear Task.vi** execute. Thus, through this programming trick, we are assured that the icons will execute in the desired sequence: **DAQmx Create Virtual Channel. vi** followed by **DAQmx Read.vi** followed by **DAQmx Clear Task.vi**.

Also, the correct manner of chaining together DAQmx icons for error reporting is shown above. If an error does occur at one point in the chain, subsequent icons will not execute and the error message will be passed to the **General Error Handler.vi**, which will display the message in a dialog box. **General Error Handler.vi** is found in **Functions>>Programming>>Dialog & User Interface**. Rather than using this dialog box approach to error reporting, one could alternately pass error information to a front-panel error cluster for viewing.

OK, we are ready to write our simple voltmeter program. Open a new VI and, using **File>>Save**, first create a folder called **Chapter 11** in the **YourName** folder, and then save this VI under the name **DC Voltmeter (DAQmx)** in YourName\Chapter 11.

Switch to the block diagram and write the following code, which continually reads and displays the voltage difference at the differential AI channel ai0 every 0.5 seconds until the front-panel **Stop Button** is pressed or an error occurs. Use **Create>>Constant** in the pop-up menu to create the wired inputs of **DAQmx Create Virtual Channel.**

vi. The **DAQmx Physical Channel Constant** 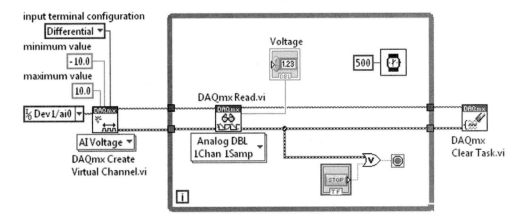 created at the **physical channels** input can be programmed by clicking on its *menu button* with the . You will be presented with a list of available analog input channels on your particular DAQ device (as determined by MAX when it was last run). Otherwise, you can highlight the interior of the **DAQmx Physical Channel Constant** and manually enter the name of the desired AI channel (e.g., *Dev1/ai0*). Also note that this diagram configures the analog input channel to accept a minimum and maximum voltage difference of $V_{min} = -10$ V and $V_{max} = +10$ V, respectively. Use values appropriate for your DAQ device on your block diagram.

If a bad wire resulted when you wired the error cluster to the input of the **Or** icon, your (older) version of LabVIEW requires that the error cluster's status value be unbundled as shown next (see Section 4.10).

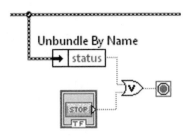

Switch to the front panel. Arrange the two objects as desired and rename the indicator **Voltage**, if you like. The voltage resolution ΔV of the *n*-bit analog-to-digital conversion process performed by your DAQ device is given by $\Delta V = (V_{max} - V_{min})/2^n$. So for a device such as the PCI-6251 with $n = 16$, $\Delta V = (20 \text{ V})/2^{16} = 0.3$ mV. For $n = 14$ (e.g., USB-6009), $\Delta V = 1$ mV. Use this information to select the proper number of **Digits of precision** on the **Voltage** indicator. For example, if $\Delta V = 0.3$ mV, **Digits of precision** should be *4*. Save your work.

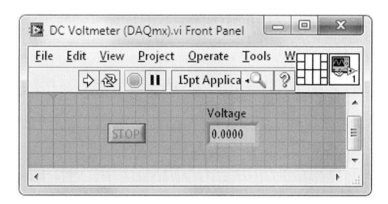

Connect a voltage difference of (approximately) +5 V to the differential analog input channel ai0 of your DAQ device. For example, on a PCI-6251 board in the differential mode, connect +5 V to pin 68 and GND to pin 34, while on a USB-6009, attach +5 V to pin 2 and GND to pin 3. On a myDAQ and NI ELVIS II, +5 V and GND connect to the pins labeled AI 0⁺ and AI 0⁻, respectively. If your voltage source is "floating," use an extra wire to connect its negative terminal to the DAQ device's AI GND pin (see discussion in Section 4.6).

Run **DC Voltmeter (DAQmx)**. Does it read the input voltage correctly?

11.3 DIGITAL OSCILLOSCOPE

With a quick change on a polymorphic VI selector and an additional DAQmx icon or two, you can morph your DC voltmeter into digital oscilloscope program. This new VI—named **Digital Oscilloscope (DAQmx)**—will acquire *N* equally spaced voltage samples of a time-varying analog input and then quickly plot the array of data values. By repeating this process over and over, we'll achieve a real-time display of the waveform input.

With **DC Voltmeter (DAQmx)** open, use **File>>Save As...** to create the new program called **Digital Oscilloscope (DAQmx)** and store it in **YourName\Chapter 11**. On the front panel, delete the **Numeric Indicator** and replace it with a **Waveform Graph**, whose *x*- and *y*-axes are labeled **Time** and **Voltage**, respectively. Also, using **Select a Control...**, locate a **Digitizing Parameter** cluster in **YourName\Controls** and place it on the front panel (if you haven't stored a **Digitizing Parameter** cluster in **YourName\Controls** previously, construct the cluster now as described in Section 3.12, "**Control and Indicator Clusters**"). Save your work.

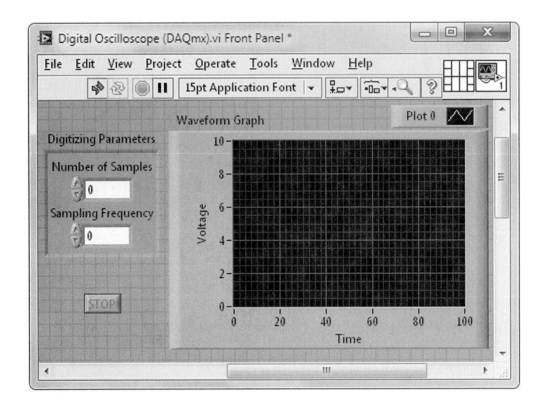

Switch to the block diagram. Remove all broken wires. Then, using the 🖑, click on the **DAQmx Read.vi**'s polymorphic VI selector and select **Analog>>Single Channel>>Multiple Samples>>Waveform**. The next illustration shows how the 🖑 is used to make this selection, which instructs **DAQmx Read.vi** to perform an *N*-sample analog input operation on one channel and output the data in the waveform data type.

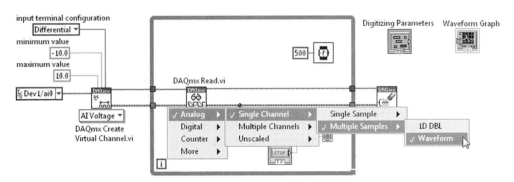

What is the *waveform* data type? Remember that previously (see Section 2.9), when using a Waveform Graph to plot an *N*-element 1D numeric array of data samples with a calibrated *x*-axis, we had to form the cluster shown below consisting of the initial *x*-axis value x_0, the *x*-axis increment Δx between neighboring data samples, and the 1D data array itself.

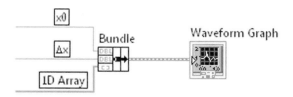

Similar to the dynamic data type used by Express VIs, the waveform data type automatically packages these items (x_0, Δx, and the 1D data array) together in a single wire. Thus, by programming DAQmx Read.vi's output to be in the waveform (rather than 1D DBL) format, its **data** output terminal can be wired directly to the Waveform Graph terminal. The resulting waveform wire will produce the same calibrated plot as would result from the "bundling" code given above.

Place the **Waveform Graph** icon terminal within the While Loop and wire it to DAQmx Read.vi's **data** output. Note the banded brown wire, which denotes the waveform data type. Also, delete the **Wait (ms)** icon and its associated **Numeric Constant** within the While Loop.

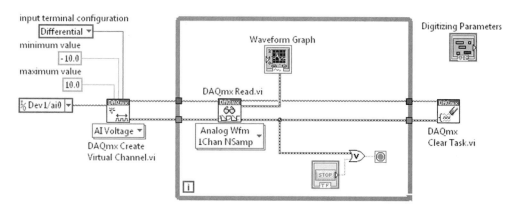

Our diagram now reads as follows: The one-time act of creating the task is done by **DAQmx Create Virtual Channel.vi**. This task is then passed into the While Loop where, during each iteration, **DAQmx Read.vi** acquires *N* data samples at a given sampling frequency f_s and then sends these data to the **Waveform Graph** for plotting. This data acquisition and presentation process repeats continuously until the While Loop ceases its execution in response to the **Stop Button** being pressed or an error having occurred. After exiting the While Loop, the one-time act of disposing of the task is done by **DAQmx Clear Task.vi**.

The unwired **Digitizing Parameters** cluster reminds us that we have a bit more work to do, namely, program the DAQ device with the desired values for N and f_s. To accomplish this feat, we use the polymorphic **DAQmx Timing.vi** icon.

DAQmx Timing.vi offers a wide array of methods to control when an incoming analog signal is digitized. From among these possibilities, we will use **Sample Clock**, a "hardware timing" method in which a digital square wave with a very precise frequency (termed the *sample clock*) controls the rate at which samples of the incoming analog signal are acquired. Each tick, which corresponds to a rising (or else, falling) digital edge of the clock, initiates the acquisition of one sample. After selecting **Sample Clock** on its polymorphic selector, the Help Window of **DAQmx Timing.vi** appears as shown next. If the **source** input is left unwired, the built-in ("on-board") sample clock on your DAQ device will be used to produce the required digital edges.

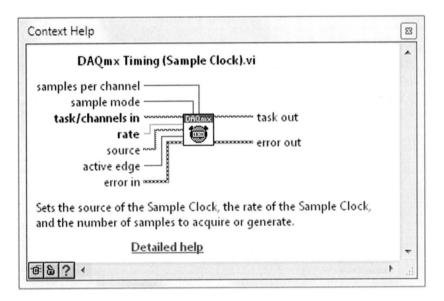

Place **DAQmx Timing.vi** on your block diagram as shown next. Since the timing information being configured by this icon becomes part of the task definition, place it outside the While Loop so that it executes one time only. Here, **Number of Samples** and **Sampling Frequency** are wired to the **samples per channel** and **rate** inputs of DAQmx Timing.vi, respectively. Using the **Create>>Constant** pop-up selection, create the ring constant wired to the **sample mode** input of DAQmx Timing.vi, and then select its **Finite**

Samples mode. In this mode, a finite number of samples (given by the value of **samples per channel**) is acquired each time DAQmx Read.vi executes. Save your work.

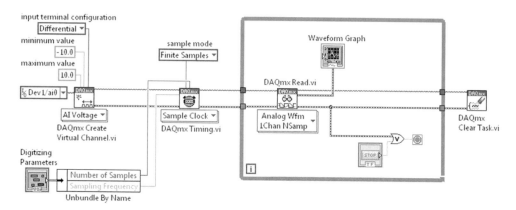

Return to the front panel and set **Number of Samples** and **Sampling Frequency** equal to *100* and *1000*, respectively. Connect the positive and negative (ground) outputs from a function generator to pins AI 0$^+$ and AI 0$^-$ on your DAQ device, and then adjust the settings on the function generator so that it is outputting an analog sine wave with amplitude less than 10 V and a frequency of about 50 Hz. Run **Digital Oscilloscope (DAQmx)**. If all goes well, you will see a plot of about five cycles of the 50 Hz sine wave.

I'm sure you're impressed by your computer-based instrument, but at the same time, concerned about an obvious flaw. Most likely, the sine-wave trace you are observing does not appear stationary but instead is moving either to the right or to the left. And, by slightly changing the function generator's sine-wave frequency, the plotted sine wave can be made to move first one direction and then the other. As explained in Chapter 4, the problem encountered here can be cured by properly *triggering* our digital oscilloscope, that is, by beginning all *N*-sample acquisitions on the same well-defined point of the incoming periodic signal's cycle. For **Digital Oscilloscope (DAQmx)** to be a useful program, we must build in this triggering capability.

In a commercially available oscilloscope, triggering is accomplished in the following manner. The input signal is monitored by an analog "*level-crossing*" circuit. The purpose of this circuit is to determine each time the incoming signal passes through a specified voltage level and then immediately trigger the scope's data acquisition process. Using knobs on the scope's front panel, the oscilloscope user sets the circuit's threshold level and specifies whether acquisition should be initiated when the level is passed through starting from above (*negative*, or *falling, slope*) or starting from below (*positive*, or *rising, slope*) .

Many National Instruments DAQ devices are capable of performing the analog level-crossing triggering procedure described above. For example, the PCI-6251 and ELVIS II have this capability; however, the lower-cost USB-6009 and myDAQ do not. So that most readers can include triggering in their **Digital Oscilloscope (DAQmx)** VI, we'll implement an alternate mode—*digital edge triggering*—in our program because this mode is available on almost all NI DAQ devices (myDAQ is the exception at the time of this writing; myDAQ owners, see Problem 8 in Chapter 4 for a software solution to this problem).

A digital signal can be in one of two possible states, termed HIGH and LOW. For the digital ports on NI DAQ devices, these states conform to the *transistor-transistor logic (TTL) standard* with HIGH (close to) 5 V and LOW (close to) 0 V. As a digital signal alternates between its two states, the transition from LOW to HIGH is called the *rising edge*, while the HIGH-to-LOW transition is the *falling edge*. To execute a data acquisition operation under digital edge triggering, one connects a digital signal to the appropriate pins on a DAQ device. Then, by choosing the proper software settings, the desired data acquisition operation can be begun ("triggered") whenever the digital signal has a rising or, alternately, falling edge.

Stop **Digital Oscilloscope (DAQmx)**, if it's running, and then switch to its block diagram. We will add triggering capability to this VI using the polymorphic **DAQ Trigger. vi** icon, which can configure a task to include digital, analog, or more novel forms of triggering. To configure this icon in its digital triggering mode, select **Start>>Digital Edge** in its polymorphic selector. After being configured thusly, the Help Window for **DAQ Trigger.vi** appears as shown below. The required **source** input specifies the DAQ-device pin (e.g., PFI0) to which the digital triggering signal is connected.

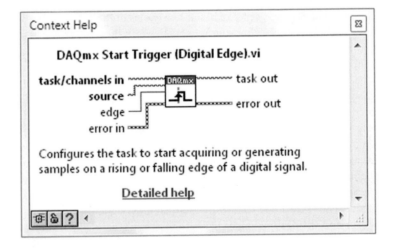

Add **DAQmx Trigger.vi** to your block diagram as shown below. Use the pop-up selection **Create>>Control** to create the **edge** front-panel control. Also use **Create>> Constant** to create a **DAQmx Terminal Constant** ⬚ wired to **source**. Clicking on its *menu button* ⬚ with the 🖑 will present you with a list of available triggering pins from which to choose. Otherwise, you can highlight the interior of the **DAQmx Terminal Constant** and manually enter the name of the desired triggering pin on your DAQ device (e.g., */Dev1/PFI0*).

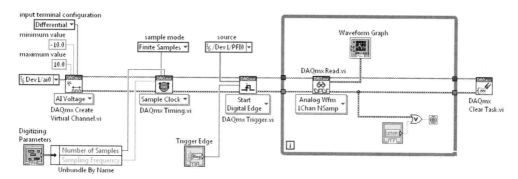

Return to the front panel. Arrange the objects to accommodate the new control and change its label from **edge** to **Trigger Edge**, if you wish. `Digital Oscilloscope (DAQmx)` is now configured for digital-edge triggering. Save your work.

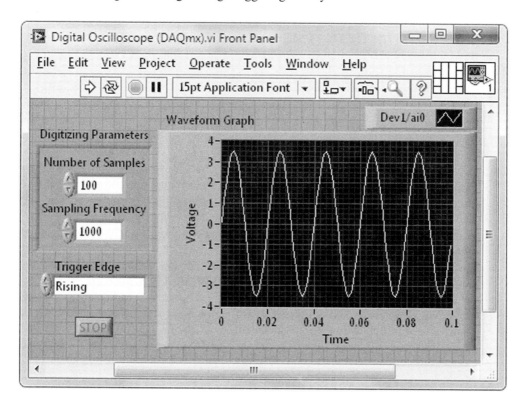

Beside producing a sine wave with frequency *f* at its *main* output, your function generator creates a TTL digital signal of the same frequency *f*, which is available at its *sync* (sometimes called *TTL* or *trigger out*) output. The digital edges of this TTL signal coincide in time with particular points on the sine-wave's cycle. For example, on many function generator models, the rising edge of the TTL signal occurs when the sine wave is at its peak, while on other units it coincides with the sine wave's positive-going zero crossing. Thus, the sync output provides a convenient digital signal with which one can control the acquisition of sine-wave data (i.e., always begin at the same equivalent point on the cycle) via digital edge triggering.

Connect the sync output from your function generator to the digital triggering pins on your DAQ device. For example, sync's positive and negative (ground) terminals would connect to PFI0 (pin 29) and GND (pin 32), respectively, on a USB-6009 device. On a PCI-6251, sync's positive and negative terminals connect to PFI0 (pin 11) and GND (pin 12), respectively. On an NI ELVIS II, the relevant pins are labeled PFI 0 and GROUND.

On the front panel, set **Number of Samples**, **Sampling Frequency**, and **Trigger Edge** equal to *100*, *1000*, and *Rising*, respectively, and set the sine-wave frequency on your function generator to approximately 50 Hz. Run **Digital Oscilloscope (DAQmx)**. You should see a "stationary" plot of about five cycles of the 50 Hz sine wave.

Stop **Digital Oscilloscope (DAQmx)**, and then change **Trigger Edge** to *Falling*. Run your VI. You should again see a "stationary" plot of about five cycles of the 50 Hz sine wave, but this time beginning on a different point on the sine-wave cycle.

11.4 EXPRESS VI AUTOMATIC CODE GENERATION

Now that we've gained some experience with the DAQmx VIs, we're in a position to appreciate the automatic code generation feature of Express VIs by programming **DAQ Assistant** to do the same task we have just coded on the block diagram of **Digital Oscilloscope (DAQmx)**.

Place the **DAQ Assistant** Express VI on a blank block diagram. When the **Create New Express Task...**dialog window opens, make the selections **Acquire Signals>>Analog Input>>Voltage>>ai0**, and then press the **Finish** button. You will recognize that these selections are the choices we programmed on the polymorphic VI selector and source input of **DAQmx Create Virtual Channel.vi** on the block diagram of **Digital Oscilloscope (DAQmx)**.

Next, when the **DAQ Assistant** dialog window opens, under the **Configuration** tab, in the **Signal Input Range** box, select **Max>>10** and **Min>>−10**. Also select **Terminal Configuration>>Differential** and, in the **Timing Settings** box, **Acquisition Mode>>N Samples**. Under the **Triggering** tab, in the **Start Trigger** box, select **Trigger Type>>Digital Edge**, **Trigger Source>>PFI0**, and **Edge>>Rising**. Again, you will recognize that these selections are the associated choices we programmed for the **DAQmx Create Virtual Channel.vi**, **DAQmx Timing.vi** , and **DAQmx Triggering.vi** icons on the block diagram of **Digital Oscilloscope (DAQmx)**.

Exit the **DAQ Assistant** dialog window by pressing the **OK** button. In the following moments, you will see a dialog window indicating that LabVIEW is building a DAQmx-based VI capable of carrying out the task that you have selected. After LabVIEW has created this (hidden) lower-level code, you will be returned to the Express VIs block-diagram icon.

To view the DAQmx-based code that LabVIEW created for the task that you configured in the dialog windows, pop up on the Express VI's icon and select **Open Front Panel** and then **Convert** in the dialog window that follows. When the front panel opens, switch to the block diagram. There, you will find DAQmx-based code that is very similar to the block diagram you wrote for **Digital Oscilloscope (DAQmx)**. If desired, you can modify, save, and run this VI just like you would with any other LabVIEW program.

11.5 LIMITATION OF EXPRESS VIs

If the DAQ Assistant Express VI can automatically write the required low-level code for any data acquisition and generation operation requested, why would one ever wish to write his or her own DAQmx-based code manually? The answer to this question is that the Express VI developers designed these high-level functions to carry out well-defined, commonly needed tasks. If your need fits these particular tasks exactly, then Express VIs are the (painless) way to go. However, if your needs deviate somewhat from the Express VIs' defined functionality or if your project demands the utmost in software efficiency, the Express VI approach may not be able to offer the flexibility and performance level you require.

As a simple demonstration of Express VI limitations, open **Digital Oscilloscope (Express)**, the DAQ Assistant–based digital oscilloscope program you wrote in Chapter 4 and stored in **YourName\Chapter 4**. As you may remember, this program is digitally triggered and expects an analog signal input at AI Channel ai0, so configure a function generator to produce a (close to) 100 Hz sine wave and input the sine wave and the generator's sync output to ai0 and PFI0 pins of your DAQ device, respectively. On the VI's front panel, select **Number of Samples** and **Sampling Frequency** equal to *100* and *1000*, respectively, and then run **Digital Oscilloscope (Express)**. You will see (about) 10 cycles of the digitized sine wave.

Now, let's say that you wanted to "zoom in" on one cycle of the sine wave, which can be accomplished by simply increasing the sampling frequency by a factor of ten, while keeping the number of samples the same. With **Digital Oscilloscope (Express)** still running, change **Sampling Frequency** to *10000* (to secure this choice, you will have click on the ✓ or press <*Enter*>). You will find that the VI does not appear to respond to this new choice for the sampling frequency.

The lack of response to an attempt at changing its sampling frequency during runtime is the result of an intrinsic feature of DAQ Assistant. DAQ Assistant is hardwired to set its various digitizing parameters (including its sampling frequency) only during its first execution after the **Run** button has been pressed. To demonstrate this feature of DAQ Assistant, halt Digital Oscilloscope (Express) by pressing its **Stop Button**. Then, with Sampling Frequency set to *10000*, press the **Run** button. You will find that DAQ Assistant has accepted the new sampling frequency, and a single cycle of the digitized sine wave will be displayed.

Unfortunately, because of DAQ Assistant's fundamental design, there is no way to modify Digital Oscilloscope (Express) to accommodate runtime changes to the Sampling Frequency control. In fact, in the interest of optimizing the performance of our VI, we programmed Digital Oscilloscope (DAQmx) to have the same behavior as Digital Oscilloscope (Express). That is, so that the digitizing parameters were not needlessly updated with each While Loop iteration, we placed all configurational DAQmx icons outside the loop.

The advantage of writing DAQmx-based code, however, is flexibility. If you now decide to make Digital Oscilloscope (DAQmx) capable of responding to runtime changes of its front-panel inputs, you can do so as follows. First, pop up on the While Loop and select **Remove While Loop**.

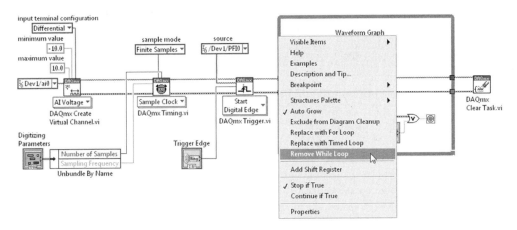

Then, modify the block diagram so that **DAQmx Timing.vi** and **DAQmx Trigger.vi** are included within a newly placed While Loop as shown next. Save your work.

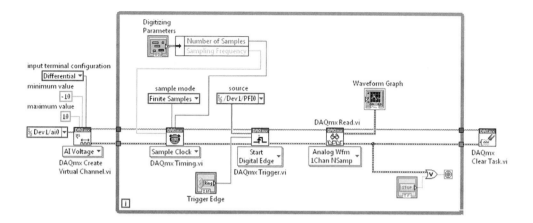

Run this modified version of **Digital Oscilloscope (DAQmx)**. Try changing the front-panel values of **Number of Samples**, **Sampling Frequency**, and/or **Trigger Edge** during runtime, and enjoy the instant response to your command.

11.6 IMPROVING DIGITAL OSCILLOSCOPE USING STATE MACHINE ARCHITECTURE

In modifying **Digital Oscilloscope (DAQmx)**, we sacrificed performance efficiency to gain runtime response. Namely, most of the DAQmx icons contained on the block diagram simply set up (**DAQmx Create Virtual Channel.vi**, **DAQmx Timing.vi**, **DAQmx Trigger.vi**) or shut down (**DAQmx Clear Task.vi**) data acquisition operations and, thus, need only be called during one, or a small subset, of the multitude of **Digital Oscilloscope (DAQmx)**'s While Loop iterations. These set-up and shut-down actions are collectively termed *overhead operations* and, as the program is now written, two of these overhead operations (**DAQmx Timing.vi**, **DAQmx Trigger.vi**) are executed each time the While Loop iterates. Since your data-taking system is not able to digitize the incoming stream of real-time data during overhead operations, needlessly including them increases the program's *dead time*, the percentage of each iteration period during which the system is not sensitive to the incoming data. With increasing dead time, more and more cycles of the incoming repetitive data will stream past your system's input undetected. In certain situations, such as when collecting a large number of data cycles with the intent of adding them together so as to average out random noise, a programming inefficiency that causes you to miss a significant percentage of the incoming data cycles can easily extend the completion time of your experiment by minutes (or sometimes even hours).

If you care to correct the above-described problem, write the following program, which only executes set-up and shut-down operations when necessary (e.g., **DAQmx**

Trigger.vi executes only when a change has been made to the front-panel **Trigger Edge** control). With **Digital Oscilloscope (DAQmx)** open, use **File>>Save As...** to create a new VI named **Digital Oscilloscope (DAQmx State Machine)** in **YourName\Chapter 11**. Consolidate all three controls in one control cluster by placing **Trigger Edge** in the **Digitizing Parameters** control cluster.

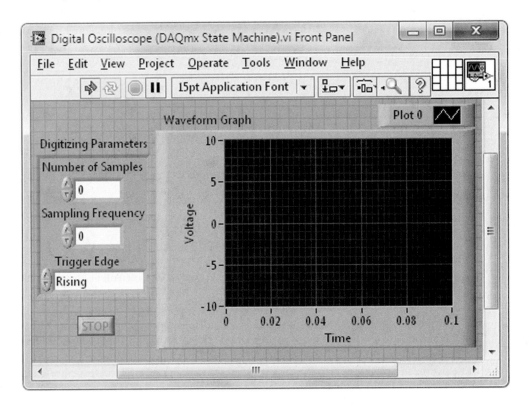

Switch to the block diagram and find a fairly large blank work area where you can build your new code. You can either delete the icons currently on the block diagram (without deleting the terminals for the front-panel objects) or stockpile these icons in some out-of-the-way place and then reuse them as you construct your new code.

The block diagram we will write is based on the *state machine* architecture. A state machine consists of a nested While Loop and Case Structure. The While Loop executes continuously until its ⏺ is set to TRUE, and with each loop iteration, one of the Case Structure cases (termed a *state*) executes. The state that executes during a particular iteration performs some operation (e.g., reads data) and, in addition, selects which state will be executed during the next iteration. The block-diagram template for a state machine is shown in the next illustration. The state selection is made using an Enum Constant, which is stored in a shift register. Information that needs to be passed from one state to another is similarly stored in other shift registers.

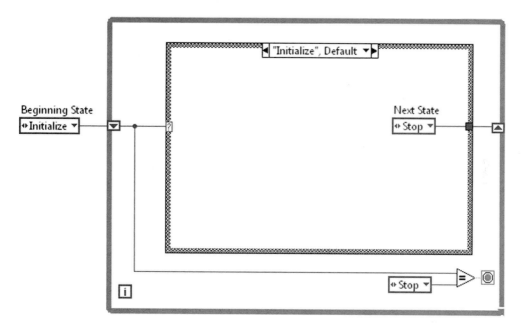

For our digital oscilloscope state machine, we require five states: *Create Task, Check Settings, Change Settings, Read Data,* and *Clear Task.* Thus, we need to create a block-diagram Enum Constant programmed with these five items.

To create the required **Enum Constant**, you can simply obtain this icon in **Functions>>Programming>>Numeric**, place it on the block diagram, select **Edit Items...** in its pop-up menu, and then program the five state names in the dialog window that appears. After closing the dialog window, the completed Enum Constant will appear on the block diagram. You can obtain multiple copies of it by cloning (pressing

<*Ctrl*> while clicking on object with the).

I'll outline an alternate method for creating the required Enum Constant that involves the **Control Editor** and includes options that can be very handy. Open the **Control Editor** as follows: First, select **File>>New...** at the top of an open VI window or the **Getting Started** window. The **New** dialog window will open. In the **Create New** box, find **Custom Control** in the **Other Files** folder and double-click on it. The **Control Editor** window will open, which appears similar to the front panel of a VI.

With the **Control Editor** window open, activate the **Controls Palette**, and then place an **Enum** (found in **Controls>>Modern>Ring & Enum**) on it.

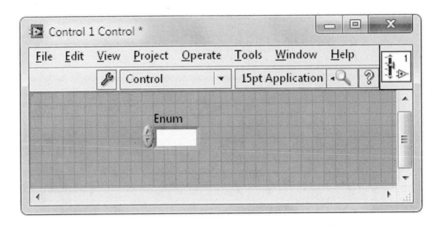

Pop up on the **Enum** and select **Edit Items...**, and then program the **Items** box with the five state names.

After programming the five states, press the **OK** button. You will be returned to the **Control Editor** window, which now contains the completed Enum. You may wish to resize the Enum to make all of the text within it visible.

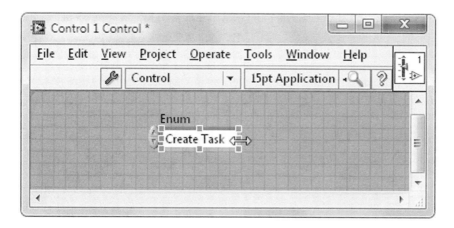

Finally, using **File>>Save**, save this customized control under the name Oscilloscope States in YourName\Controls (if this folder doesn't already exist, create it). The extension **.ctl** will be added automatically. Then close the **Control Editor** window.

For future reference, the above method describes saving our customized object in the **Control** mode. In this mode, the **Enum Constant** we place on the block diagram is an independent copy of the Oscilloscope States file in YourName\Controls. One can, instead, select to save the customized object in the **Type Definition** (or, the related **Strict Type Definition**) mode by selecting **Type Def.** as shown below before saving the customized control.

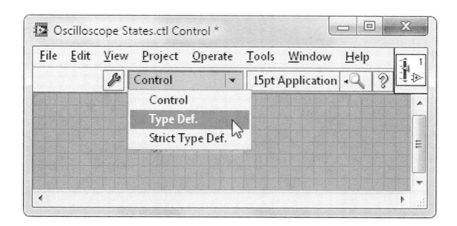

When the **Oscilloscope States** file in **YourName\Controls** is saved in the **Type Definition** mode, it becomes a master file that is linked to every related Enum Constant placed on the block diagram. Then, any change made to the **Oscilloscope States** file will be transmitted to every related Enum Constant. This feature can be quite labor saving, say, when writing a state machine and finding well into the project that an additional state must be added.

Begin coding the block diagram of **Digital Oscilloscope (DAQmx State Machine)** by constructing the **Create Task** state as shown in the next diagram. To obtain a copy of your customized Enum, click on **Select a VI...** in the Functions Palette. In the dialog window that opens, navigate to the **YourName\Controls** folder and double-click on the filename **Oscilloscope States.ctl**. You will be returned to the block diagram, where you can place the customized **Enum Constant**.

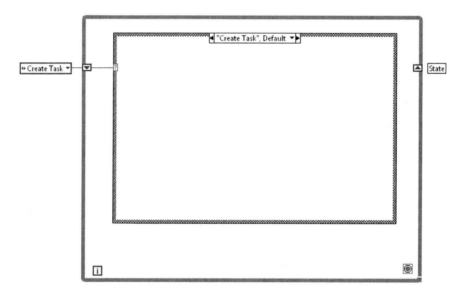

When the **Enum Constant** is connected to the Case Structure's selector terminal, two cases will be activated that are associated with the first two items programmed on the **Enum Constant** (**Create Task** and **Check Settings**). Pop up on the case selector label and choose **Add Case for Every Value**. The Case Structure will then create a case for each of the five states.

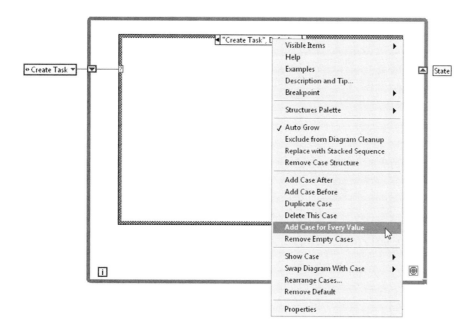

Place a **DAQ Create Virtual Channel.vi** icon in the **Create Task** state, configure it for **Analog Input>>Voltage**, and complete the rest of the code shown in the next illustration. Once on the block diagram, the desired item of the **Enum Constant** can be selected by clicking on it with the 🖑.

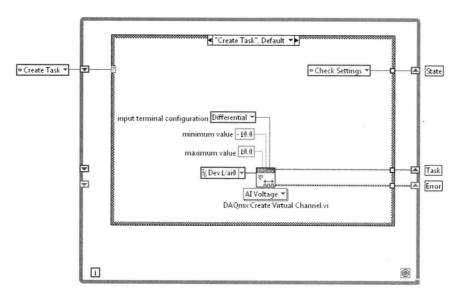

Switch to the **Check Settings** state and place the **Digitizing Parameters** icon terminal here. Complete the code shown, which reads the current cluster values and stores them in a shift register for use in subsequent iterations.

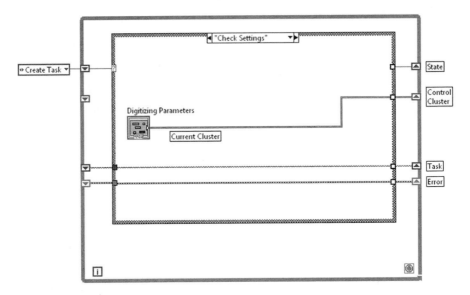

Code the **Check Settings** state, which compares the current **Digitizing Parameters** cluster with that from the previous iteration to determine whether a change has occurred in the front-panel settings. Because we are checking whether the whole current cluster equals the whole previous cluster, pop up on the **Equal?** icon and select **Comparison Mode>>Compare Aggregates**. In this mode, the **Equal?** output is a single Boolean value (in the **Compare Elements** mode, each pair of associated elements in the two clusters is compared and the output is an array of Boolean values). If no change has occurred (**Equal?** is TRUE), **Take Data** is chosen as the next state to execute. Alternately, if a change has occurred (**Equal?** is FALSE), **Change Settings** is chosen as the next state. The **Select** icon used here is found in **Functions>>Programming>>Comparison**. Also, to create the **Cluster Constant**, which initializes the shift register, pop up on the shift register and select **Create>>Constant**.

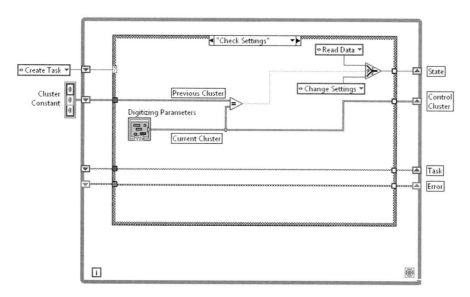

Switch to the **Change Settings** state and code it as shown. The order of **Unbundle By Name**'s output terminals is chosen (using the 🖑) to avoid crossed wires.

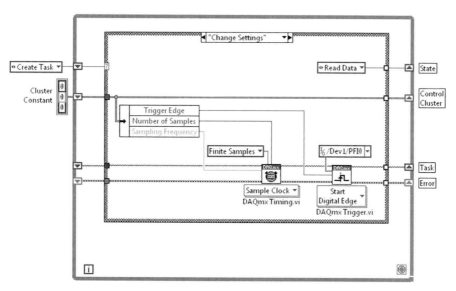

Next, code the **Read Data** state. Here, if the **Stop Button** has been pressed (so that its value is TRUE), **Clear Task** is chosen as the next state to execute. Alternately, if the **Stop Button** has not been pressed (so that its value is FALSE), **Check Setting** is chosen as the next state (which commences another cycle of data taking).

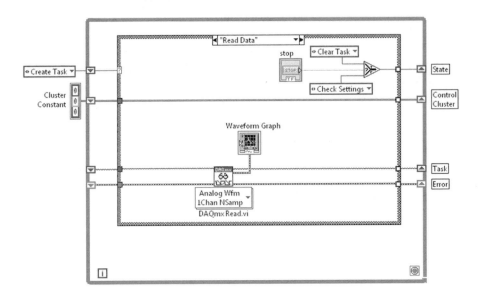

Finally, switch to the **Clear Task** state and complete the diagram as shown. With this code, after the **Clear Task** state executes, the While loop will cease execution on its next iteration because an **Enum Constant** with a value of **Clear Task** will be passed to the **Equal?**'s upper terminal, causing a TRUE to be output to the **Or** icon, hence passing a TRUE to the ⊙. If an error occurs at any time during the execution of the state machine, this too will cease execution of the VI.

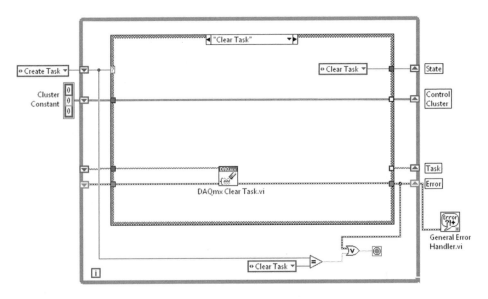

Note that two of the Case Structure's output tunnels are not solidly colored, indicating that these terminals are not wired in all five cases. These unfilled tunnels each constitute a programming error that must be rectified before the VI can run.

In clicking through the five cases, we find that the *task*-related output tunnel is unwired in only the **Clear Task** state. Since there is no task output for **DAQmx Clear Task.vi**, we can simply wire this tunnel as show next.

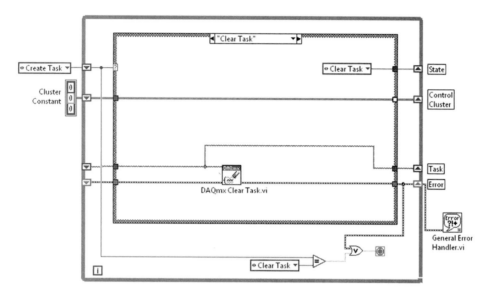

For the *cluster*-related output tunnel, the **Create Task** state is the culprit. Resolve this problem by including the wire shown.

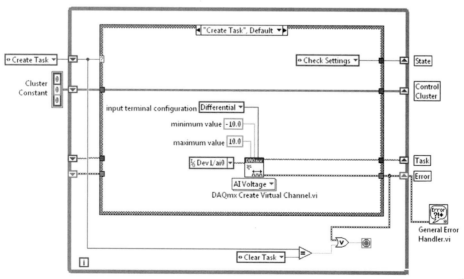

The program is finished now. Return to the front panel and save your work. Run the VI and see if it works.

11.7 ANALOG OUTPUT OPERATIONS

By simply reconfiguring the DAQmx icons with which we are now familiar, analog output operations can be carried out. Let's first write a program that simply outputs a requested voltage value. On a new VI, place a **Numeric Control** labeled Voltage and a **Stop Button**. If you'd like, hide the Stop Button's label (toggle it off via **Visible Items>>Label**) and make the Voltage control's **Digits of precision** appropriate for the accuracy of your particular DAQ device. Save this VI in the YourName\Chapter 11 folder under the name DC Voltage Source (DAQmx).

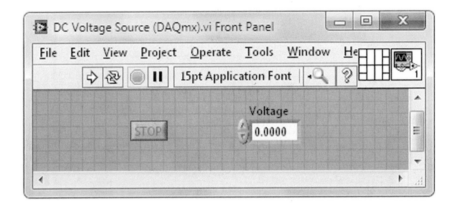

Switch to the block diagram and write the following code, which will output the requested analog voltage at AO channel ao0 until the **Stop Button** is pressed or an error occurs. The values for **minimum value** and **maximum value** in the given diagram are appropriate for a USB-6009 DAQ device. If using an *M* series DAQ device such as the PCI-6251, you can choose **minimum value** and **maximum value** to be *–10* and *10*, respectively. Using the polymorphic VI selectors, configure **DAQmx Create Virtual Channel.vi** and **DAQmx Write.vi** to be in the **Analog Output>>Voltage** and **Analog>>Single Channel>>Single Sample>>DBL** mode, respectively. In the **RSE** output terminal configuration, the voltage output at the ao0 pin is referenced to the DAQ device AO GND.

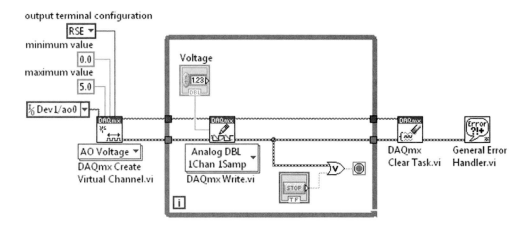

Return to the front panel and save your work. Connect the positive and negative terminals of a voltmeter or oscilloscope to AO 0 (pin 14 on a USB-6009; pin 22 on a PCI-6251) and AO GND (pin 13 on a USB-6009; pin 54 on a PCI-6251), respectively, of your DAQ device. On a myDAQ and NI ELVIS II, AO 0 is labeled as such; AO GND is labeled as AGND and GROUND, respectively. Run the VI with various values for **Voltage**. You will now find that when **Voltage** is set to outside of the range allowed by **minimum value** and **maximum value**, the program ceases operation.

When you are finished, run **DC Voltage Source (DAQmx)** one more time with **Voltage** set to *0*, so the channel ao0 is left in the zero-voltage state.

11.8 WAVEFORM GENERATOR

If your DAQ device supports hardware-timed analog output operations (PCI-6251, myDAQ, and ELVIS II do; the USB-6009 doesn't), try building the following hardware-timed waveform generator called **Waveform Generator (DAQmx)**. This VI is based on the **DAQmx Write.vi** icon configured in its **Analog>>Single Channel>>Multiple Samples>>1D DBL** mode. When in this mode, the Help Window for **DAQmx Write.vi** is as shown in the following illustration. Here, a 1D array of *N* double-precision floating-point numbers describing a few cycles of a periodic waveform is input at **data**. The icon then writes this *N*-element array to a memory buffer. If **autostart** is set to TRUE, the icon then instructs the DAQ device to begin outputting the sequence of analog voltage values as dictated by the 1D array within the memory buffer.

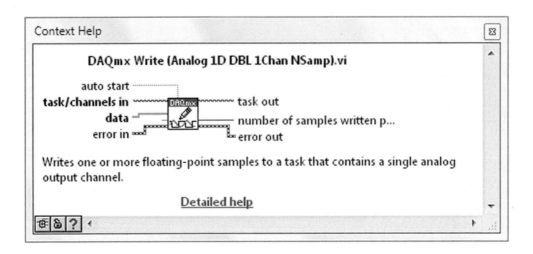

In **Waveform Generator (DAQmx)**, the "waveform-describing" array will be created using **Waveform Simulator,** whose operation is controlled by the **Digitizing Parameters** and **Waveform Parameters** control clusters. Starting with a blank VI, build the following front panel. **Digitizing Parameters** and **Waveform Parameters** are found in **YourName\Controls**. Save this VI under the name **Waveform Generator (DAQmx)** in **YourName\Chapter 11**.

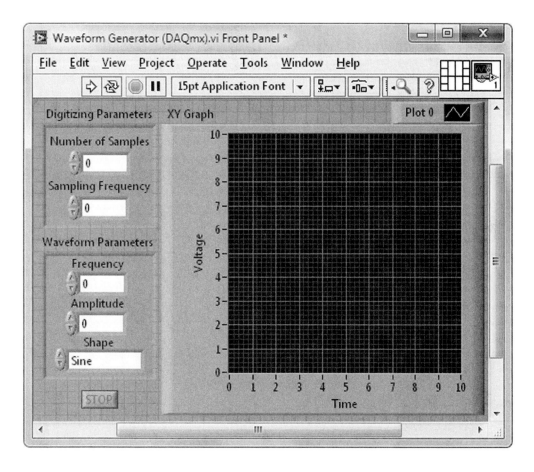

Switch to the block diagram and code it as shown next. Wire Waveform Simulator's **Displacement** array to the **data** input of DAQmx Write.vi. Also, wire **Sampling Frequency** to the **rate** input of DAQmx Timing.vi and configure this icon's **sample mode** to be **Continuous Samples**.

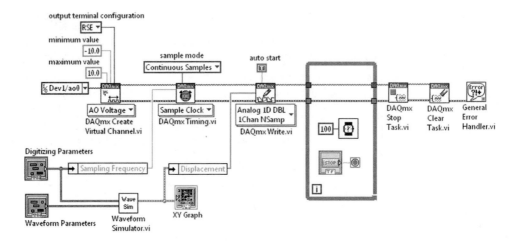

When this diagram is run, **DAQmx Write.vi** will write the N-element **Displacement** array into the memory buffer associated with the defined AO task, and the array will also be plotted on the front-panel **XY Graph** for the user to view. The continuous AO operation will then be started. In this operation, the DAQ device will output the prescribed sequence of voltages at the selected analog output channel in a circular fashion (i.e., after the last array value is output, the next output is the first array value). The rate at which the output voltage is changed ("updated") from one value to the next is controlled by **Sampling Frequency**. While this continuous analog output operation executes, the value of the front-panel **Stop Button** is checked every 100 ms. When the button is pressed, the AO task is stopped and cleared.

Return to the front panel and save your work. Then, connect the positive and negative (ground) inputs of an oscilloscope to the AO 0 and AO GND pins of your DAQ device, respectively.

Program **Waveform Generator (DAQmx)** to create two cycles of a 5 V, 100 Hz sine wave, where each cycle is defined by 100 samples. That is, in the **Waveform Parameters** control cluster, set **Frequency Amplitude**, and **Type** equal to *100, 5,* and *Sine,* respectively; in the **Digitizing Parameters** control cluster, set **Number of Samples** and **Sampling Frequency** to be *200* and *10000,* respectively. Run the VI. If your oscilloscope is triggered properly, you will observe a 100 Hz sine wave produced continuously until you press the VI's **Stop Button**.

What happens if you don't write a whole number of sine-wave cycles to the memory buffer? Stop the VI. Then program it to write 1½ (rather than two) cycles of the 100 Hz sine wave to the memory buffer. That is, keep all of the front-panel settings as before, except change **Number of Samples** to *150*. Run the VI. Can you explain the waveform observed on the oscilloscope based on your knowledge of the circular fashion in which the memory buffer values are sequenced?

DO IT YOURSELF

Build a (crude) DAQmx-based frequency meter VI called **Frequency Meter (DAQmx Count Edges)**, which will determine the frequency of a digital signal input to your DAQ device. Construct the instrument as follows.

1. Configure a counter input task that uses *counter 0* on your DAQ device (e.g., *Dev1/ctr0*) to count falling digital edges at its input (e.g., *PFI0*).
2. Within a While Loop, write code that executes the following procedure for each iteration: start the counting task, wait for 100 milliseconds, read the number of counts N accumulated during the 100 milliseconds, calculate the frequency $f = N/(0.1s)$ and send it to the front panel for display, and then stop the task.
3. Configure the While Loop so that its operation ceases when a front-panel Stop Button is pressed or an error occurs.

After you have completed **Frequency Meter (DAQmx Count Edges)**, connect the sync output of a function generator to the input of counter 0 (and GROUND) on your DAQ device, and then use it to measure the frequency of this digital signal.

(OPTIONAL) All DAQ devices, including the USB-6009, are equipped to count falling edges of digital signals; thus, everyone can carry out the above project. If you have a more advanced DAQ device (such as the PCI-6251), construct another frequency meter VI called **Frequency Meter (DAQmx Frequency)** in which the DAQmx task is configured as **Counter Input>>Frequency** (as opposed to **Counter Input>>Count Edges**). Then, rather than measuring time with a Wait (ms) icon, the time measure can be carried out by a much more accurate clock on board your DAQ device.

PROBLEMS

1. Observe the resolution ΔV of an analog input signal digitized by your DAQ device and verify that your observation is consistent with the prediction of Equation [1] in Chapter 4, that is, $\Delta V = V_{span}/2^n$. Attach a slowly varying analog signal (e.g., 1 Hz triangle wave) and the sync output from a function generator to the analog input channel and digital triggering pins programmed into **Digital Oscilloscope (DAQmx)**. Run **Digital Oscilloscope (DAQmx)** with a fast sampling rate and fairly small number of samples so that the input signal is effectively constant over the trace that you obtain. Stop the VI, and then by changing the y-axis scaling, zoom in on the acquired data (alternately, you can use a **Probe** to view the numerical values of the data samples). Under "high magnification," you should find that the data samples of this "effectively constant" input signal are distributed in discrete levels (because of electronic noise). Measure the spacing ΔV between two of these adjacent levels. Does your value for ΔV agree with the prediction of Equation [1]?
2. By passing an analog signal through a low-pass filter prior to inputting it to a digitizer, the aliasing effect can be suppressed. To demonstrate this procedure,

attach an analog sine-wave signal and the sync output from a function generator to the analog input channel and digital triggering pins programmed into **Digital Oscilloscope (DAQmx)**.

On the front panel of **Digital Oscilloscope (DAQmx)**, set **Number of Samples** and **Sampling Frequency** equal to *100* and *1000*, respectively, and set the sine-wave frequency *f* on your function generator to approximately 100 Hz. Run **Digital Oscilloscope (DAQmx)**. If all goes well, you will see a "stationary" plot of about ten cycles of the 100 Hz sine wave. Turn off autoscaling on the *y*-axis.

Next, set *f* on your function generator to approximately 1900 Hz, and then fine-tune *f* until **Digital Oscilloscope (DAQmx)** displays an aliased waveform of about 100 Hz. Use Equation [2] in Chapter 4 to explain why the input with *f* = 1900 Hz appears as a 100 Hz sine wave when digitized because of aliasing.

Finally, without changing the frequency setting on your function generator, pass the *f* = 1900 Hz sine-wave voltage through the low-pass op-amp filter shown in Figure 11.1 with *R* = 15 kΩ and C = 0.1 *μ*F, prior to inputting it to the DAQ device. When **Digital Oscilloscope (DAQmx)** is now run, by what factor is the amplitude of the aliased signal attenuated in comparison to the unfiltered situation? What is the 3-dB frequency $f_{3dB} = 1/2\pi RC$ of the low-pass filter? If you wanted the aliased signal to be attenuated even further, what component in the filter would you change? (This process can be optimized using a low-pass filter with a steeper roll-off than the simple first-order circuit used here.)

3. Build a two-channel digital oscilloscope. With **Digital Oscilloscope (DAQmx)** open, use **File>>Save As...** to create a new VI named **Dual Digital Oscilloscope (DAQmx)**. On the block diagram, you must change the single physical channel selected (e.g., *Dev1/ai0*) to a list of two channels. Use the Help Window of **DAQmx Create Virtual Channel** to learn the proper syntax for making such a list. Beside programming this list of physical channels, there is only one other change you must make on the block diagram to complete your program.

Once the VI is finished, attach a voltage waveform and the sync out digital signal from a function generator to the two AI channels of your DAQ device that are programmed on the block diagram. Additionally, attach the sync out digital

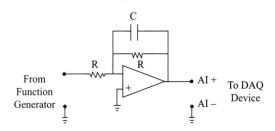

FIG. 11.1 Low-pass filter for Problem 2.

signal to the PFI0 channel of your DAQ device. Run **Dual Digital Oscilloscope (DAQmx)** and verify that it simultaneously displays traces of the two inputs.

4. Thermocouples are widely used as temperature sensors. A thermocouple is constructed by joining the ends of two dissimilar metals, for example, a copper and a constantan wire for a type T thermocouple. This junction produces a millivolt-level voltage, which has a well-documented temperature dependence, where the temperature is measured relative to a "cold junction" reference temperature. Conveniently, this cold junction can be provided by a compact electronic device called a Cold Junction Compensator (CJC), which effectively makes the reference temperature equal to 0 °C.

   Connect a thermocouple to a CJC and then connect the plus and minus output of the CJC to the AI 0$^+$ and AI 0$^-$ pins of your DAQ device. Using **DC Voltmeter (DAQ)** as a guide, write a program called **Thermocouple Thermometer (DAQmx)** that, every 250 milliseconds until a Stop Button is pressed, reads the thermocouple voltage, converts this value to the corresponding temperature in Celsius, and then displays this temperature in a front-panel indicator. When using a Cold Junction Compensator, **CJC Source** and **CJC Value** are *Constant Value* and *0*, respectively. Program **Thermocouple Type** for your particular type of thermocouple (e.g., T).

   Run **Thermocouple Thermometer (DAQ)** and use it to measure room temperature as well as the temperature of your skin.

5. Use **Digital Oscilloscope (DAQmx)** to measure the time constant of a *RC* circuit. Attach a TTL digital square wave from the sync output of a function generator to the PFI0 channel of your DAQ device as well as across the *RC* series circuit as shown in Figure 11.2 with $R = 4.7$ kΩ and $C = 0.1$ μF. When the square wave transitions from its LOW-to-HIGH (HIGH-to-LOW) state, the capacitor will be charged (discharged) through the resistor. Set the square wave frequency to a low frequency (say, 10 Hz) to allow the capacitor to fully charge and discharge after each transition. To observe the capacitor discharge, attach the analog input channel for which **Digital Oscilloscope (DAQmx)** is configured across the capacitor. Then, choosing appropriate values for **Number of Samples**, **Sampling Frequency**, and **Trigger Edge**, observe the discharge of the capacitor and accurately measure its time constant τ. By definition, τ is the time taken for the capacitor's voltage to discharge to $e^1 = 0.37$ of its initial value. Compare the value you measure to the theoretically expected value of $\tau = RC$.

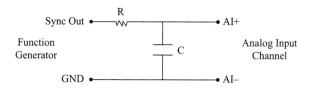

FIG. 11.2 *RC* circuit for Problem 5.

To assist in measuring voltages along the discharge curve, you may wish to activate a cursor by popping up on the Waveform Graph and selecting **Visible Items>>Cursor Legend**. When the cursor legend appears, pop up on it and select **Create Cursor>>Single Plot**. Once activated, by moving the cursor to particular data points on the plot, the x- and y-coordinates of this point of interest are displayed in the cursor legend.

6. Write a program called **Waveform Generator (DAQmx State Machine)** that allows the parameters of the analog waveform being output by your DAQ device to be changed while the program is running. Use the state machine architecture for your block diagram. When a change in front-panel setting is detected, the task must be first stopped before changes are made to the sampling frequency.

7. Using the information given in Chapter 9's **Do It Yourself** project, build a digital thermometer with a DAQmx-based (rather than DAQ Assistant Express VI–based) program called **Digital Thermometer (DAQmx)**.

8. Using the information given in Chapter 10's **Do It Yourself** project, build a spectrum analyzer with a DAQmx-based (rather than DAQ Assistant Express VI–based) program called **Spectrum Analyzer (DAQmx)**.

9. In this problem, you will create a task using the Measurement & Automation Explorer (MAX). This approach provides an alternate method to that used within the chapter text, where tasks were created on the block diagram. Use MAX to create a digital oscilloscope task similar to your **Digital Oscilloscope (DAQmx)** VI through the following steps:

    With MAX open, right-click on **Data Neighborhood** and select **Create New...>>NI-DAQmx Task**. In the dialog windows that follow, choose selections **Acquire Signals>>Analog Input>>Voltage>>ai0**, name your task *Oscilloscope*, and then press the **Finish** button. Within the MAX window, you will then find the **Configuration** tab displayed for your task. There, select **Acquisition Mode>> N Samples**, **Samples to Read>>100**, and **Rate (Hz)>>1k**. Next, click on the **Triggering** tab, and select **Trigger Type>>Digital Edge**. Finally, click on the **Save** button near the top of the window. The creation of your task is now complete.

    (a) To implement your task, open a blank VI and save it under the name **Digital Oscilloscope (MAX Task)**. On the block diagram place a **DAQmx Task Name**, found in **Functions>>Measurement I/O>>DAQmx**. Using the 👆, click on its menu button and select *Oscilloscope*. Then, complete the following diagram.

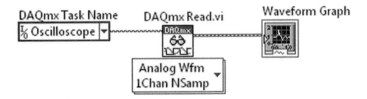

Using a function generator, attach a 100 Hz sine-wave voltage waveform and sync out digital signal to the AI 0 and PFI0 channels of your DAQ device. Run Digital Oscilloscope (MAX Task) and verify that it obtains a single $N = 100$ sample trace.

(b) Alternately, the DAQmx Task Name can generate code for you automatically. To implement this feature, open a blank VI and save it under the name Digital Oscilloscope (MAX Task Autocode). On the block diagram place a **DAQmx Task Name**, using the 🖑, click on its menu button and select *Oscilloscope*. Then, pop up on the DAQmx Task Name and select **Generate Code>>Configuration and Example**. A **DAQmx Read.vi** icon wired to a configurational subVI will be automatically created. Open the subVI and you will find code that is familiar. With the 100 Hz sine-wave and digital-sync signal inputs from the function generator, run Digital Oscilloscope (MAX Task Autocode) and verify that it obtains a single N sample trace. With some simple modifications, you can transform this VI into a program that is equivalent to Digital Oscilloscope (DAQmx), if you wish.

10. In this problem, you will create a "virtual channel" using the Measurement & Automation Explorer (MAX). This approach provides an alternate method to that used within the chapter text, where virtual channels were created on the block diagram using the **DAQmx Create Virtual Channel.vi** icon. Use MAX to create an analog input voltage virtual channel such as that used in your Digital Oscilloscope (DAQmx) VI through the following steps:

With MAX open, right-click on **Data Neighborhood** and select **Create New...>>NI-DAQmx Global Virtual Channel**. In the dialog windows that follow, choose selections **Acquire Signals>>Analog Input>>Voltage>>ai0**, name your virtual channel *AI Voltage*, and then press the **Finish** button. Within the MAX window, you will then find the **Configuration** tab displayed for your virtual channel where you can choose parameters related to the virtual channel such as the Signal Input Range and Terminal Configuration. If any changes are made from the default values, then click on the **Save** button near the top of the window. The creation of your virtual channel is now complete.

To implement your virtual channel, open Digital Oscilloscope (DAQmx) and, using **File>>Save As...**, create a new VI named Digital Oscilloscope (MAX Virtual Channel). On the block diagram, delete the **DAQmx Create Virtual Channel.vi** icon (and the various items wired to it) and replace it with a **DAQmx Global Channel**, found in **Functions>>Measurement I/O>>DAQmx**. Using the 🖑, click on its menu button and select *AI Voltage*. Then, simply modify the diagram as shown next.

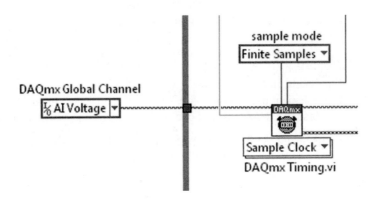

Using a function generator, attach a 100 Hz sine-wave voltage waveform and sync out digital signal to the AI 0 and PFI0 channels of your DAQ device. Run **Digital Oscilloscope (MAX Virtual Channel)** and verify that it performs exactly like **Digital Oscilloscope (DAQmx)**.

CHAPTER 12

PID Temperature Control Project

In this chapter, you will apply your LabVIEW programming skills in the construction of a feedback-based system that controls the temperature of an aluminum block with great precision. This feat will be accomplished by controlling the current through a *thermoelectric (TE) device*, which performs as a "heat pump" between the aluminum block and a large temperature reservoir. When current flows in one direction through the TE device, the device pumps heat from the block to the reservoir, hence cooling the block. When current flows in the other direction, the TE device pumps heat from the reservoir to the block, heating the block. Your job will be to write a top-level program called **PID Temperature Controller** that uses the *Proportion-Integral-Derivative (PID) control algorithm* to produce the current flow required to maintain the aluminum block at a desired "*set point*" *temperature* $T_{set\text{-}point}$. If your program is written correctly, **PID Temperature Controller** should be able to maintain the block's temperature at $T_{set\text{-}point}$ with fluctuations less than about 0.05 °C.

12.1 VOLTAGE-CONTROLLED BIDIRECTIONAL CURRENT DRIVER FOR THERMOELECTRIC DEVICE

Build the circuit in Figure 12.1. It is designed to provide the rather large bidirectional current flow necessary to power both the heating and the cooling capabilities of the TE device. The voltage level V_{in} serves as a "selector code word," which dictates the magnitude and direction of the current flow through the TE module. If the power source that supplies the ±10 V in your circuit has built-in ammeters, use them to monitor the current flow through the TE device, eliminating the need to insert the ammeter shown in the diagram. See Appendix I for a discussion of important practical issues regarding this circuit, especially the need (and method) for dissipating the prodigious amount of heat it produces.

Use **DC Voltage Source (Express)** or **DC Voltage Source (DAQmx)** found in **YourName\Chapter 4** and **YourName\Chapter 11**, respectively, to apply various voltages

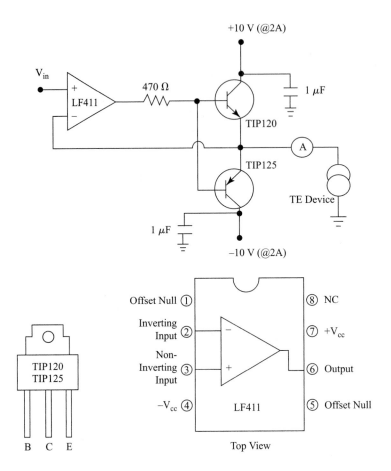

FIG. 12.1   Voltage-controlled bidirectional current driver.

at V_{in}. First, test that when V_{in} is positive (negative), an electric current flows in the proper direction through the TE device to heat (cool) the aluminum block in your experimental setup. If a positive V_{in} cools the block, reverse the connections to the TE device within your circuit. Second, find the positive and negative voltages V_{sat}, where V_{sat} is the minimum voltage at V_{in} for which the current to the TE device becomes saturated. The value of this saturation current is determined by the capability of the power supplies attached to the TIP transistors. We then define the range of acceptable values for V_{in} to be between $\pm V_{sat}$.

12.2 PID TEMPERATURE CONTROL ALGORITHM

Build a digital PID temperature controller that can control the temperature of an aluminum block to within less than 0.05 °C of a given set-point temperature. The set-point temperature can be chosen anywhere within the range of 0 to 40 °C.

An algorithm for such a controller is as follows:

1. Read the block's temperature T_{block} using a thermistor as the temperature sensor.
2. Compare T_{block} with the desired set-point temperature $T_{set-point}$.
3. Based on this comparison, decide what value of V_{in} will most appropriately command the TE device to provide the heating or cooling needed to bring T_{block} closer to $T_{set-point}$.
4. Apply this voltage at the V_{in} input of the voltage-controlled bidirectional current driver circuit.
5. Repeat this process continuously to obtain the desired temperature control of the aluminum block.

The *Proportional Control* method offers a simple procedure for deciding what voltage should be applied at V_{in} in your temperature-control algorithm. In this method, the difference between the desired set-point temperature $T_{set-point}$ and the actual sample temperature T_{block} is defined to be the error $E \equiv T_{set-point} - T_{block}$. Then the control voltage V_{in} is simply taken to be directly proportional to E:

$$V_{in} = AE \qquad \text{(Proportional Control)} \qquad [1]$$

where A is a constant called the *gain*. The value for the gain is chosen empirically such that, when $T_{set-point}$ is changed, the optimal value for A will cause the system to ramp to the new set-point and then stabilize near there (see the following discussion) quickly.

Here is a practical consideration you'll confront when trying to implement Equation [1]: When the sample temperature is far from the set-point, the error E may become large enough to generate a calculated value for V_{in} that falls outside its acceptable range of $\pm V_{sat}$. The solution to this problem is to truncate the expression given in Equation [1] so that the magnitude of V_{in} is never allowed to be greater than V_{sat}. The graphical representation for the Proportional Control algorithm then is as shown in Figure 12.2.

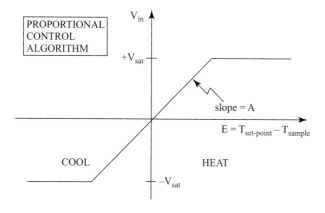

FIG. 12.2

Although pleasingly simple, the Proportional Control method is intrinsically flawed. Here's why. Assume the sample is initially at room temperature and that you select $T_{set\text{-}point}$ to be above room temperature. Turning on the Proportional Control algorithm, the initial error E will be positive, resulting in a command to the TE device to heat the sample. So far so good. As time goes on, the algorithm will issue the proper heating instructions to bring the sample closer and closer to the desired set-point temperature. When T_{block} nears $T_{set\text{-}point}$, the positive error E will become small, causing the Proportional Control to ease off on the applied heating power so that $T_{set\text{-}point}$ is gently approached. Then, at the decisive moment when T_{block} equals $T_{set\text{-}point}$, the error E becomes zero and the Proportional Control turns off power to the TE device. There is the flaw. Unfortunately, when at an elevated temperature, the sample will constantly be losing heat to its surrounding (room-temperature) environment through the heat-transferring processes of conduction, convection, and radiation. Thus, to maintain a sample at a set-point above room temperature, heat must constantly flow into the sample to counteract the heat it is losing to its environment. Since the Proportional Control turns off the heating when T_{block} equals $T_{set\text{-}point}$, the sample will never be able to stabilize at the desired temperature. Rather, the sample will stabilize at some temperature $T_0 < T_{set\text{-}point}$. The positive error E produced at the magic temperature T_0 commands just the right amount of heating by the TE device to cancel out the heat being lost by the sample to the environment at that temperature. By similar reasoning, a set-point below room temperature will result in the sample stabilizing at a temperature $T_0 > T_{set\text{-}point}$.

Luckily, there is an easy remedy to the above-described defect in the Proportional Control algorithm—simply include a constant term V_0 on the right-hand-side of Equation [1]:

$$V_{in} = AE + V_0 \tag{2}$$

When this expression is implemented, the proportional term will ramp the sample to the set-point temperature as before. Then, once at the set-point (where $E = 0$), V_0 will instruct the TE module to provide the constant heating (or cooling) necessary to counteract heat losses (or gains) to the environment, thereby stabilizing the sample at $T_{set\text{-}point}$. Of course, the value of V_0 must be precisely chosen so that the environmental influences on the sample, when at $T_{set\text{-}point}$, are perfectly neutralized.

Now the best news: There is an elegant way to build intelligence into the control algorithm so that the proper value for V_0, appropriate to the chosen set-point, will be found automatically. In the *Proportional-Integral (PI) Control* method, rather than defining V_0 as a fixed constant, an integral term is used to construct the correct constant during runtime according to the following expression:

$$V_{in} = AE + B \int E \, dt \qquad \text{(PI Control)} \tag{3}$$

Here, the integral keeps a running sum of all the error values that occur over the entire execution time of the algorithm. During the times when T_{block} is below $T_{set\text{-}point}$, a positive contribution will be made to the sum. When T_{block} is above $T_{set\text{-}point}$, a negative contribution will be made. Because of the self-correcting manner in which these contributions are made, the second term in Equation [3] will eventually converge to the constant V_0 that allows the sample to stabilize at the set-point $T_{set\text{-}point}$. From that point on, the error E will equal zero and the value of the integral (and thus constant V_0) will no longer change.

Inclusion of a derivative term to damp out oscillations provides one further refinement to the control algorithm. The expression for this so-called *Proportional-Integral-Derivative (PID) Control* algorithm then is given by

$$V_{in} = A\,E + B\int E\,dt + C\frac{dE}{dt} \qquad \text{(PID Control)} \qquad [4]$$

where A, B, and C are constants. For the experimental situation of discrete data sampling, where the error E is determined every Δt seconds, the value of the control voltage V_{in} after the nth sampling can be approximated by

$$V_{in} = A\,E_n + B\,\Delta t \sum_{m=0}^{n} E_m + \frac{C}{\Delta t}\{E_n - E_{n-1}\} \qquad \text{(Discrete-Sampling PID)} \qquad [5]$$

where the summation $\sum_{m=0}^{n} E_m$ is over *all* of the error values determined since the control algorithm was turned on.

12.3 PID TEMPERATURE CONTROL SYSTEM

Start construction of your digital temperature control system by rebuilding the circuit of the thermistor-based digital thermometer described in Chapter 9's **Do It Yourself** project. The thermistor, which is embedded in the aluminum block, will be used to monitor T_{block}. If needed, add the 1 μF capacitor as shown in Figure 12.3 to shunt high-frequency noise signals to ground and the unity-gain buffer to prevent the voltage-sensing circuitry from loading the thermistor circuit at low temperatures (when the thermistor resistance is large). Then connect V_{out} and GND to the inputs of an analog input channel of your LabVIEW system.

Next, create a folder called **Chapter 12** within the **YourName** folder and save a VI called **PID Temperature Controller** in **YourName\Chapter 12**. Code **Temperature Controller** according to the following guidelines.

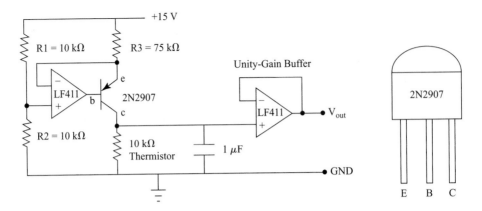

FIG. 12.3   Hardware circuit for a digital thermometer. A constant current of 0.1 mA flows through thermistor.

12.3.1 Block Temperature Determination

Noise-free temperature measurements of the block's temperature (assumed equal to the embedded thermistor's temperature) will be a "must" for your temperature-control algorithm to function properly. To average out random fluctuations when reading the thermistor temperature, you might try the following: At each determination of the temperature, quickly sample the thermistor temperature N times, and then analyze these data to determine the mean. Given the Steinhart–Hart coefficients for your thermistor, the **DAQ Assistant** or **DAQmx** VIs make acquiring the array of N temperature samples simple. To do the job of finding the mean from this data set, look at the VIs available in **Functions>>Mathematics>>Probability & Statistics**.

12.3.2 Control Algorithm

To determine the proper heating or cooling needed by the TE device, implement the Discrete-Sampling PID Control algorithm in the software. Remember that Δt in Equation [5] is the time between successive determinations of the error E; think carefully about what this time is in your particular VI. Also make sure to include a truncation feature; that is, if Equation [5] yields a value of V_{in} with magnitude greater than V_{sat}, then truncate the magnitude of V_{in} to V_{sat}. You will have to determine the optimum value for the constants A, B, and C empirically. If your setup is similar to the one described in Appendix I, try $A = 7$, $B = 1$, and $C = 0.5$.

12.3.3 Output to the TE Device

Use **DAQ Assistant** to output the PID-determined value for V_{in} over one of LabVIEW's analog output channels to the input of the TE device's voltage-controlled bidirectional current driver circuit.

12.3.4 Front Panel

Use your creativity here. Remember you have a wide array of controls, indicators, and graphs available. A front-panel control that resets the summation $\sum_{m=0}^{n} E_m$ to zero when a button is pressed is an optional feature that you might find handy.

When you get the temperature controller to work, show it off to your friends and instructor!

CHAPTER 13

Control of Stand-Alone Instruments

13.1 INSTRUMENT CONTROL USING VISA VIs

In previous chapters, you have used LabVIEW software to transform a personal computer (connected to an appropriate National Instruments DAQ device) into several handy laboratory instruments. In particular, you programmed this system to become a DC voltmeter, digital oscilloscope, spectrum analyzer, waveform generator, and digital thermometer. Pause to consider the following tantalizing prospect: Perhaps the only instrument required in a modern-day laboratory is a DAQ device–equipped computer controlled by LabVIEW software. That is, by simply writing a collection of appropriate VIs, it might be possible for you—the cutting-edge experimentalist—to satisfy all of your laboratory instrumentation needs with this single LabVIEW-based data acquisition and generation system. This system's tremendous flexibility would then obviate the need to purchase an expensive collection of stand-alone electronic equipment such as power supplies, function generators, picoammeters, spectrum analyzers, and oscilloscopes.

The functioning VIs that you have written in previous chapters demonstrate that the above "tantalizing prospect" can, at least in certain situations, be realized. But don't discard your stand-alone instruments just yet. The timing, speed, sensitivity, and simultaneous data-taking requirements of many contemporary research experiments are beyond the capabilities of your DAQ device. For instance, although the LabVIEW-based digital oscilloscope we constructed worked well for observing audio-range frequencies (less than 20 kHz), it would prove miserably inadequate at displaying the several nanosecond-wide voltage pulses emanating from a photomultiplier tube. In this latter situation, a stand-alone digital scope with a very fast analog-to-digital converter (on the time scale of several gigasamples per second) would do the job nicely. Thus, stand-alone instruments play a central role in state-of-the-art research, and so it might not surprise you to find that they, too, fall under the scope of LabVIEW.

Over the past few decades, a message-based communications standard has evolved by which stand-alone instruments can be software controlled using a personal computer. In this communications scheme, a particular instrument obeys an array of manufacturer-defined ASCII character commands that represent all the possible ways of manually pressing buttons, turning dials, and viewing output data on its front panel. Although the hardware conduit (called an *interface bus*) through which these ASCII messages are passed between the PC and laboratory instrument can take on various guises (including RS-232, GPIB, Ethernet, and USB), there is a single set of LabVIEW icons available to control this communication process. This icon set is named *VISA* (short for Virtual Instrument Software Architecture) and is found in **Functions>>Instrument I/O>>VISA**.

In this chapter, you will learn how to use VISA icons to control the message-based communication between a stand-alone instrument and your computer. You will explore generic features of this communication process such as the Standard Commands for Programmable Instruments (SCPI) language and various synchronization methods while writing code that controls a particular stand-alone instrument—the *Agilent 34410A Digital Multimeter*—using two interface buses, the *General Purpose Interface Bus* and the *Universal Serial Bus*.

13.2 THE VISA SESSION

When using VISA icons to facilitate message-based communication between a computer and a particular stand-alone instrument, the instrument is termed a *VISA resource* and the communication activity is called a *VISA session*. To "*query*" a VISA resource (i.e., send it a command and then receive back its response), the required VISA session consists of the following four steps: *open the session, write the command message to the resource, read the response from the resource, close the session*. In **Functions>>Instrument I/O>>VISA** (and its subpalette **VISA Advanced**), the following four icons are available to perform the four given steps—**VISA Open**, **VISA Write**, **VISA Read**, and **VISA Close**. To understand how to wire these four icons together to query an instrument, we will first briefly describe the function of each individual icon. In the Help Windows that follow, remember that required inputs are given in boldface text; recommended and optional inputs are in plain and dimmed text, respectively.

The Help Window for **VISA Open** is shown in the following illustration. The job of this icon is to begin a VISA session between your computer and the resource defined at its **VISA resource name** input. The **VISA resource name** consists of text that specifies the interface type being used (e.g., GPIB or USB), followed by resource information (in the case of GPIB, the instrument's address, a number we'll discuss in a few minutes), and ending with the resource type. For our work, the resource type will be a stand-alone instrument denoted by *INSTR*. To pass the **VISA resource name** to other VISA icons, this quantity is available at the **VISA resource name out** output terminal.

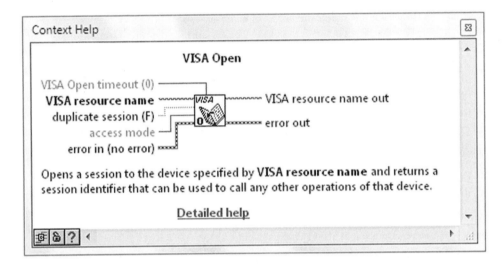

VISA Write, whose Help Window follows next, performs the actual ASCII message transfer from your computer to the stand-alone instrument. Once presented with the open session's **VISA resource name**, this icon writes the ASCII string at its **write buffer** input to the instrument. This string is one of the commands recognized by the instrument and, when received by the instrument, configures it properly for a desired data-taking measurement. Additionally, the VISA resource name is available at the **VISA resource name out** output terminal.

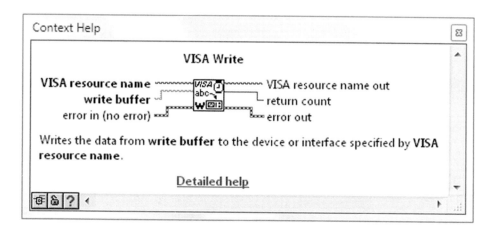

Next, the Help Window for **VISA Read** is shown. This icon transfers the response message from the stand-alone instrument into your computer's memory. When given the open session's **VISA resource name**, this icon receives the ASCII response string consisting of (a maximum of) **byte count** number of bytes and outputs this string at its **read buffer** terminal. This string typically contains the results of a data-taking measurement performed by the instrument. Additionally, the VISA resource name is available at the **VISA resource name out** output terminal.

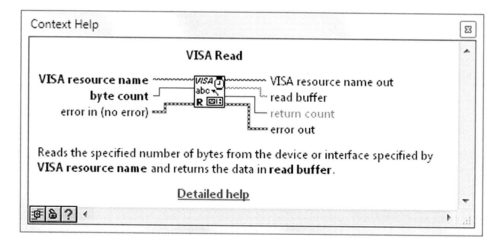

Finally, **VISA Close**'s Help Window is given below. This icon closes the VISA session specified at its **VISA resource name** input.

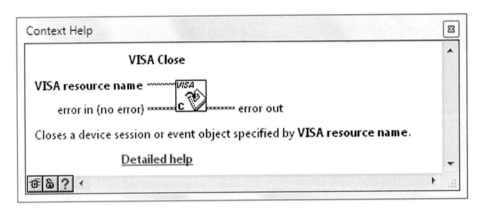

Note that all four of these VISA icons include error reporting via an error cluster, which appears at the **error in** and **error out** terminals.

The four-step VISA session to query an instrument is accomplished by wiring these four VISA icons together as follows.

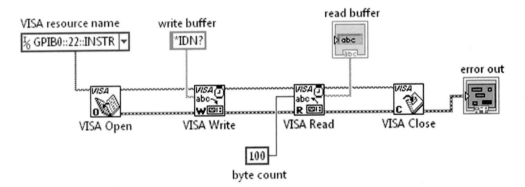

In this example, the message *IDN?* is sent over a GPIB interface to an instrument at address 22. *IDN?*, a command recognized by most instruments, instructs the instrument to identify itself. After receiving this command, the instrument's response (which is an ASCII string consisting of identification information) is received by the computer over the GPIB and displayed in the **read buffer** front-panel string indicator.

Similar to the File I/O and DAQmx icons that you have studied previously, the wiring scheme of VISA icons takes advantage of the principle of LabVIEW programming called *data dependency*. Simply stated, data dependency means that an icon cannot execute until data are available at *all* of its inputs. In the previous diagram, all of **VISA Open**'s required inputs (there is only one—**VISA resource name**) are wired. So, when this diagram is run, **VISA Open** executes immediately. Upon completion, **VISA Open** outputs a VISA resource name at its **VISA resource name out** terminal, which is passed through the wiring to VISA Write's **VISA resource name** input. Because of data dependency, **VISA Write** cannot execute until it receives the VISA resource name from **VISA Open.** In a similar way, **VISA Read** cannot execute until **VISA Write** completes, and so on. Thus, through this programming scheme, we are assured that the icons will execute in the desired sequence: **VISA Open** followed by **VISA Write** followed by **VISA Read** followed by **VISA Close**.

Also, the correct manner of chaining together VISA icons for error reporting is shown above. If an error does occur at one point in the chain, subsequent icons will not execute and the error message will be passed to the **error out** indicator cluster.

Because VISA-based programming is so robust, you can write highly dependable data-taking programs with just the information already presented. However, with a bit more grounding in the message-based communication scheme, you'll be able to create programs in which you can have near-total confidence. The following paragraphs will take you to the next level of sophistication in stand-alone instrument control.

13.3 THE IEEE 488.2 STANDARD

When remote control of laboratory instruments first became possible, there was a chaotic period during which, more or less, each instrument manufacturer defined its own communications protocol through a unique blend of parallel and serial modes, positive and negative polarities, and assorted handshaking signals. In 1965, Hewlett Packard (now named Agilent) ended this cacophony by designing a universal instrument interface called the Hewlett Packard Interface Bus (HP-IB) and offered it as the only option on all of its new computer-programmable instruments. Because of its high transfer rates, HP-IB quickly gained popularity with other instrument manufacturers and, in 1975, was accepted as an industry-wide standard known as IEEE-488 or, more commonly, the General Purpose Interface Bus (GPIB). In 1987, an improved version of this standard called IEEE-488.2 was adopted, which enhanced and strengthened message-based communication by specifically defining an instrument's minimally required communication capabilities, a protocol for message exchange, a generic set of commonly needed commands, and a status reporting system. Today, most computer-controlled laboratory instruments are IEEE 488.2 compliant, even those that communicate over interface buses other than the GPIB (for example, Ethernet and USB).

13.4 COMMON COMMANDS

One important innovation of the IEEE 488.2 standard was the introduction of a standardized set of "*common commands*" for the many generic operations that all instruments must perform. The mnemonics for these common commands begin with asterisks to delineate them from the other device-specific commands recognized by a particular

TABLE 13.1   Common Commands for IEEE488.2 Compliant Instruments

Mandatory Common Commands	Function
*IDN?	Reports instrument identification string.
*RST	Resets instrument to known state.
*TST?	Performs self-test and reports results.
*OPC	Sets operation complete (OPC) bit in SESR upon completion of the command.
*OPC?	Returns "1" to the output buffer upon completion of the command.
*WAI	Waits until all pending operations complete execution.
*CLS	Clears status registers.
*ESE	Enables event-recording bits in SESR.
*ESE?	Reports enabled event-recording bits in SESR.
*ESR?	Reports value of SESR.
*SRE	Enables a SBR bit to assert the SRQ line.
*SRE?	Reports SBR bits that are enabled to assert the SRQ line.
*STB?	Reports the contents of the SBR.

instrument. All IEEE 488.2 compliant instruments, at the very least, are required to recognize the subset of 13 common commands given in Table 13.1. Many of these commands are related to the reporting of events using two status registers called the *Status Byte Register* (SBR) and *Standard Event Status Register* (SESR), which will be described in detail starting in the next paragraph.

13.5 STATUS REPORTING

Another IEEE 488.2 innovation is a standardized scheme for *status reporting*. This status report system is available to inform you of significant events that occur within each instrument connected to an interface bus. In this scheme, an instrument is equipped with two status registers called the *Standard Event Status Register (SESR)* and the *Status Byte Register (SBR)*. Each bit in these registers records a particular type of event that may occur while the instrument is in use, such as an execution error or the completion of an operation. When the event of a given type occurs, the instrument sets the associated status register bit to a value of one, if that bit has previously been enabled (as described in the following paragraphs). Thus, by reading the status registers, you can tell what events have transpired.

The Standard Event Status Register, which is schematically shown here, records eight types of events that can occur within a data-taking instrument.

Standard Event Status Register (SESR)

7	6	5	4	3	2	1	0
PON	URQ	CME	EXE	DDE	QYE	RQC	OPC

The eight events associated with the eight bits of the SESR are described in Table 13.2. In our work, the OPC bit will be most useful.

TABLE 13.2   Eight Standard Event Status Register (SESR) Events

Bit	Associated Events of SESR
7 (MSB)	**PON** (Power On): Instrument was powered off and on since the last time the event register was read or cleared.
6	**URQ** (User Request): Front-panel button was pressed.
5	**CME** (Command Error): Instrument received a command with improper syntax.
4	**EXE** (Execution Error): Error occurred while the instrument was executing a command.
3	**DDE** (Device Error): Instrument is malfunctioning.
2	**QYE** (Query Error): Attempt was made to read the instrument's output buffer when no data were present, or a new command was received before previously requested data had been read from the output buffer.
1	**RQC** (Request Control): Instrument requests to be controller.
0 (LSB)	**OPC** (Operation Complete): All commands prior to and including an *OPC command have been executed.

The SESR exists as an event-signaling tool for you to use in your programs. However, this status register completely lacks initiative and will not perform any work unless you request it to do so. Thus, when initiating communications with an instrument, one of the messages that you may wish to send is an instruction that activates the subset of event-reporting SESR bits that are of interest to you. For instruments that conform to the IEEE 488.2 standard, this activation process is accomplished via the *ESE* (Event Status Enable) command. For example, suppose you wish the QYE bit to be activated and thus record any execution errors in the SESR's bit 2. Since $00000100_2 = 4_{10}$, the QYE bit can be activated by performing a VISA Write of the ASCII command *ESE 4* to the instrument. In our work to come, we will activate the OPC bit with the command *ESE 1*.

The Status Byte Register, which is schematically shown next, records whether data are available in the instrument's output buffer, whether the instrument requests service, and whether the SESR has recorded any events.

Status Byte Register (SBR)

7	6	5	4	3	2	1	0
—	RQS	ESB	MAV	—	—	—	—

The functions of the eight bits of the SBR are described in Table 13.3. The SBR bits are studious, performing their status reporting duties without need of a request from you.

TABLE 13.3   Functions of Eight Standard Byte Register (SBR) Bits

Bit	Function of SBR Bit
7 (MSB)	May be defined for use by instrument manufacturer.
6	**RQS** (Request Service): The instrument has asserted a SRQ because it requires service from the GPIB controller.
5	**ESB** (Event Status Bit): An event associated with an enabled SESR bit has occurred.
4	**MAV** (Message Available): Data are available in the instrument's output buffer.
3–0	May be defined for use by the instrument manufacturer.

An instrument can be configured to assert a *Service Request (SRQ)* in response to either of two events—an event detected by the Standard Event Status Register or the presence of previously requested data in the instrument's output buffer (that is, the assertion of the Event Status Bit (ESB) or Message Available (MAV) bit, respectively). This configuration process is accomplished on IEEE 488.2 instruments using the *SRE* (Service Request Enable) command. For example, if you wish an event detection by the SESR to trigger a request for service by the instrument, initialize the instrument by writing the ASCII command *SRE 32* to the instrument. Since $32_{10} = 00100000_2$, the setting of the SBR's fifth (ESB) bit will then be the criterion for the instrument asserting a SRQ. If, instead, you wish the presence of data in the output buffer (signaled by the MAV bit being set) to trigger a SRQ, then write *SRE 16*. Finally, *SRE 0* will disable the instrument's ability to assert a SRQ.

The relationship between the Standard Event Status Register, the instrument's output buffer, and the Status Byte Register (along with the common commands that configure and query each) is illustrated in Figure 13.1.

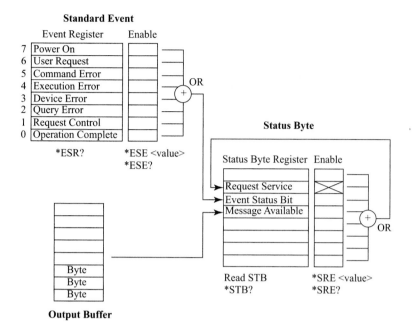

FIG. 13.1 Relationship between the Standard Event Status Register (SESR) and the Status Byte Register (SBR).

As an alternative to the use of the SRQ, *serial polling* is a common method for determining the status of an instrument. In a serial poll process, the interface bus queries an instrument and the instrument responds by returning the value of the bits in its Status Byte Register. A serial poll is easily accomplished in LabVIEW using **VISA Read STB**, found in **Functions>>Instrument I/O>>VISA**. This icon's Help Window is shown next.

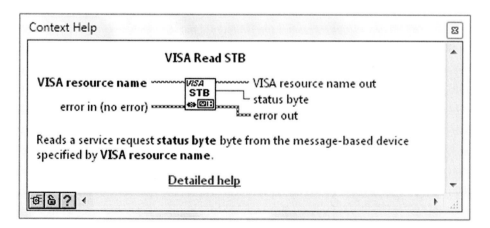

13.6 DEVICE-SPECIFIC COMMANDS

Finally, each stand-alone instrument is designed for a specialized purpose and has its own idiosyncratic methods for accomplishing its objectives. Thus, every programmable instrument comes with a set of *device-specific commands* that allow the user to control its functions remotely and to transfer the information it produces into a computer's memory. The array of device-specific commands for an instrument is listed in its user manual. This set of commands is defined by the instrument's maker . . . and therein lies a problem. When surveying the user manuals for programmable instruments of varying models and manufacturers, you will find a great diversity in the style of the various command sets. Some (especially those associated with older model instruments) are an alphabetized collection of cryptic one- or two-character strings (the designers' thinking was obviously "shorter commands yield quicker and, therefore, better computer-instrument communication"). At the other extreme are the user-friendly sets, with similar commands logically grouped, each represented by an easy-to-read-and-remember mnemonic.

As programmable instruments have come into wider use, it has become apparent that development costs and unscheduled delays can be diminished markedly by simplifying the instrument programmers' task whenever possible. Thus, user-friendly device-specific command sets are the rule, rather than the exception, for instruments currently being manufactured. Commonly, these command sets are organized in a hierarchical *tree structure*, similar to the file system used in computers. Each of an instrument's major functions, such as **TRIG**ger, **SENS**e (alternately, **MEAS**ure), **CALC**ulate, and **DISP**lay, defines a *root* and all commands associated with that root form its *subsystem*. So, for example, to configure the Agilent 34410A Digital Multimeter to measure a DC voltage whose value is expected to fall within the range of ±10 V (an action within its **SENS**e subsystem), the appropriate command is as follows:

SENSe:VOLTage:DC:RANGe 10

Here, *SENSe* is the root keyword and colons (:) represent the descent to the lower-level *VOLTage*, then *DC*, then the lowest-level *RANGe* keywords. Finally, *10* is a parameter associated with *RANGe*. Although the full command mnemonic given here can be sent to the instrument, it is only absolutely necessary to send the capitalized characters.

In 1990, a consortium of equipment manufacturers defined the *Standard Commands for Programmable Instruments (SCPI)* in an effort to standardize the device-specific command sets of computer-controlled instrumentation. Although this standard has not been universally adopted, it is not uncommon to discover that your post-1990 instrument is SCPI compliant. As a means of categorizing generally applicable command groups, the SCPI standard posits the following model for a generic programmable instrument. An instrument that performs measurements on an input signal is assumed to have the root functions shown in Figure 13.2. Here, for example, **SENS**e includes any action involved in the actual conversion of an incoming signal to internal data, such as setting the range, resolution, and integration time, while **INP**ut consists of actions that condition the signal prior to its conversion, such as filtering, biasing, and attenuation.

SIGNAL MEASUREMENT INSTRUMENT

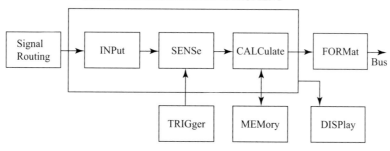

FIG. 13.2

Alternately, an instrument that generates signals is modeled by Figure 13.3.

SIGNAL GENERATION INSTRUMENT

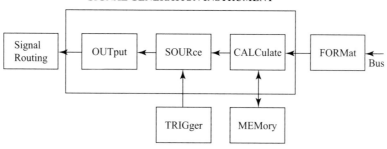

FIG. 13.3

The SCPI command set is organized in a hierarchical tree structure using the syntax illustrated above by the *SENSe:VOLTage:DC:RANGe 10* command. You'll learn more about the SCPI command syntax as you work your way through this chapter. But maybe now is a good time to dive in and actually control a stand-alone instrument.

13.7 SPECIFIC HARDWARE USED IN THIS CHAPTER

In designing this chapter, I faced the following problem. There are thousands of computer-controllable stand-alone instruments available for purchase from the myriad of worldwide scientific instrument makers. Each of these instruments communicates using one (or many times, several) of the available interface buses. I have a small subset of these instruments in my laboratory and I can use them there to practice the art of message-based communication. You also, hopefully, have a small subset of such instruments available to practice with in your own laboratory. What's the problem? Well, because of the high cost and specialized nature of such equipment, the probability that my subset and your subset have some common instrument is most likely very small. The unfortunate thing about this situation is that each stand-alone instrument is designed to take specialized measurements and understands its own unique set of ASCII commands (which are defined by its maker and are listed in its user manual). Thus, before attempting to control a particular instrument using an interface bus, the programmer must have a detailed understanding of the measurement that the instrument is designed to take, the procedure that it implements in doing its work, and the command list that it recognizes. All of these considerations greatly constrain the writing of a set of generic laboratory exercises that everyone can perform.

That said, I still was faced with the fact that I had to choose a particular instrument and interface bus to work with in this chapter's exercises. With regard to the interface bus, the GPIB is a must. It is the interface you will almost certainly encounter in your computer-based laboratory work and is currently (by far) the most widely used interface bus for laboratory equipment. With an estimated 10 million GPIB-equipped instruments in use in research and industry worldwide, the GPIB most likely will retain its popularity for years to come. However, nearly every new computer comes equipped with USB and Ethernet ports, making these interfaces both familiar and virtually cost free for users. Hence, USB and Ethernet are gaining in popularity with scientific instrument makers, so I want to demonstrate the use of at least one of these newer interfaces as well. Thankfully, most of the latest models in laboratory instruments are equipped to communicate over multiple interface buses and so a single such device can be used to demonstrate the use of both the GPIB and, say, the USB interfaces.

For the following reasons, the Agilent 34410A Digital Multimeter (DMM) has many features that make it ideal for our work and so I have chosen it as our "device to practice with" in this chapter's exercises. First, this instrument measures voltage, current, and resistance—vanilla-flavored quantities that require no specialized knowledge to understand (unlike, for instance, the control of grating angle and slit size in a spectrometer). Second, this instrument is equipped to communicate via three possible interfaces—GPIB, USB, and Ethernet (we will use the first two of these in this chapter). Third, for such a high-quality instrument standardly equipped with several interfaces, its price tag of approximately $1300 makes it affordable (at least by scientific instrument standards). Every lab should have one and many do! Fourth, this instrument is both IEEE 488.2 and SCPI compliant and so we will be able to use it to gain experience with generic features of widely adopted communication schemes such as SCPI command syntax and IEEE 488.2 status reporting. Fifth, the Agilent 34410A is the successor to the longtime industry-standard Agilent 34401A DMM. Thanks to Agilent's interest in maintaining backward compatibility between these two instruments, the commands I have chosen to use with the newer 34410A DMM can also be used by 34401A owners (of which there are many) to carry out this chapter's exercises on their instrument's GPIB interface (the 34401A has no USB or Ethernet capabilities).

In the best circumstance, an Agilent 34410A (or 34401A) DMM is already available for your use or, with a modest investment, you can purchase this worthwhile instrument. Then, without need for modification, you can straightforwardly work your way through the given exercises to learn the basics of message-based communication. If, instead, you have some other interface-equipped instrument available, try reading the following pages to understand the generic issues being investigated. Then, by consulting the user manual, it may be fairly easy, for instance, to use the interface-appropriate VISA resource name and substitute an ASCII command string here and there to adapt the exercises to your particular instrument and interface bus. If neither of the above describes your situation, simply read through the following pages. I believe you will learn some valuable features of instrument control that will serve you well in future work.

13.8 MEASUREMENT & AUTOMATION EXPLORER (MAX)

To carry out the exercises in this chapter, you must have a stand-alone instrument equipped with a GPIB and/or USB interface. To communicate over the GPIB, you must also have a National Instruments GPIB device (e.g., PCI-GPIB) connected properly to your computer, which is in turn connected to the stand-alone instrument by a GPIB cable. To use the USB, simply connect an appropriate USB cable between the stand-alone instrument's USB connector and a USB port on your computer. Additionally, the National Instrument's NI Device Drivers software (which is included on your LabVIEW installation disks) must be correctly installed. To verify that these conditions are met, we will use the handy utility *Measurement & Automation Explorer*, which is nicknamed *MAX*.

To open MAX, either select **Tools>>Measurement & Automation Explorer...** (if you have an open VI or **Getting Started** window) or double-click on MAX's desktop icon (if available). After MAX opens, double-click on **Devices and Interfaces** under the **My System** heading at the left. This action will command MAX to determine all of the National Instrument devices present within your computing system.

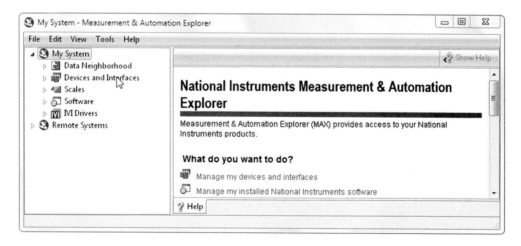

MAX will list the findings of its device survey in hierarchical tree fashion as shown next. If a GPIB device is connected correctly to your computer, a folder labeled **GPIB0** will appear in the resulting list (your system may have a different number than *0* in the folder's label). To find all of the stand-alone instruments properly connected to this device, right-click on the **GPIB0** folder and select the **Scan for Instruments** option. Alternately, you can click on the **Scan for Instruments** button in the toolbar near the top of the window.

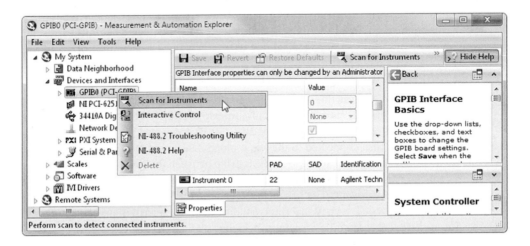

In a few moments, MAX will complete the scan. To view its results, double-click on the **GPIB0** folder. For the case shown below, one stand-alone instrument was found and information about it is stored in the folder labeled **Instrument 0**.

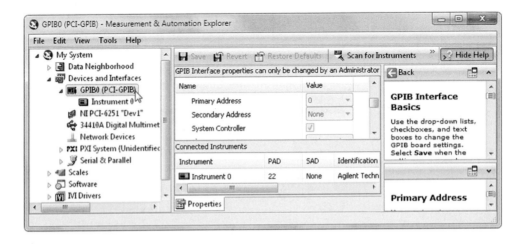

Because up to 15 instruments can be connected to a single GPIB device, each instrument has an identification number called its *GPIB address*. A GPIB address can be any integer between 0 and 30 and is typically defined via a hardware DIP switch setting within the instrument or a sequence of button pressing and/or knob turning on its front panel. The instrument's user manual will describe the method for setting its address.

The GPIB address of the instrument found in the **Scan for Instruments** operation is viewed by double-clicking on the **Instrument 0** folder. After the double-click, in the box associated with the **Attributes** tab, we find that the instrument's GPIB address is 22 (the address of your instrument may be different). A text description identifying the instrument also appears.

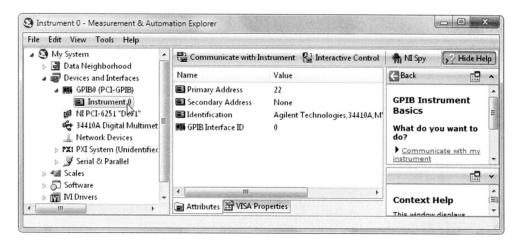

Clicking on the **VISA Properties** tab, we find that the correct **VISA resource name** for this instrument is *GPIB0::22:INSTR* and also are told (under **Device Status**) that the instrument is working (i.e., communicating) properly.

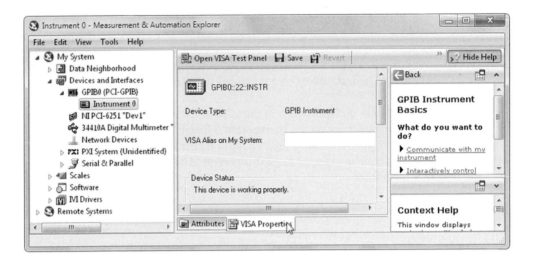

To verify that the instrument is indeed properly communicating over the GPIB, click on the **Open VISA Test Panel** button in the toolbar near the top of the window.

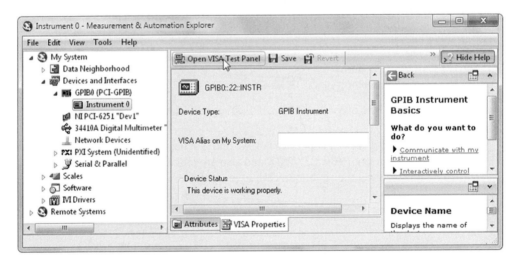

An interactive dialog panel will appear. You might wish to take a few moments to explore some of the interface-related information available in this panel by selecting its various windows (e.g., **Configuration**) and their associated tabs. After exploring, select the **Input/Output** window by clicking on its name.

Once in the **Input/Output** window, select its **Basic I/O** tab. Here, after typing an ASCII message in the **Enter or Select Command** box, a mouse-click on the **Query** button will carry out a write-then-read action. That is, the message in **Enter or Select Command** will be written over the GPIB to the addressed instrument, and then the instrument's ASCII response will be read back over the GPIB to the computer and displayed in the large rectangular box at the window's bottom left. When the **Input/Output** window opens, its **Enter or Select Command** box is preloaded with *IDN?*, the IEEE 488.2 common command for an instrument to identify itself (the backslash message terminator \n is also included; more on that later).

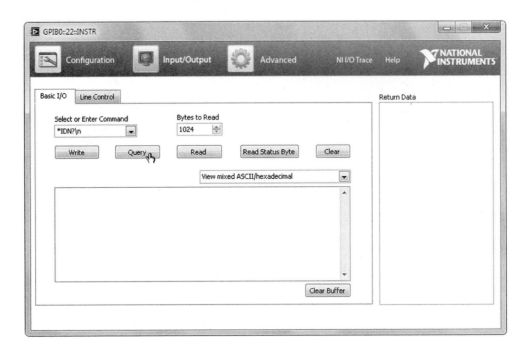

Click on the **Query** button. If your GPIB communication is configured properly, the identification string received from the instrument will appear in the large rectangular box. For the Agilent 34410A DMM used here, this identification string identifies the instrument's manufacturer and model information followed by some integers, which denote the version numbers of installed firmware that controls the multimeter's internal microprocessors. Note that this identification string is 60 bytes long.

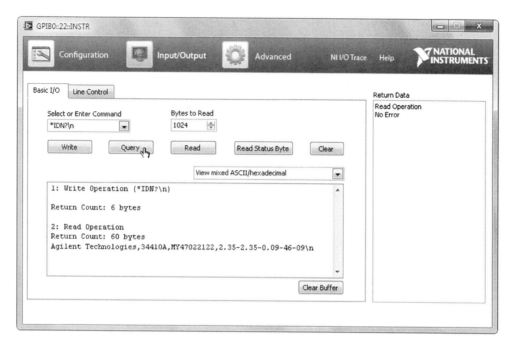

For future reference, this interactive dialog window is a handy tool for use in determining correct command syntax when developing message-based communication programs for a new instrument.

Alternately, if your instrument is set up to communicate over the USB, after the device survey initiated by the double-click on **Devices and Interfaces**, its name will appear in the resulting list. As seen below, included next to its name, the Neptune's Trident symbol 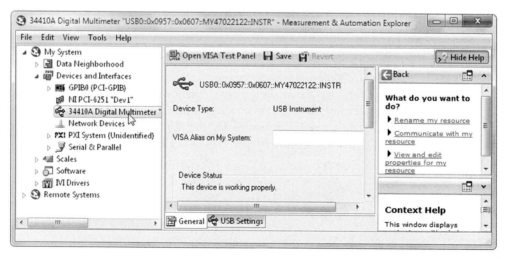 denotes that this instrument is communicating over the USB interface.

Following similar steps to those described above for the GPIB, the USB-related **VISA resource name** for your instrument can be found (as seen below, the DMM on my system is named *USB0::0x0957::0x0607::MY47022122::INSTR*). After opening a **VISA Test Panel**, try querying your instrument for its identification string over the USB. You should, of course, get the same result as when querying over the GPIB.

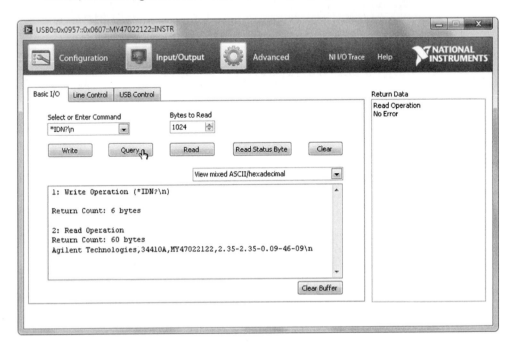

After you are finished, exit this dialog window and then close MAX.

13.9 SIMPLE VISA-BASED QUERY OPERATION

Let's begin by writing a VISA-based program that carries out the query (i.e., write-then-read) action that you just completed using MAX.

On the front panel of a blank VI, place a **String Control** and a **String Indicator** (found in **Functions>>Programming>>String**), and then label them **Command** and **Response**, respectively. As shown in the next diagram, you'll want to resize these objects so that they can display strings much larger than their default sizes allow. Using **File>>Save**, first create a new folder named **Chapter 13** within the **YourName** folder, and then save this VI under the name **Simple VISA Query** in **YourName\Chapter 13**.

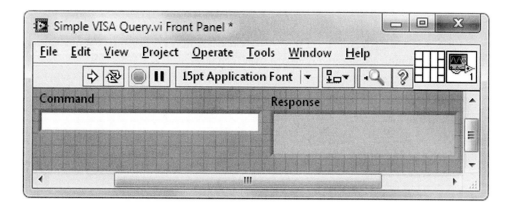

Switch to the block diagram and place a **VISA Open** icon (found in **Functions>> Instrument I/O>>VISA>>VISA Advanced**) there. Pop up on its **VISA resource name** input and create a **VISA Resource Name Constant** using **Create>>Constant**.

You must now load the **VISA Resource Name Constant** with the **VISA resource name** of the instrument with which you wish to communicate. By clicking on the Constant's *menu button* with the , you will be presented with the list of VISA resources that MAX found when it performed the **Scan for Instruments** operation (or after you select the menu button's **Refresh** option) as shown next.

You can then simply choose the desired **VISA resource name** from this list. Alternately, you can manually enter the appropriate **VISA resource name** for your instrument (as found using MAX) into the . The syntax for a **VISA resource name** is *Interface Bus Name::Resource Information::Resource Type.*

Here, from the menu button's list, I have chosen the **VISA resource name** appropriate for communicating with my DMM over the GPIB.

If, instead, I wished to communicate with my DMM over the USB, I would choose this available **VISA resource name** for the menu button's list.

Here is the almost-too-good-to-be-true fact: The VISA icons we will use to write our program are interface independent. That is, once the appropriate **VISA resource name** is wired to **VISA Open**, the rest of our VISA-based block diagram will be the same, regardless of whether we are communicating over the GPIB, the USB, or several other possible interface buses (including Ethernet, RS-232, PXI, VXI). All of the low-level details related to the specific interface bus used are taken care of "under the hood" by the VISA icons. We are left to simply direct the communication between the instrument and PC at a generic high level, for example, orchestrating what, and at what time, commands and data are written and read. For the block diagrams in the rest of this chapter, my selected **VISA resource name** will be appropriate for GPIB communication. As just described, however, it is a simple matter to change the **VISA resource name** choice to switch to communication over the USB, if desired.

Complete the block diagram as shown next using the VISA icons found in **Functions>>Instrument I/O>>VISA** (and its subpalette **VISA Advanced**). Wire the **Command** control terminal to VISA Write's **write buffer** input and the **Response** indicator terminal to VISA Read's **read buffer** terminal. When executed, **VISA Read** will read up to N bytes from the selected resource, where N is equal to the integer wired to its **byte count** input. In a moment, we will read the Agilent 34410A identification string. Using the VISA Test Panel a few moments ago, we saw that this identification string is 60 characters long. Thus, wire the **byte count** input to an integer (**U32**) greater than or equal to 60 (I used *100*) as shown. Create the **error out** indicator cluster using the pop-up menu option **Create>>Indicator**.

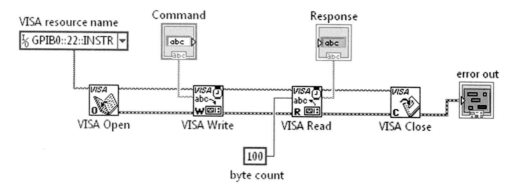

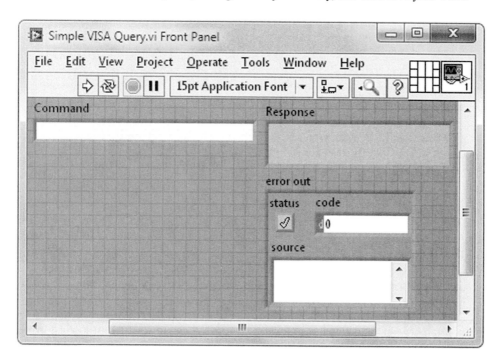

Return to the front panel, arrange the objects nicely, and then save your work.

Enter *IDN?* into **Command**, and then run your VI. If all goes well, **Response** will display the instrument's identification string upon completion of the VI execution, as shown.

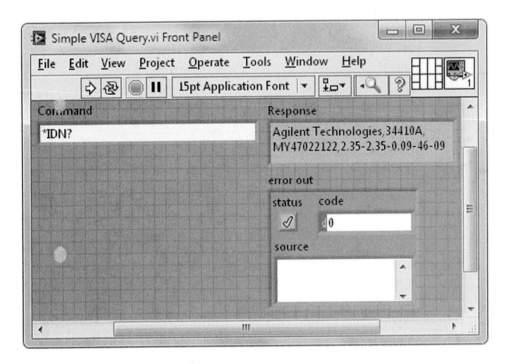

Simple VISA Query will leave the multimeter in *remote mode* with its triggering circuitry "idled." You can return to *local* mode, which continuously "triggers" measurements, by pressing the instrument's front-panel Shift/Local key.

13.10 MESSAGE TERMINATION

At the conclusion of a message-transfer process, some method must be used to signal that the complete message has been passed. The IEEE 488.2 standard appoints the ASCII LF (line feed, also called new line) as its special *end of string* (EOS) character. That is, when receiving a message string, the LF character is always interpreted by the receiver as the last byte of a message. Thus, appending LF to a command string is one method of signaling message termination in IEEE 488.2 communication. Alternately, the IEEE 488.2 standard allows the assertion of an *end or identify* (EOI) while the last character in the string is being passed as another acceptable termination method. The EOI is a digital signal on a dedicated wire within the GPIB cable. When using VISA icons to control an IEEE 488.2 compliant instrument, message termination is taken care of automatically, allowing you to remain blissfully ignorant of this lower-level activity.

If you'd like to view an example of this (usually invisible) message termination activity, pop up on the **Response** indicator on the front panel of **Simple VISA Query**, and then select '****' **Code Display**. The \n character you see at the end of the identification string is the backslash code for LF. The multimeter appended this termination character to its identification string to the alert the receiver (in this case, the GPIB device) that the message has ended.

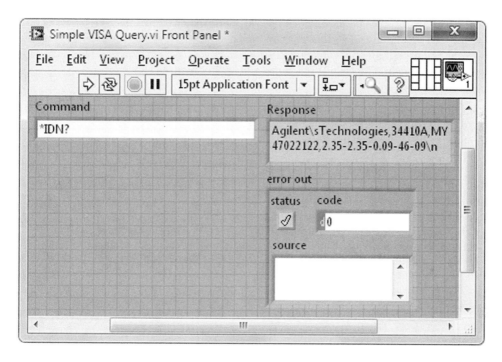

To deactivate backslash coding, pop up on **Response** and select **Normal Display**.

13.11 GETTING AND SETTING COMMUNICATION PROPERTIES USING A PROPERTY NODE

In addition to message termination, there are other low-level functions connected with message-based communication. Many of these low-level functions have an associated parameter setting, which is termed a *VISA property*. VISA assigns default values for these properties and, as long as the VISA-based programs that you write fall within the scope of these default settings, the VISA icons will automatically take care of these low-level functions without any programming effort needed by you (as demonstrated by the message termination example shown above). At times, however, you will most likely write programs that fall outside the scope of the default VISA property settings and so will need to assign nondefault values to these quantities. Reading ("getting") and writing ("setting") VISA property values can be done within your programs using a *Property Node* (and also can often be done in MAX).

As a concrete example of a VISA property, consider **VISA Read**'s *timeout*, which is a fail-safe feature of **VISA Read** that prevents a program from running endlessly if an error occurs. If, for example, an instrument that is being queried doesn't seem to be responsive (perhaps a nameless experimenter forgot to flip on the instrument's power switch), **VISA Read** will only wait for the instrument's response for a certain number of milliseconds (given by the value of the VISA property named **Timeout**) before aborting the read operation and issuing an error message.

The default timeout value for VISA Read on your system can be determined using a **Property Node**. The Help Window for a **Property Node** is shown next.

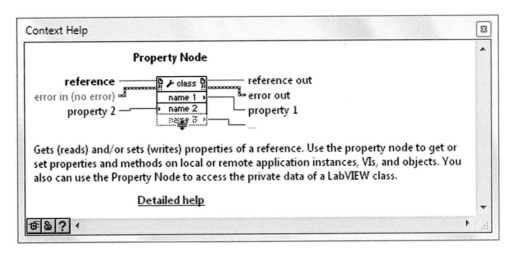

Write the following VI, which reads the current **Timeout** value on your system. Open a new VI, and save it under the name **Get Timeout Value** in **YourName\Chapter 13**. Switch to the block diagram and place a **VISA Property Node** (found in **Functions>>Instrument I/O>>VISA>>VISA Advanced**) there. Then, using **Create>>Constant**, wire the **VISA resource name** for your instrument to the Property Node's **reference** input as shown.

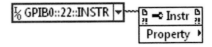

Next, using the 🖑, click on the **Property** terminal and select **General Settings>> Timeout Value**. You might explore what other Properties appear in this menu, many of which are specific to a particular interface bus.

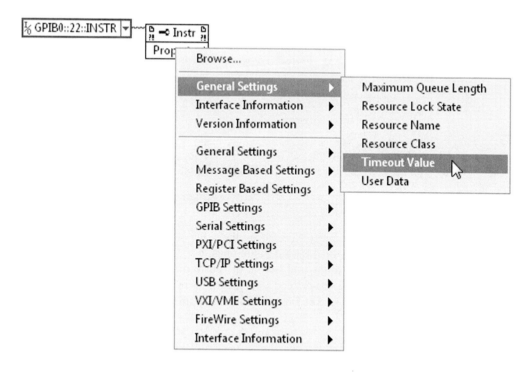

Note that, within the **Timeout** terminal, a small arrow at the right points outward from the terminal's interior. This outward-directed arrow indicates that the **Timeout** terminal is configured as an indicator, that is, it reads ("gets") the current **Timeout** value. Using **Create>>Indicator**, create a front-panel indicator to display the value of **Timeout**.

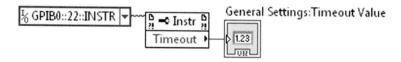

Switch to the front panel, change the indicator label to **Timeout Value (ms)**, and then save your work. Run the VI. As shown below, the (default) **Timeout** value for my system is 3000 ms = 3 seconds.

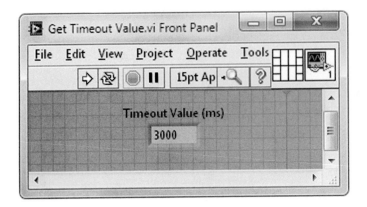

You can also use a Property Node to set the value of a VISA property. To demonstrate this procedure, with **Get Timeout Value** open, use **Save As...** to create a new program called **Set Timeout Value**, and store it in **YourName\Chapter 13**. When run, **Set Timeout Value** will change VISA Read's **Timeout** to a value input from its front panel.

On the block diagram of **Set Timeout Value**, pop up on the **Timeout** terminal and select **Change To Write**.

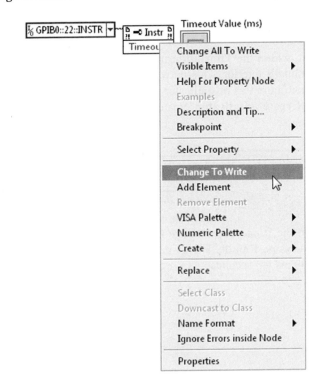

Note that, within the **Timeout** terminal, the small arrow is now at the left pointing inward toward the terminal's interior. This inward-directed arrow indicates that the **Timeout** terminal is configured as a control, that is, it writes ("sets") the **Timeout** value. Delete the Timeout Value (ms) indicator terminal and then, using **Create>>Control**, create a front-panel control labeled **Timeout Value (ms)**.

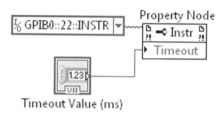

Timeout Value (ms)

Return to the front panel and save your work.

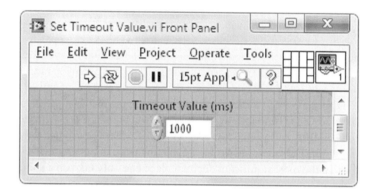

Set Timeout Value (ms) to be *1000*, and then run Set Timeout Value. Next, run Get Timeout Value. Is **Timeout** now equal to 1000 ms = 1 s? Try setting Timeout Value (ms) equal to *2500*. You will find that only certain values for **Timeout** are allowed. LabVIEW takes the value you input to Set Timeout Value as a suggestion (rather than an order) and sets **Timeout** to the nearest allowed value. When finished, run Set Timeout Value one last time, setting **Timeout** equal to *3000*.

13.12 PERFORMING A MEASUREMENT OVER THE INTERFACE BUS

Now that Simple VISA Query has given us a template for the VISA query process, let's try controlling a real measurement. Hook up some known DC voltage difference, say 5 or 6 V, between the HI and LO Voltage Inputs of the Agilent 34410A (or 34401A) DMM. This instrument's user manual instructs us that delivering the following sequence of ASCII commands will result in one DC voltage sample being acquired and then loaded into the instrument's output buffer (which is part of its interface circuitry):

CONF:VOLT:DC<Space>10,0.00001
INIT
FETC?

Here is the meaning of this secret code. First, the Agilent 34410A can be programmed to perform 14 different types of measurement functions, including DC voltage, AC voltage, DC current, AC current, resistance, frequency, capacitance, and temperature. Given these options, the first command instructs the instrument that we desire to take a DC voltage measurement. The full command is *CONFigure:VOLTage:DC <Space><Range>,<Resolution>* (this command actually executes a collection of commands drawn from the Agilent 34410A's **INP**ut, **SENS**e, **TRIG**ger, and **CALC**ulate root subsystems). The command mnemonic *CONFigure:VOLTage:DC* is constructed in the hierarchical tree structure, typical of SCPI-compliant instruments. *CONFigure* is the root-level keyword and colons (:) represent the descent to the lower-level *VOLTage* and then the lowest-level *DC* keywords. Although the full command mnemonic can be sent to the instrument, it is only absolutely necessary to send the capitalized characters. Separated from the command mnemonic *CONF:VOLT:DC* by a *<Space>*, the numerical values for two measurement parameters—*<Range>* and *<Resolution>*—are specified. *<Range>* selects among the instrument's five available voltage measurement scales. Each scale offers a different sensitivity with *<Range>* giving the maximum measurable value on a particular scale. The five available ranges are 100 mV, 1 V, 10 V, 100 V, and 1000 V. In our situation of measuring a signal of approximately 5 V, the 10 V scale is appropriate. *<Resolution>* specifies the precision of the measurement, with eight levels of accuracy available on the 34410A DMM. For our programs to be compatible with both the 34410A and the older-model 34401A DMM, we will use only two resolution levels in our work. These resolutions have 5 1/2 and 6 1/2 digits of precision (where the 1/2 digit means that the most significant decimal place can only take on a value of "1" or "0"), and so, on the 10 V scale, voltages can be resolved at the level of either 0.0001 or 0.00001 V, respectively. The trade-off in requesting higher accuracy is that the measurement takes a longer time. In the command sequence above, the highest resolution of 6 1/2 digits is selected by setting *<Resolution>* equal to *0.00001* when *<Range>* equals *10*. Note that the syntax of the *CONF* command obeys the conventions of the SCPI language: a *comma* (,) separates the parameters from each other and a *<Space>* separates the command mnemonic from the parameters.

Once the multimeter has been configured for the desired measurement function as described in the previous paragraph, the data-taking process is begun by sending the *INITiate* command (from the **TRIG**ger root subsystem). Upon receipt of *INIT*, the multimeter will acquire the requested voltage sample and then store this value in its internal memory. Finally, the *FETCh?* command (from the **MEM**ory root subsystem) instructs the instrument to transfer the reading in its internal memory to its interface-related output buffer.

We'd like now to place this command sequence into **Simple VISA Query**. Since there are three commands to be sent, it appears that we must modify the VI to include a sequence of three successive implementations of **VISA Write**. Although you are free to do so, a much easier solution is available. The SCPI language allows the programmer to concatenate several commands together into one long multi-command string that can be sent in a single **VISA Write** statement. The syntax for this concatenation process is as follows:

- Use a semicolon (;) to separate two commands within the string.
- Begin a command with a colon (:) if it has a different root-level than the command preceding it. The first command in the concatenated string and IEEE 488.2 common commands (which begin with an asterisk) do not require a leading colon.

Since each of our three commands has a different root level, applying the above rules results in the following concatenated string:

CONF:VOLT:DC<Space>10,0.00001;:INIT;:FETC?

Type this command into the **Command** control on the front panel of **Simple VISA Query** as shown next. Run the VI. Your computer will instruct the multimeter to acquire a 6 1/2 digit voltage reading on the 10 V range, retrieve this value, and then display it on the front-panel in the **Response** indicator. Cool, eh?

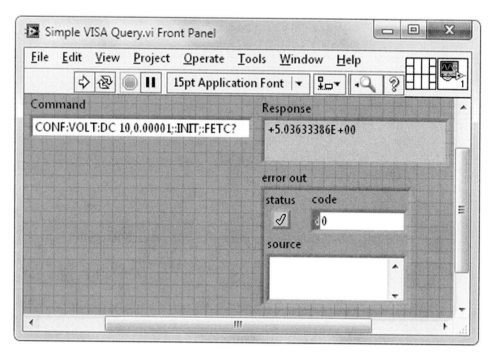

Note that, although extra digits are displayed, the value within **Response** is only accurate to the fifth decimal place.

As shown above, the Agilent DMM reports its data samples in the form of an ASCII character string using the exponential format SD.DDDDDDDDESDD, where S is a positive or negative sign, D is a numeric digit, and E is an exponent. For future reference, note that the string that represents a data sample is 15 bytes long. If you want to use this reading as input to a mathematical calculation (a common situation), you will need to convert the string representation into a numerical format. Such conversion operations can be easily accomplished in LabVIEW using the collection of conversion icons found in **Functions>>Programming>>String**. In the present case, use **Fract/Exp String To Number** in **Functions>>Programming>>String>>String/Number Conversion**. The Help Window for this icon is given next.

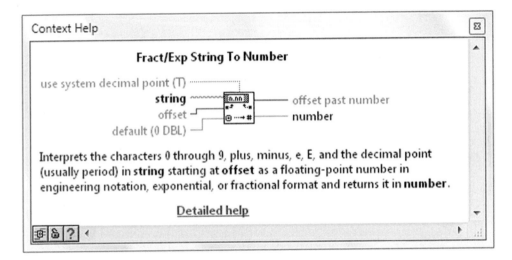

Place a **Number Indicator** on the front panel of **Simple VISA Query** and label it **Numeric Voltage**. Use **Display Format...** in this indicator's pop-up menu to make its **Digits of precision** equal to *5* and disable **Hide trailing zeros**. Then modify the block diagram as follows.

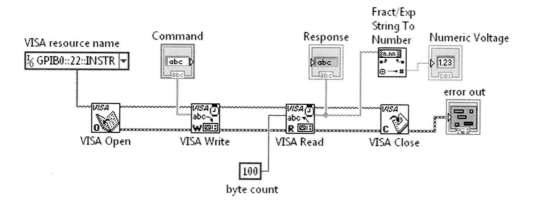

Run the VI to verify that the string-to-number conversion icon performs as expected.

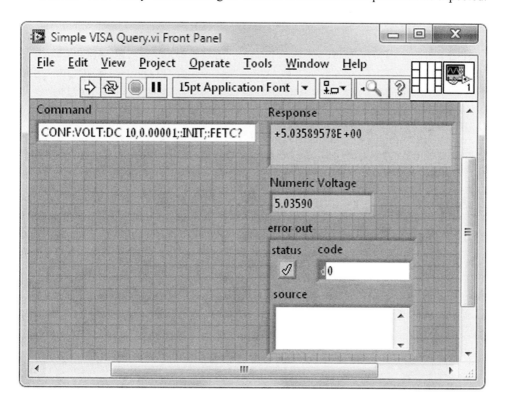

13.13 SYNCHRONIZATION METHODS

Although most ASCII commands are completed quickly after being received by a programmable instrument, some commands start a process that requires a significant amount of time (such as acquiring a large amount of data or moving an object from Point *A* to Point *B*). The time required for such processes must be taken into account when writing a data acquisition program, or else, upon execution, the program may request data before they are available, induce undesirable motion, or cause some other chaotic outcome.

As an example, in its default configuration, the Agilent 34410A multimeter acquires one data sample after receipt of the *INITiate* command and then stores this measured value in its internal memory. However, through use of the *SAMPle:COUNt <Space><Value>* command, the multimeter can be instructed to take and store multiple data samples upon receiving *INITiate*. The Agilent 34410A is configured to acquire 1500 DC voltage samples with 6 1/2-digit resolution via the following concatenated string of commands:

CONF:VOLT:DC<Space>10,0.00001;:SAMP:COUN<Space>1500;:INIT;:FETC?

The *FETCh?* command will load the 1500 acquired samples from the multimeter's internal memory (which, by the way, can hold up to a maximum number of 50000 measured values; the 34401A can store only 512 values) into the instrument's interface-related output buffer.

Let's write a VI that uses the given command string to gather a sequence of 1500 voltage samples. Open **Simple VISA Query**, and then use **Save As...** to create a new VI called **Simple VISA Query (Long Delay)**. Delete **Numeric Voltage** from the front panel and enlarge **Response** so that it can display a very long string (which is the concatenation of 1500 voltage values) and activate its scrollbar by selecting **Visible Items>>Vertical Scrollbar** in its pop-up menu. Type the command given above into the **Command** control (Agilent 34401A owners, your DMM performs more slowly, so configure it to acquire 15, rather than 1500, i.e., use *SAMP:COUN<Space>15* in your command string). Once entered into **Command**, you can keep this command permanently loaded there by selecting **Edit>>Make Current Values Default**.

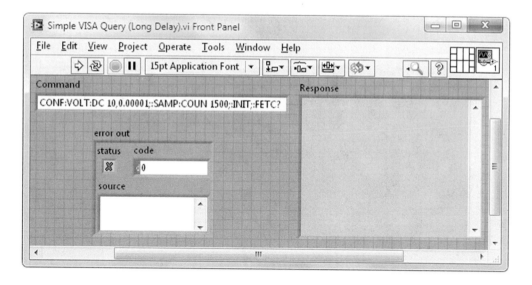

Switch to the block diagram. Delete **Fract/Exp String To Number**. After the given command string is written to the multimeter by **VISA Write**, **VISA Read** will receive a string containing of the 1500 voltage samples. Since each voltage sample is reported as a 15-bytes string and a (single) delimiting ASCII character will be needed to separate each sample, this 1500-sample string is expected to be about $(1500 \times 15) + 1500 = 24000$ bytes long. Input an integer larger than 24000 to **byte count** as shown.

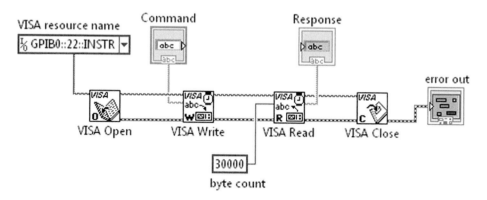

Run **Simple VISA Query (Long Delay)**. Count down the seconds, 3...2...1... Disappointingly, you will find that your VI outputs only a subset of the expected 1500 values and produces an error.

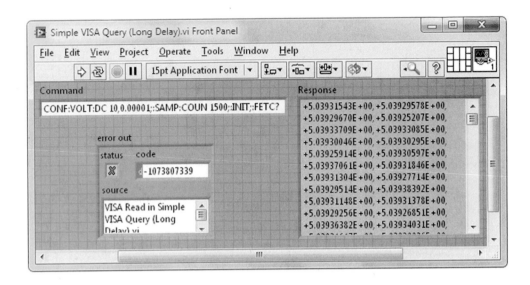

To find out what produced this error, pop up on the **code** indicator within the **error out** cluster and select **Explain Error**.

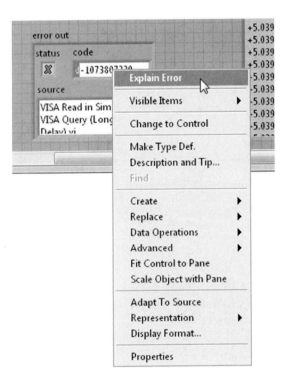

A dialog window appears, where we are told that a "timeout expired" at **VISA Read** before the requested operation (i.e., take 1500 data samples) could be completed.

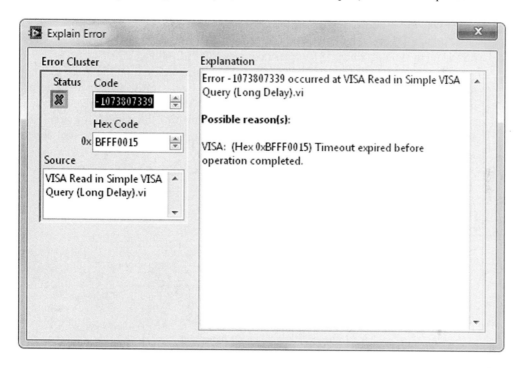

After some head scratching and checking of the Agilent 34410A user manual, the following explanation then emerges for the error we observed when running **Simple VISA Query (Long Delay)**. Simply stated, voltage sampling takes time. When configured for 6 1/2 digit resolution, it takes the Agilent 34410A multimeter 0.2 power line cycles (PLC) for each voltage sample. Additionally, the Agilent 34410A has an optional autozero feature, which operates as follows: After each voltage measurement, the multimeter internally disconnects the input signal and takes a zero reading. The instrument then subtracts the zero reading from the preceding measured value to prevent offset voltages in the multimeter's internal circuitry from affecting measurement accuracy. Since the zero reading also takes the same number of PLC as a regular voltage measurement, in comparison to when autozeroing is not activated, each complete voltage sample by the multimeter takes twice as long. Consulting the DMM user manual, we find that the autozeroing feature is turned off for the resolution we have chosen, so each voltage sample takes 0.2 (not 0.4) PLC. Assuming this instrument is plugged into a 60 Hz power source (that is, 60 PLC per second), 1500 voltage samples will take about

$$1500 \times \left(\frac{0.2 \text{ PLC}}{60 \text{ PLC/s}} \right) = 5 \text{ s}$$

There's the problem! A few moments ago, we found that the default timeout value for **VISA Read** is 3 seconds, but the measurement we have initiated takes about 5 seconds. Thus, before all of the requested data are available, **VISA Read** terminates the execution of Simple VISA Query (Long Delay).

For 34401A users, each data sample by your DMM takes 10 PLC and autozeroing is turned on by default. Thus, 15 samples take $15 \times (20 \text{ PLC}/60 \text{ PLC/s}) = 5$ s.

There are a couple of crude solutions to this dilemma. First, on the block diagram of Simple VISA Query (Long Delay), you can insert a single-frame **Sequence Structure** into the VISA execution chain that simply contains a **Wait (ms)** icon, wired to produce a delay of greater than 5 seconds between the issuance of the data-taking command and the order to read the gathered data samples. The resulting diagram would appear as follows.

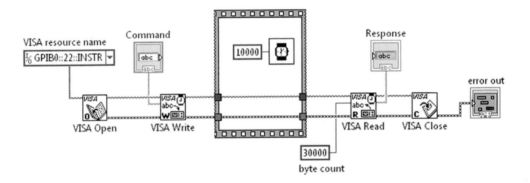

Second, for a slightly more elegant fix, you can use a **Property Node** to change the **Timeout** value for **VISA Read** from its default value (on my system) of 3000 ms to something larger than 5 seconds. Use this approach to modify the block diagram of Simple VISA Query (Long Delay) as shown below. Here, the **Timeout** value is chosen to be 10 seconds.

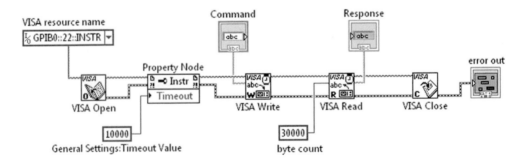

Return to the front panel of **Simple VISA Query (Long Delay)**. Turn your DMM off and then on again to clear the unread data in its output buffer from the previous unsuccessful run. Then, with the command to perform 1500 samples programmed into **Command**, run the VI (34401A owners, request 15 samples, not 1500). About 5 seconds later you should see something like the following front panel. Note that the delimiter used by the Agilent DMM to separate neighboring data values is a comma.

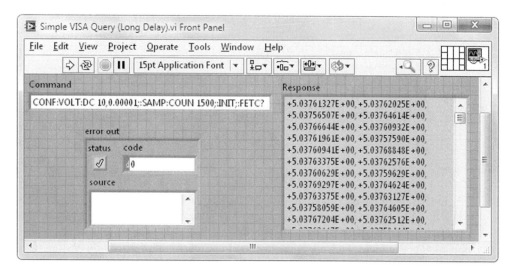

In the preceding example, we found that with a detailed knowledge of the measurement process being implemented, it was possible to troubleshoot a malfunctioning VISA-based VI. Please note that lack of communication (in particular, the interface bus not correctly knowing when the instrument's data will be available) is the root problem that led to the malfunction.

Fortunately, powerful tools exist that allow one to monitor the status of tasks being performed by a programmable instrument. For IEEE 488.2-compliant instruments, these tools are the Standard Event Status Register (SESR) and Status Byte Register (SBR) that were discussed at the beginning of this chapter. With proper use of the SESR and SBR, many potential data-taking glitches, such as the one just experienced, can be avoided.

The status reporting capabilities of the SESR and SBR can be employed in several ways. We will explore two commonly used techniques—the Serial Poll and Service Request Methods. The core operation for both of these methods is the same—the completion of an assigned task triggers the Operation Complete (OPC) bit in the Standard Event Status Register to be set, which in turn sets the Event Status Bit (ESB) of the Status Byte Register.

In the Serial Poll Method, the setting of ESB is detected by directly checking the Status Byte Register, whose state is obtained by serial polling the instrument. The complete step-by-step process of this method is shown in Figure 13.4.

Serial Poll Method

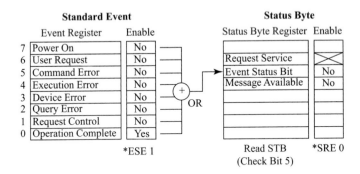

FIG. 13.4

In the Service Request Method, the Status Byte Register is configured such that, when its ESB is set, the Request Service bit is induced to be set also. This action then causes the instrument to assert a SRQ, which alerts the interface bus that the assigned operation is complete. This method is pictured in Figure 13.5.

Service Request Method

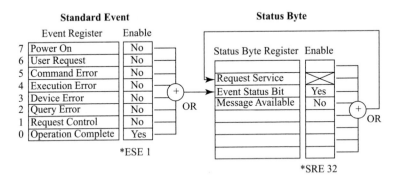

FIG. 13.5

We'll write VIs that implement both of these approaches to status reporting.

13.14 MEASUREMENT VI BASED ON THE SERIAL POLL METHOD

Let's try the Serial Poll Method first. To configure the Agilent 34410A (or 34401A) DMM for status reporting using the Serial Poll Method, write the following VI called

Status Config (Serial Poll) and save it in **YourName\Chapter 13**. First, code the VI's block diagram as shown next. Use the autocreation feature in pop-up menus to create all of the constants, controls, and indicators.

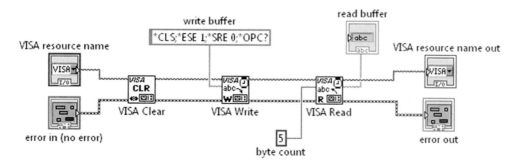

Switch to the front panel and arrange the object logically. Design an icon and assign the connector pane's terminals consistent with the Help Window shown.

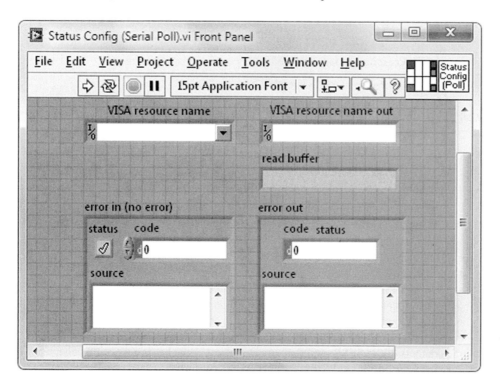

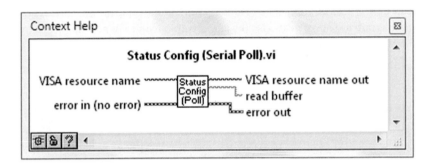

Here's how the VI works, assuming that the instrument referenced by **VISA resource name** is IEEE 488.2 compliant. Within the chain of VISA icons, **VISA Clear** (found in **Functions>>Instrument I/O>>GPIB**) executes first. The Help Window for this icon is shown below.

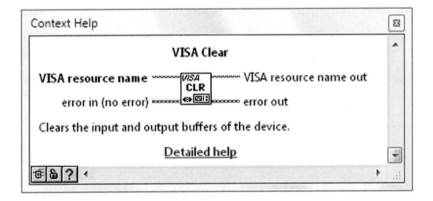

Although not an absolute necessity for inclusion in **Status Config (Serial Poll)**, this VI performs the precautionary action of "clearing" the instrument. **VISA Clear** instructs the instrument to abort all measurements in progress, disable its triggering circuitry, clear its interface-related output buffer, and prepare to accept a new command string.

Next, **VISA Write** sends the concatenated command string *CLS;*ESE 1;*SRE 0;*OPC?* to configure the instrument for status reporting using the Serial Poll Method. Note that since the component strings are all IEEE 488.2 common commands, leading colons are not required in the concatenation. In this sequence of commons, *CLS* clears the contents of the SESR and SBR. As described in the beginning of this chapter, *ESE 1* enables the SESR's OPC bit to set the ESB in the Status Byte Register and *SRE 0* disables the instrument from asserting a SRQ. Then, *OPC?* requests the instrument to return a "*1*" to the instrument's output buffer after this command is completed. This last command is included simply as a method of checking that the entire sequence of commands has been executed.

Finally, **VISA Read** reads the contents of the instrument's output buffer. If all goes well, there should be a single ASCII character *1* read into the computer.

Test drive your VI as follows. Click on **VISA resource name** control's menu button with the 🖑.

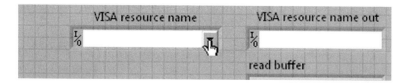

Then, from the list presented, select the **VISA resource name** for your computer-controlled instrument.

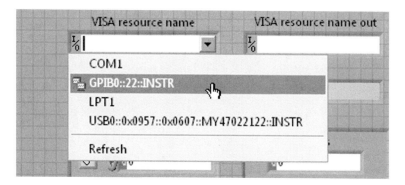

Then, run **Status Config (Serial Poll)**. Upon completion, does the **Buffer Reading** string indicator display an ASCII character *1*?

Next, construct a VI called **Serial Poll**, which continuously reads the Status Byte Register of an instrument until a given bit is set. A suggested coding of **Serial Poll** is shown in the following diagrams, and explanations of the unfamiliar icons are in the subsequent paragraphs. Save **Serial Poll** in **YourName\Chapter 13**.

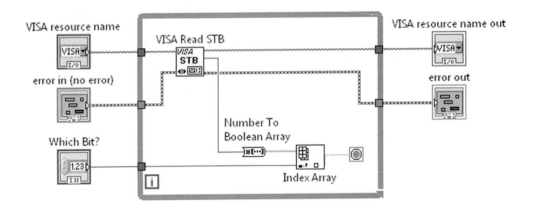

VISA resource name

VISA Read STB

VISA resource name out

error in (no error)

error out

Which Bit?

Number To
Boolean Array

Index Array

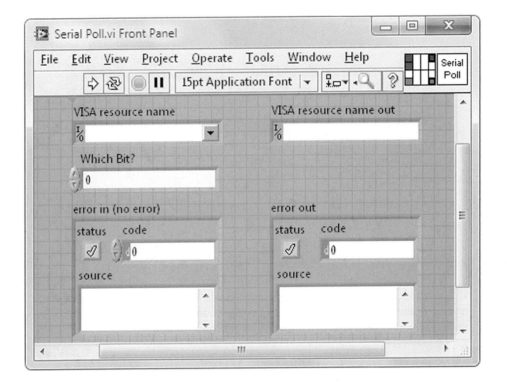

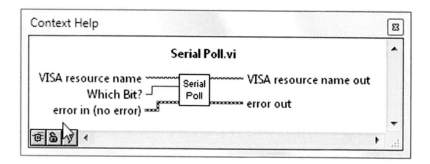

VISA Read STB, found in **Functions>>Instrument I/O>>VISA**, is the workhorse of this VI. With each iteration of the While Loop, its **status byte** output returns the current values of the SBR's eight bits in the form of an integer. For example, if the SBR's fifth bit (ESB) is set, then **status byte** outputs the integer 32, since $00100000_2 = 32_{10}$. The Help Window for **VISA Read STB** is shown next.

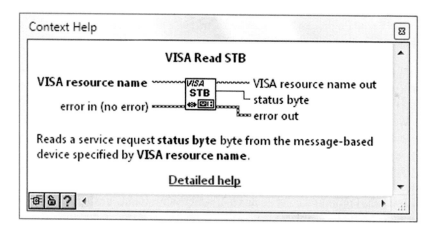

The individual bits of **status byte** can be checked through the use of **Number To Boolean Array** (found in **Functions>>Programming>>Boolean** with its Help Window shown next). This VI creates an array of TRUE and FALSE values that mirror the sequence of zeros and ones (starting from the least-significant bit) in the binary representation of the integer input **number**. For example, if **number** equals the decimal integer 48, then the **Boolean array** output will be [F, F, F, F, T, T, F, F], since $48_{10} = 00110000_2$. **Index Array** can then be used to ascertain the value of a particular element in this array. Serial Poll's While Loop will continue to iterate until **Which Bit?** becomes TRUE.

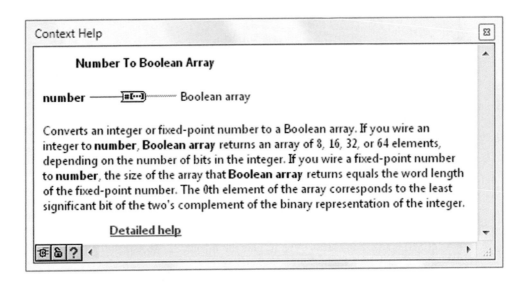

Run **Serial Poll** under **Highlight Execution** (which is activated by clicking the light bulb button in the toolbar) and, through your observations, gain a better understanding of its operation. Remember to input values for **VISA resource name** and **Which Bit?** on the front panel. When run in this isolated manner, the VI will most likely never be able to exit the While Loop, so you'll have to stop it using the **Abort Execution** button in the toolbar. Also, turn off **Highlight Execution** by clicking the light bulb button.

We're finally ready to write **VISA Query (Serial Poll)**. This top-level program implements serial polling to synchronize the interface bus activities necessary in acquiring 1500 voltage samples using an Agilent 34410A multimeter.

Open **Simple VISA Query (Long Delay)**, and then use **Save As...** to create **VISA Query (Serial Poll)**. The front panel can remain unchanged. Type the following command into the **Command** control, much of which may already be there by default; be sure to include the *OPC (on the older 34401A model DMM, use *15* rather than *1500*).

*CONF:VOLT:DC<Space>10,0.00001;:SAMP:COUN<Space>1500;:INIT;*OPC;:FETC?*

Switch to the block diagram and modify it as shown below with **Status Config (Serial Poll)** and **Serial Poll** used as subVIs.

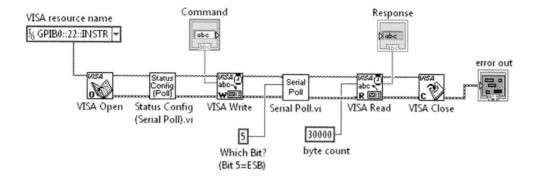

Here is how this diagram works: The concatenated command string is sent to the instrument by **VISA Write**. After configuring the multimeter for the desired DC Voltage measurement function, the acquisition process is begun by the *INIT* command. The succession of 1500 samples is acquired and temporarily stored in the multimeter's internal memory. After the 1500th sample is obtained, *OPC* instructs the instrument to set its SESR's OPC bit (which, in turn, sets the SBR's ESB), and then *FETC?* loads the contents of the internal memory into the instrument's output buffer. At that point, **Serial Poll** detects the setting of ESB, which then triggers the instrument's output buffer to be read by **VISA Read**. One might be tempted to write the concatenated command with *OPC* after *FETC?*, rather than sandwiched between *INIT* and *FETC?*, as above. It is best, however, to avoid sending *OPC* after a query (a query is a command like *FETC?* that ends in a question mark) because such commands cause a message to be loaded into an instrument's output buffer. If the message exceeds the finite size of the output buffer, the query must be immediately followed by **VISA Read** as the program executes to read the long message string over the bus successfully.

Return to the front panel, save your work, and then run **VISA Query (Serial Poll)**. Does the VI obtain the requested 1500 DC voltage samples successfully? If so, try running it again with **Highlight Execution** activated for both **VISA Query (Serial Poll)** and its subVI **Serial Poll**. This exercise will illustrate the weakness of the Serial Poll Method, namely, the large volume of interface bus traffic required by this technique. During the 5 seconds while the 1500 data samples are being gathered, the instrument is polled numerous times by the interface bus so that its status can be continuously monitored. Although effective, the Serial Poll Method is rather inefficient because of its excessive use of the interface bus and processor time.

13.15 MEASUREMENT VI BASED ON THE SERVICE REQUEST METHOD

The Service Request Method provides status reporting with a minimum of interface bus activity. To configure an IEEE 488.2 compliant instrument for status reporting using the Service Request Method, open **Status Config (Serial Poll)**, and then create **Status Config (SRQ)** using **Save As...** and save it in **YourName\Chapter 13**. The front panel and terminal assignments can remain as is, but the icon should be redesigned as shown here.

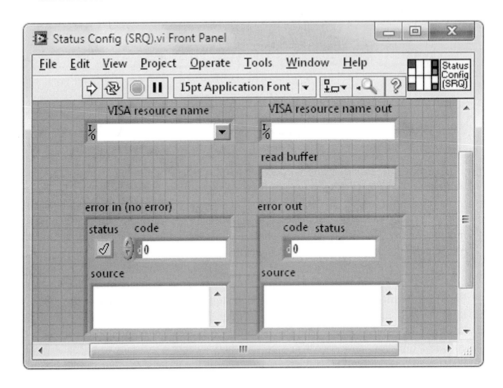

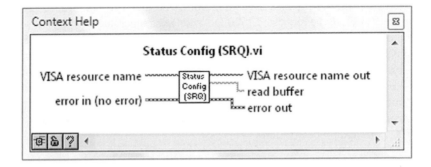

Only two modifications of the block diagram are needed. First, by changing *SRE 0* to *SRE 32* in the command string sent to the instrument, the instrument will assert the SRQ line when the SBR's fifth (ESB) bit is set. The already present *ESE*

1 command configures the instrument to set the ESB in response to the SESR's OPC (Operation Complete) bit being set. Second, for VISA icons to detect service request (SRQ) events during this VISA session, **VISA Enable Event**, with **Service Request** wired to its **event type** input, must be included in the diagram as shown. **VISA Enable Event** is found in **Functions>>Instrument I/O>>VISA>>VISA Advanced>>Event Handling.**

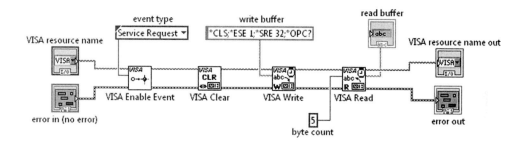

Save your work as you close this VI.

Now, let's write a program to acquire 1500 voltage samples from an Agilent 34410A DMM using the Service Request Method. Open **VISA Query (Serial Poll)**, then use **Save As...** to create a new VI named **VISA Query (SRQ)**, and store it in **YourName\ Chapter 13**. The front panel is fine as is.

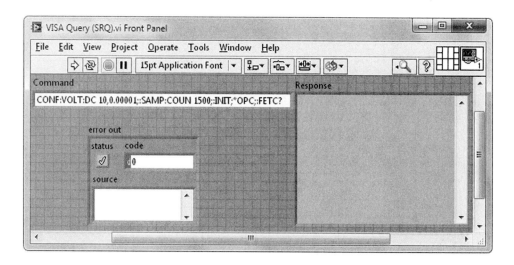

Switch to the block diagram and modify it as shown next.

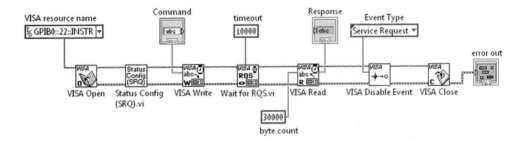

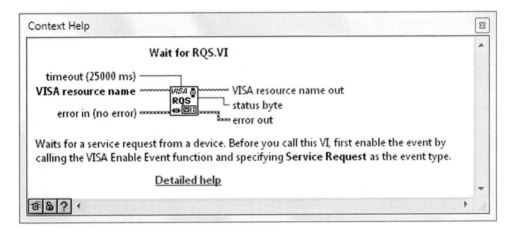

Here, **VISA Disable Event**, found in **Functions>>Instrument I/O>>VISA>>VISA Advanced>>Event Handling**, must be included to disable VISA servicing of SRQ events before the VISA session is closed.

Wait for RQS.vi, also found in **Functions>>Instrument I/O>>VISA>>VISA Advanced>>Event Handling** (Help Window shown below), sits idly until the instrument denoted by **VISA resource name** asserts a SRQ. However, there is a limit to the patience of this icon. It will only wait up to a total time of **timeout**, with a default value of 25000 ms = 25 seconds. Because our measurement requires only 5 seconds, we can go with this default value by keeping the **timeout** input unwired, as shown in the above diagram.

Save your work, and then run **VISA Query (SRQ)**. Does it successfully acquire the requested 1500 DC voltage samples from your DMM? Do you understand the operation of this program and how the Service Request Method manages to work with a minimum of interface bus activity?

To simplify the block diagram of **VISA Query (SRQ)**, you might consider packaging **VISA Disable Event** and **VISA Close** together in a subVI called **Close (SRQ)**, since both of these icons are involved in closing down the service request–based VISA session.

To accomplish this feat easily, simply create a highlighting box around the two icons using the ⬉.

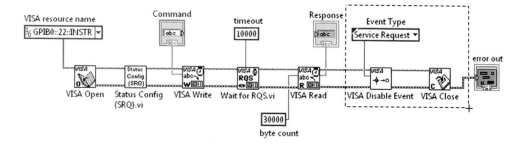

Then select **Edit>>Create subVI**. A new subVI icon will appear wired on your diagram.

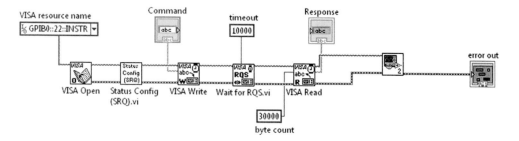

Double-click on this new icon to open it. Then, relabel the front-panel objects appropriately, design an icon, and assign the connector pane's terminals consistent with the Help Window shown (using the $4 \times 2 \times 2 \times 4$ pattern). Save this VI under the name **Close (SRQ)** in YourName\Chapter 13.

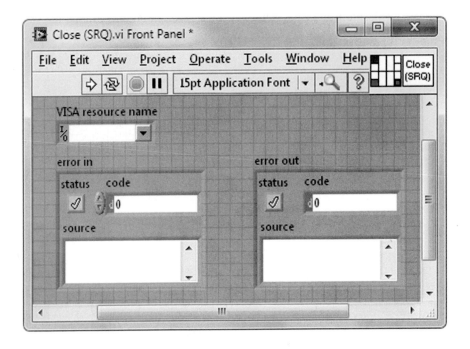

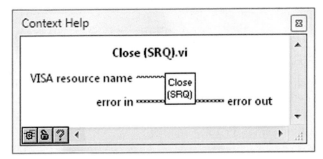

Switch to the block diagram of **Close (SRQ)**. It should appear as follows.

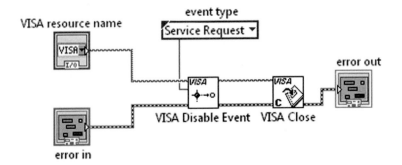

Close **Close (SRQ)**, and return to the block diagram of **VISA Query (SRQ)**. You may have to delete the originally created subVI and load a new copy of **Close (SRQ)** there using **Functions>>Select a VI...** After that, the finished block of **VISA Query (SRQ)** will appear as shown next. Try running this VI to verify that it functions correctly.

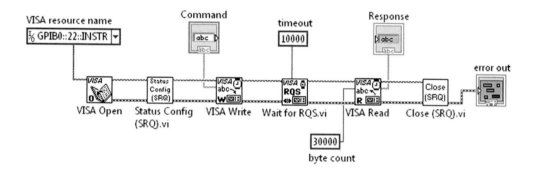

13.16 CREATING AN INSTRUMENT DRIVER

An instrument driver is a collection of modular software routines that perform the operations required in the computer control of a programmable instrument. These operations include configuring, triggering, status checking, sending commands to, and receiving data from, the instrument. Above, **Status Config (Serial Poll)** and **Status Config (SRQ)** are examples of configuration VIs that would be useful to include as part of the Agilent 34410A instrument driver. You will now write another configuration VI, this time one that prepares the multimeter for taking a desired measurement.

The Agilent 34410A is capable of implementing 14 types of measurement functions, including DC and AC voltage, DC and AC current, 2- and 4-wire resistance (2-wire is the "normal" method for measuring resistance; the more involved 4-wire technique is necessary only when measuring very small resistance samples), frequency and period of an AC signal, continuity, diode check, capacitance, and temperature. To gain experience with some of the LabVIEW tools available for developing instrument drivers, let's write a driver that offers the choice of configuring the Agilent 34410A for a DC voltage, AC voltage, or 2-wire resistance measurement. You, of course, can be more ambitious and write your VI to control up to all 14 possible measurement functions.

Referring to the Agilent 34410A user manual, we find that our driver must allow a user to select one of the following three possible commands to configure the instrument for the desired measurement function:

CONFigure:VOLTage:DC <Space> <Range>, <Resolution>
CONFigure:VOLTage:AC <Space> <Range>, <Resolution>
CONFigure:RESistance <Space> <Range>, <Resolution>

Here, the possible values of *<Range>* for both the DC and the AC voltage measurements are 0.1, 1, 10, and 100 V (we ignore the highest range because it's different—1000 vs. 750—for DC and AC voltage). For the resistance measurement, the allowed *<Range>* values are 100, 1k, 10k, 100k, 1M, 10M, and 100M ohms (here, we ignore the 1 Gohm range because it's not available on the older 34401A model). In all cases, the measurement precision may be 5 1/2 or 6 1/2 digits, which corresponds to *<Resolution>* being 10^{-5} or 10^{-6} times the *<Range>* value, respectively.

We will write two programs called **Range and Resolution Decoder** and **Command String**, which will allow a user to construct the desired command string using front-panel controls. On **Range and Resolution Decoder**, given range and resolution choices from a user-friendly front-panel listing of the multimeter's available offerings, the program will convert these choices to the double-precision floating-point numeric format needed in **Command String**. **Command String** will construct the appropriate ASCII command string to be sent to the Agilent DMM, based on selections made on its front-panel controls.

Create a new VI named **Range and Resolution Decoder** and save it in **YourName\ Chapter 13**. Place four **Emun** controls (found in **Controls>>Modern>>Ring & Emun**) on the front panel and label them **Function, Voltage Range, Resistance Range**, and **Resolution**, respectively. Pop up each **Enum**, select **Edit Items...**, and then program it with the items, in the given order, shown in the following list.

> **Function**: *DC Voltage, AC Voltage, Resistance*
> **Voltage Range**: *100 mV, 1 V, 10 V, 100 V*
> **Resistance Range**: *100 ohm, 1 kohm, 10 kohm, 100 kohm, 1 Mohm, 10 Mohm, 100 Mohm*
> **Resolution**: *5 1/2 Digits, 6 1/2 Digits*

Then place these four **Enum** controls in a cluster shell (found in **Controls>> Modern>>Array, Matrix & Cluster**) labeled **Function Parameters** as shown next.

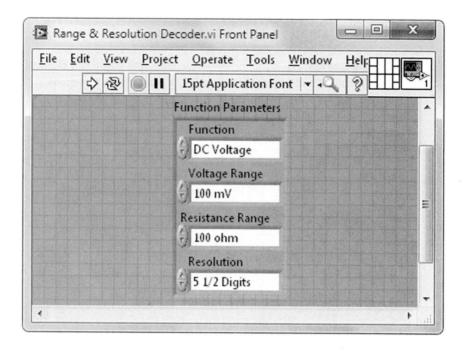

Switch to the block diagram, place a **Case Structure** there, and complete the code shown. Pop up on the Case Structure and select **Add Case for Every Value**, and then verify that it has three cases labeled **DC Voltage**, **AC Voltage**, and **Resistance**.

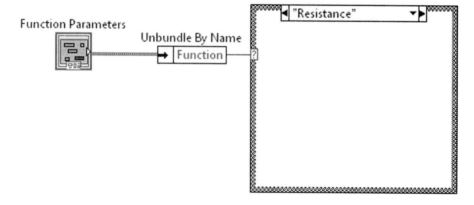

Select the **DC Voltage** case, and then place an **Index Array** icon within it. Pop up on Index Array's **n-dimension array** input and select **Create>>Constant** to create an **Array Constant** and label it **Voltage Ranges**. Next, program the index-0 through index-3 elements of this **Array Constant** as *0.1, 1.0, 10.0,* and *100.0,* respectively. The **Array Constant** then will serve as a *look-up table* of the multimeter's allowed voltage ranges, given as double-precision floating-point numbers. Complete the code for the **DC Voltage** case shown below. Here, the integer associated with a selected **Voltage Range** on the front-panel Enum control provides the index of the desired look-up table element. This element is then output by **Index Array**.

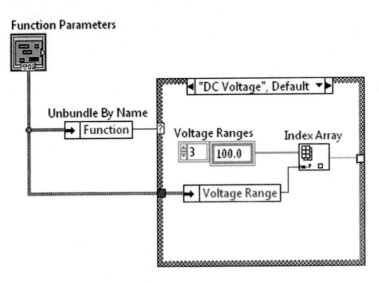

Clone the **Voltage Ranges** Array Constant (mouse-click while holding down *<Ctrl>*), and place the copy somewhere on the block diagram. Then, switch to the **AC Voltage** case and (using your cloned **Voltage Ranges**) write the code shown.

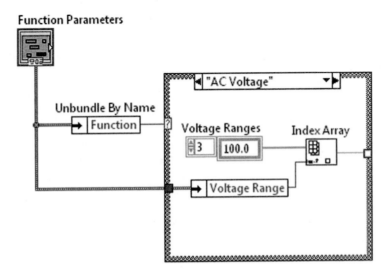

Finally, switch to the **Resistance** case, and program it as shown. Here, the index-0 through index-6 elements of the **Resistance Ranges** Array Constant are *1.0E2, 1.0E3, 1.0E4, 1.0E5, 1.0E6, 1.0E7,* and *1.0E8,* respectively.

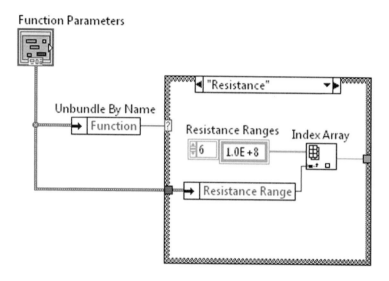

Add a second **Case Structure** and complete the diagram as shown next. The **Range & Resolution** indicator cluster is created by popping up on **Bundle** and using **Create>>Indicator**.

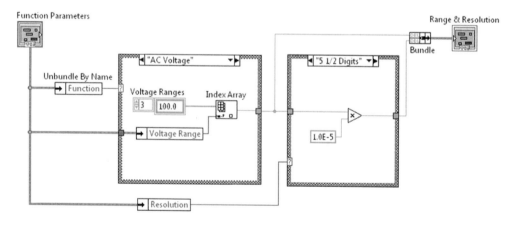

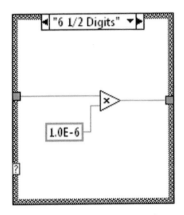

Return to the front panel. Within the **Range & Resolution** indicator cluster, name the owned labels of the top and bottom arrays **Numeric Indicator** as **Range** and **Resolution**, respectively, by popping up on each array's index display and selecting

Visible Items>>Label (do not carry out the labeling using the 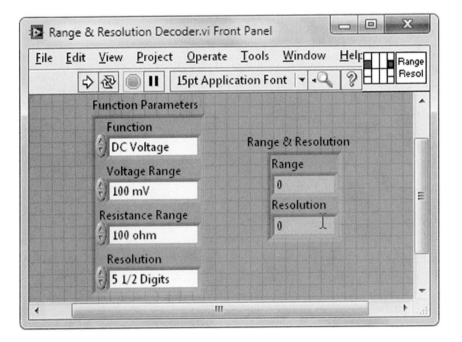 as this will create free, rather than owned, labels). Design an icon and assign the connector pane's terminals as shown. Save your work.

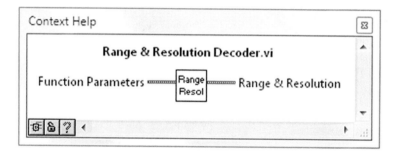

Run **Range and Resolution Decoder** and verify that it functions properly. For example, with **Function, Voltage Range,** and **Resolution** equal to *DC Voltage, 10 V,* and *6 1/2 Digits,* respectively, **Range** and **Resolution** should equal *10.0* and *0.00001.*

Next, open a blank VI, and save it under the name **Command String** in **YourName\ Chapter 13.** Switch to the block diagram and write the following code, which constructs the desired ASCII command string. The **Function Parameters** control cluster and **output string** string indicator are made using the autocreation feature in the pop-up menus.

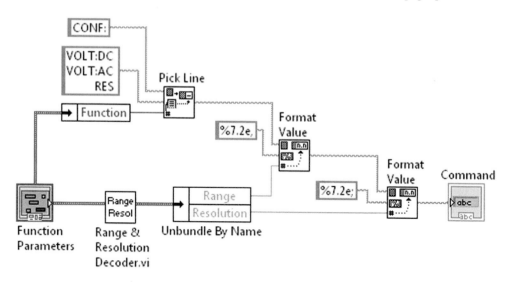

This diagram constructs the desired command string in a three-step process. First, all three possible commands begin with the keyword *CONF:,* so this sequence of ASCII characters is wired to the **string** input of **Pick Line** (found in **Functions>> Programming>>String>>Additional String Functions** with its Help Window given next). The value of the **line index** input (an integer given by the front-panel **Function** Enum control) then selects which of the three possible lines programmed into the **String Constant** wired to **multi-line string** is to be appended to *CONF:.* Create the three lines in this **String Constant** by the following sequence of keystrokes: *VOLT:DC<Space>* *<Enter>VOLT:AC<Space>* *<Enter>RES<Space>*. Be sure to include the *<Space>* character at the end of each command string. You can make the invisible space and line feed characters visible by popping up on the **String Constant** and selecting "****" **Codes Display**. The correct entry will then appear as *VOLT:DC\s\nVOLT:AC\s\nRES\s*, where *\s* and *\n* are backslash codes for space and line feed, respectively.

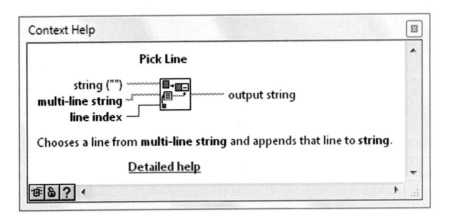

Format Value, from **Functions>>Programming>>String>>String/Number Conversion** (Help Window shown next), then is used to attach two more string fragments, each with embedded ASCII-coded numbers that program the *<Range>* and *<Resolution>* settings of the multimeter. The **Format Value** icon takes the number at its **value** input and converts it to an ASCII string representation with the format defined at its **format string** input. This ASCII string is appended to **string** and presented at **output string**. In the above diagram, the scientific notation format *%7.2e* (see Section 5.6) is used for both *<Range>* and *<Resolution>* parameters. Note that a comma (,) and semicolon (;) follow *<Range>* and *<Resolution>*, respectively.

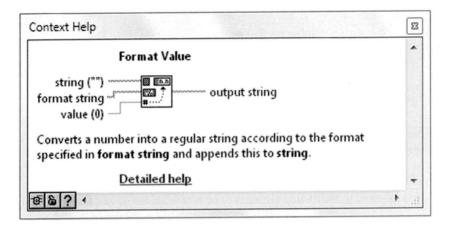

Switch to the front panel and change the label of the String Indicator from **output string** to **Command**. Run the VI with a given choice of the controls within **Function Parameters**, and verify that the correct command string appears in the **Command** indicator. Save your work.

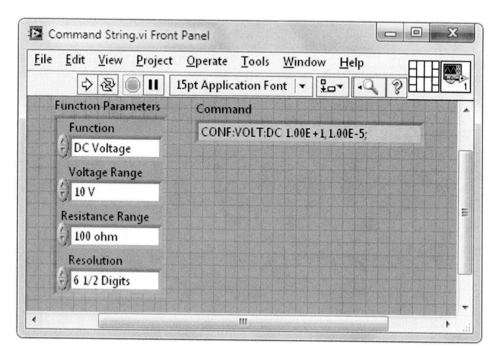

We'll next add the ability to control the multimeter's autozeroing feature and to program the desired number of data samples to be taken. Add a **Push Button** (found in **Controls>>Modern>>Boolean**) and a **Numeric Control** to the front panel and label them **Autozero** and **Sample Count**, respectively. Change the representation of **Sample Count** to **U16**.

Switch to the block diagram, and then include the autozero and sample count code shown below. The *%5d* format in the *SAMPle:COUNt* command specifies a five-place decimal integer because the maximum allowed value for *SAMPLe:COUNt* (according to the Agilent 34410A user manual) is 50000. The format string entry for this command should be *:SAMP:COUN<Space>%5d*.

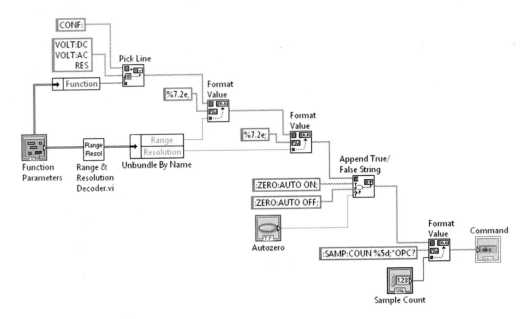

Autozero can be either turned on and off with the following commands:

$$:ZERO{:}AUTO{<}Space{>}ON$$
$$:ZERO{:}AUTO{<}Space{>}OFF$$

Append True/False String (found in **Functions>>Programming>>String Additional String Functions**), whose Help Window follows, provides an easy way to choose which of these two choices is concatenated to the command string. Remember to include the leading colon and final semicolon in the **false string** and **true string** entries to assure proper command concatenation.

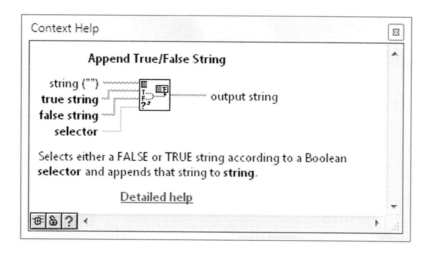

To guarantee that the instrument fully processes the sent command string before exiting the configuration VI (which you will write in a moment), the command string concludes with *OPC?.

Return to the front panel. Run the VI with a given choice of the front-panel controls to verify that the correct command string appears in the **Command** indicator. Then design an icon and assign the connector pane consistent with the following Help Window. Save your work as you close the VI.

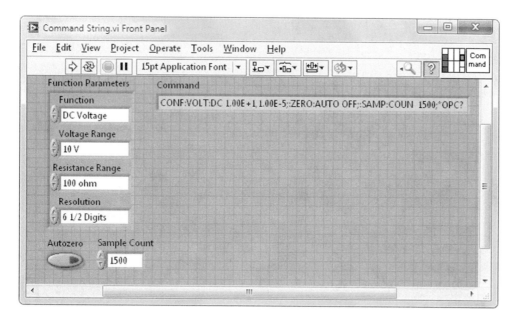

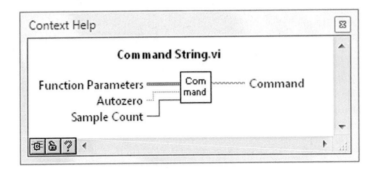

Finally, create a VI named **Measurement Config** and save it in **YourName\Chapter 13**. Switch to the block diagram and code it as shown. This VI will write the command string to the instrument. When this diagram runs, the **read buffer** indicator will display an ASCII "*1*" if the command string was successfully read by the instrument.

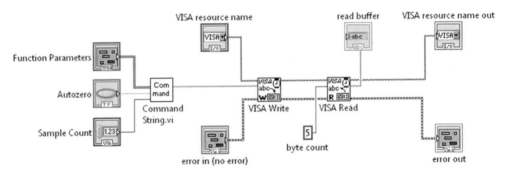

Switch to the front panel, and arrange the objects there as you wish. Then design an icon and assign the connector pane's terminals consistent with the Help Window shown below. Make the default value for the **Sample Count** control equal to *1* using **Edit>>Make Current Values Default**. Save your work.

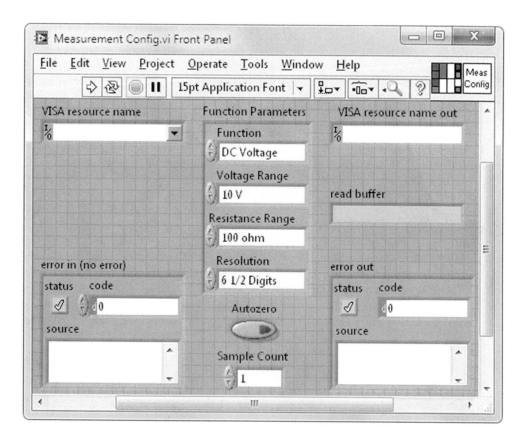

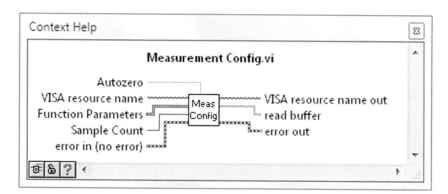

Input the VISA resource name for your instrument into the **VISA resource name** control, and then run **Measurement Config** with front-panel control settings shown above so that the following command is sent to the instrument:

$$CONF{:}VOLT{:}DC{<}Space{>}1.00E{+}1,1.00E{-}5,{:}ZERO{:}AUTO{<}Space{>}$$
$$OFF{;:}SAMP{:}COUN{<}Space{>}\ 1;{*}OPC?$$

If the command is successfully sent over the interface bus, an ASCII "*1*" will appear in **output buffer**. If the Agilent DMM beeps, there is most likely an error in the sent command. Open the front panel of **Command String**, and then run **Measurement Config** again. Check that the concatenated command in the **Command** indicator on **Command String**'s front panel has a form given above; make sure all of the colons, semicolons, and spaces are included. If there is an error, correct it on the block diagram of **Command String**.

After running **Measurement Config** the multimeter will be left in *remote mode*. You can switch to *local mode* by pressing the instrument's front-panel Shift/Local key. The Agilent 34410A can then be triggered (equivalent to sending the INIT command over the interface bus) with the Trigger button. A star (*) annunciator will blink on the instrument's front-panel display as it acquires each voltage sample. Run your VI with **Sample Count** equal to *5*, then in local mode press the Trigger button. Does the annunciator blink **Sample Count** times after the Trigger button is depressed?

Save **Measurement Config** as you close it.

Write a final modular VI for your Agilent DMM instrument driver called **Take Data** as shown below, and save it in **YourName\Chapter 13**. The leading **CLS* command assures that all bits in the SESR and SBR register are set to zero, prior to each data-taking process.

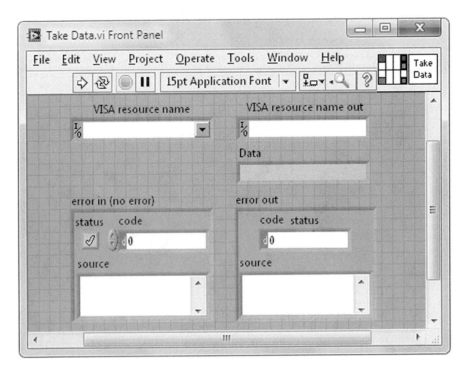

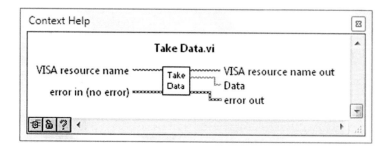

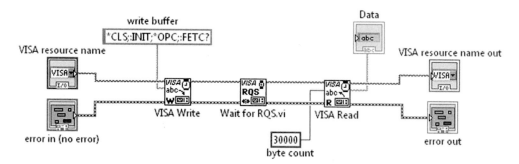

Note: **Take Data** cannot be run independently without generating an error. However, if you first run one of the other VIs that you have written (can you figure out which one?), then **Take Data** can be run successfully.

13.17 USING THE INSTRUMENT DRIVER TO WRITE AN APPLICATION PROGRAM

Ultimately, the merit of an instrument driver is measured by the ease with which you can use it to write an *application program* to fulfill some specialized need in your laboratory work. Let's quickly write an application program called **Data Sampler** that can be configured to take a multi-sample voltage or resistance measurement.

With **VISA Query (SRQ)** open, use **File>>Save As...** to create **Data Sampler** and save it in **Your Name\Chapter 13**. Rewrite the block diagram using your modular driver software as shown below.

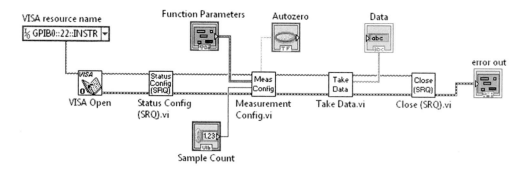

Return to the front panel and arrange the object there as desired. Save your work.

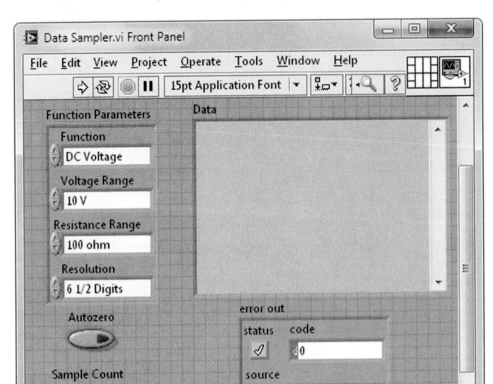

Run **Data Sampler** with various choices of front-panel settings (or interface buses!), and then pat yourself on the back for a job well done.

Now that you know some of what goes into writing an instrument driver, here's some very good news. In many cases, the LabVIEW instrument driver you will need for a particular instrument in your laboratory has already been written and is available for your use free of charge. National Instruments provides an extensive library of downloadable instrument drivers at *http://www.natinst.com/idnet/*. You can also access this resource within LabVIEW by selecting **Tools>>Instrumentation>>Find Instrument Drivers...** Most of these drivers are written using VISA icons and so, using the interface-appropriate VISA resource name for your instrument, can be used to communicate over various interface busses—RS-232, GPIB, Ethernet, and USB. If interested, try using **Tools>>Instrumentation>>Find Instrument Drivers...** to download the Agilent 34410A DMM instrument driver. After exiting LabVIEW and then restarting it, you should find the driver in **Functions>>Instrument I/O>>Instrument Drivers>>Agilent 34410**. Take a look at some of the icons in this palette and see if you can decipher them.

Finally, a useful Agilent DMM instrument driver utility would perform the following task: Take the instrument's **Data** string (data samples delimited by commas and the string terminated by a LF character) and convert it to a numeric array and a spreadsheet format. One manifestation of that utility called **Reformat Data String** is shown next. On the front panel, the string control and string indicator have been resized and scrollbars have been activated by selecting **Visible Items>>Vertical Scrollbar** in the pop-up menus. The **Single Numeric Sample** indicator is included for convenience (converting the array format to a single DBL numeric) when **Sample Count** is equal to *1*.

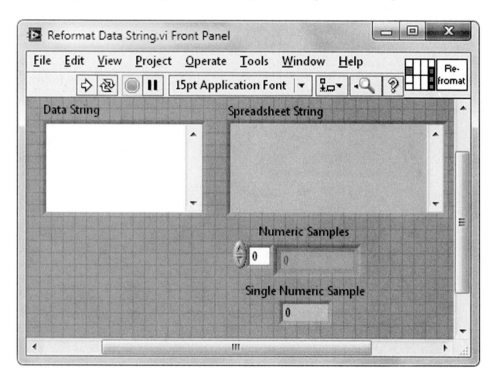

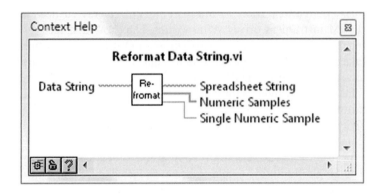

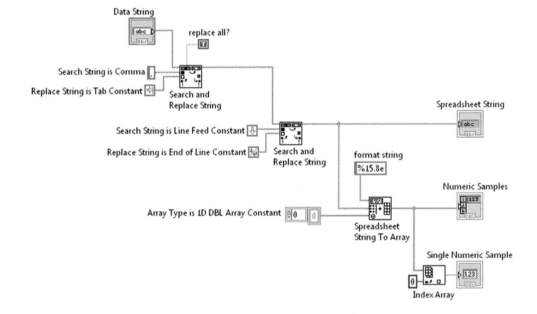

Write **Reformat Data String**. This VI implements the **Search and Replace String** icon found in **Functions>>Programming>>String** (Help Window shown below) to coerce the original **Data** string into the spreadsheet format (by replacing comma delimiters and the LF terminator with tabs and an EOL, respectively). Do you understand how it works?

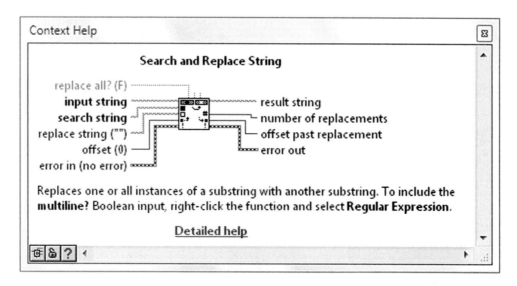

Finally, return to **Data Sampler**. Include **Reformat Data String** as a subVI on its block diagram and modify the front panel as shown.

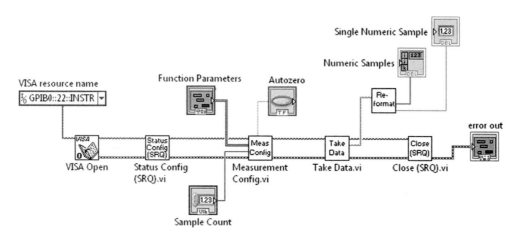

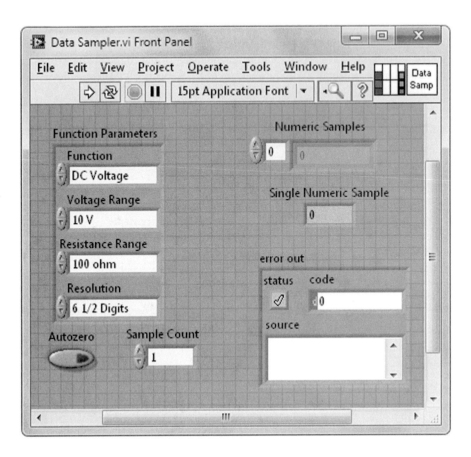

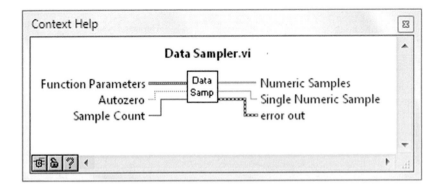

Once written, run **Data Sampler** and watch it perform its magic.

DO IT YOURSELF

Assume that you have a widget in your laboratory that is providing you with some interesting information about X, where X might be the position of an object or the intensity of a light source. Additionally, say, the widget provides this information about X in the form of a "voltage code," that is, it produces an output voltage V that is some known function of X. Then, with an Agilent 34410A multimeter and an appropriate application VI (called it **Time Evolution of X**), you can monitor X (via measurement of V) as a function of time.

Using your Agilent 34410A instrument driver programs as subVIs, write **Time Evolution of X**. When run, this top-level VI continuously obtains a single DC voltage sample every **Wait Time** seconds (where **Wait Time** is given by the value on the similarly named front-panel control) until the front-panel **Stop Button** is pressed. While running, the VI provides real-time graphing of *Voltage* vs. *Time* data on a **Waveform Chart** with the Chart's *Time* axis properly calibrated. The front panel also provides the option of storing all of the accumulated data in a spreadsheet file with the *Time* and *Voltage* data in the spreadsheet's first and second columns, respectively. Define *Time* = 0 at the moment that the first voltage sample is acquired.

The front panel of **Time Evolution of X** should appear as shown next. All needed parameters without a front-panel control should be input on the block diagram. After building this VI, run it to observe a time-dependent voltage input (e.g., from a function generator) to the multimeter and save the resulting *Voltage* vs. *Time* data in a spreadsheet. Determine the minimum value allowed for **Wait Time (second)** using your knowledge of how long it takes the DMM to acquire a single voltage sample.

A helpful tip: The *Time* axis of the Waveform Chart can be calibrated using a Property Node. Pop up on the Waveform Chart's icon terminal and select **Create>>Property Node>>X Scale>>Offset and Multiplier>>Multiplier**. Then set the **Multiplier** property appropriately.

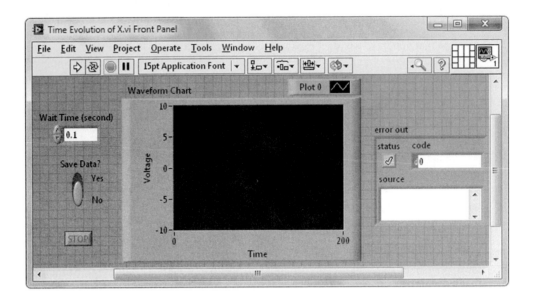

PROBLEMS

1. Thermocouples are widely used as temperature sensors. A thermocouple is con-
    structed by joining the ends of two dissimilar metals, for example, a copper and
    a constantan wire for a type T thermocouple. This junction produces a millivolt-
    level voltage, which has a well-documented temperature dependence, where the
    temperature is measured relative to a "cold junction" reference temperature.
    Conveniently, this cold junction can be provided by a compact electronic device
    called a Cold Junction Compensator (CJC), which effectively makes the reference
    temperature equal to 0 °C.

    Connect a thermocouple to a CJC and then connect the plus and minus out-
    put of the CJC to the HI and LO Voltage Inputs of the Agilent 34410A Multimeter.
    Then, write a program called **Thermocouple Thermometer (VISA)** that, every 250
    milliseconds until a Stop Button is pressed, reads the thermocouple voltage, con-
    verts this value to the corresponding temperature in Celsius, and then displays this
    temperature in a front-panel indicator.

    To convert the thermocouple voltage to its corresponding temperature,
    use **Convert Thermocouple Reading.vi**, which is found in **Programming>>
    Numeric>>Scaling**, with its **CJC Voltage** input wired to *0* (the **CJC Sensor** and
    Type of Excitation inputs can be left unwired). Program **Thermocouple Type** for
    your particular type of thermocouple (e.g., T).

    Run **Thermocouple Thermometer (VISA)** and use it to measure room tem-
    perature as well as the temperature of your skin.

2. As written in this chapter, **Serial Poll** is flawed in that, if the bit being monitored in the Status Byte Register is never set, this VI will loop endlessly. With **Serial Poll** open, use **Save As...** to create a new VI called **Serial Poll with Timeout**. Then modify the block diagram so that, if the bit being monitored is not set within 10 seconds, the While Loop is stopped.

3. Use **Tools>>Instrumentation>>Find Instrument Drivers...** to download the Agilent 34410A Digital Multimeter instrument driver. After closing and restarting LabVIEW, this driver software will be found in **Functions>>Instrument I/O>> Instrument Drivers>>Agilent 34410**. Use the icons from this "built-in" driver to write a program called it **Time Evolution of X (Built-In Driver)**, which carries out the task described in this chapter's **Do It Yourself** project. The icon VI Tree gives a helpful overview of the "built-in" driver.

4. Regardless of the chosen resolution, the Agilent 34410A Multimeter always reports data sample values with eight digits to the right of the decimal point. Thus, some of these decimal-place values are not significant. With **Take Data** open, use **File>>Save As...** to create a new VI called **Take Data (Accurate Resolution)**. Add a **Function Parameters** front-panel control to this new VI (so that the selected resolution setting can be input) and then modify the block diagram so that the **data** output reports values with the actual resolution selected (e.g., 5 1/2 digits if that is the selected resolution).

5. Use the **Instrument I/O Assistant** Express VI to query the Agilent 34410A Multimeter. Place an **Instrument I/O Assistant** (found in **Functions>> Express>>Input**) on the block diagram of a VI called **Simple VISA Query (Express)**. When this Express VI's dialog window opens, select the desired instrument, and then click on **Add Step**. In the **Add Step** dialog window that appears, double-click on **Query and Parse**. In the Enter a command box, type

CONF:VOLT:DC 10,0.00001;:INIT;:FETC?

and then click **Run this step**. The command will be sent to the Agilent 34410A Multimeter and its string response will be displayed. Click the **Auto parse** button to convert the response string to numeric format and then close the dialog box by clicking the **OK** button. When returned to the block diagram, simply create an indicator for the icon's **token** output terminal.

Run **Simple VISA Query (Express)** and demonstrate that it successfully obtains a DC Voltage sample from the Agilent 34410A Multimeter.

APPENDIX I

Construction of Temperature Control System

To perform the project in Chapter 12, you will need access to an apparatus that controls the temperature of a small object through the use of a thermoelectric (TE) device. There is, of course, a multitude of ways to construct the required gadget. In this Appendix, I offer a design (total cost approximately $100) that has worked successfully for me.

First, I use a rectangular aluminum block of dimensions $2'' \times 1.5'' \times 5/16''$ ($1'' = 2.54$ cm) as the object whose temperature is to be controlled. Hereafter, I'll simply call this object "the block." So that accurate temperature measurements can be taken on the block, a small hole of diameter $3/16''$ is drilled into one of its sides to a depth of approximately $1/2''$. A thermistor (preferably coated with thermal grease; see below) can then be inserted into this hole and held securely by the addition of a $1/4''$ long #6–32 nylon set screw. For the thermistor, I use an inexpensive $10 \, k\Omega$ model from Epcos (Part No. B57863S0103F040, available from Digikey P/N 495–2149-ND, http://www.digikey.com, $3) with soldered-on $12''$ lead wires for easy connection to a constant-current circuit. Finally, two #8–32 clearance holes are drilled (using a #19 drill bit) into the block as shown in Figure A.1. The $1.450''$ spacing is chosen so that a 30-cm wide TE module will fit between these holes.

SIDE VIEW

TOP VIEW

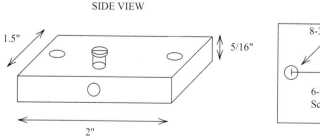
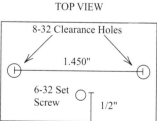

FIG. A.1   Aluminum block design

The required heating or cooling of the block is facilitated by placing it in contact with one side of a thermoelectric module. A TE module is a compact solid-state device that, via the Peltier Effect, acts as a heat pump. When current is caused to flow through the TE device in one direction, heat will be absorbed from the contacting block, cooling it, as shown in Figure A.2.

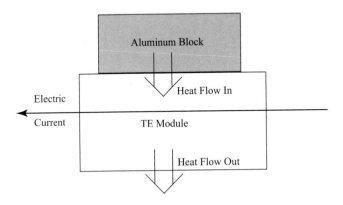

FIG. A.2   Thermoelectric module cools the block.

When current is made to flow through the TE module in the opposite direction, heat is pumped into the block, heating it as in Figure A.3.

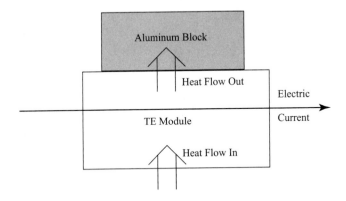

FIG. A.3   Thermoelectric module heats the block.

Note that, because the TE device is a heat pump, a heat reservoir (not pictured) must be in contact with the bottom side of the TE module in both Figures A.2 and A.3. With insignificant change to its temperature, this heat reservoir accepts (provides) the heat pumped through the TE module when the block is being cooled (heated).

Laird Technologies (http://www.lairdtech.com) manufactures over 200 standard models of TE devices. The choice of a particular TE model is dictated by the heat-pumping capacity necessary and electric power supply limitations in your application. For our project, I use a Laird Technologies Model CP 1.0–127–05L TE device (available from Mouser Electronics, http://www.mouser.com, $25 each). Within the relatively small range of current flow between ±2 Amps, this TE module has enough heat-pumping ability to change the temperature of a $2'' \times 1.5'' \times 5/16''$ aluminum block significantly away from room temperature (including cooling it to below 0 °C and icing it up!).

How can we construct the required heat reservoir? Remember that a heat reservoir is simply an object that can accept or produce an amount of heat (up to a maximum value determined by your particular application) without significantly changing its temperature. A big slab of aluminum (for example, $5'' \times 3'' \times 3''$) possesses a large "thermal mass" and thus can perform as a fairly decent heat reservoir. By tapping two #8–32 screw holes 1.450" apart into such a slab, a TE module can be sandwiched between the block and the slab. Then using two 3/4" long #8–32 (insulating) nylon screws, you can secure the three items together and, in the process, obtain good thermal contact between contacting surfaces. To guarantee efficient heat flow, coat the TE device's heat-pumping surfaces with thermal grease (available, for example, from Laird Technologies and Jameco) before sandwiching the three items. This method for realizing a heat reservoir is inexpensive, but imperfect. If the TE module is run over moderately long time periods, the temperature of the aluminum slab will begin to significantly change from room temperature because of its rather poor efficiency at transferring heat with the surrounding room.

A much more stable heat reservoir can be constructed using a finned aluminum heat sink. Laird Technologies and Aavid Thermalloy (http://www.aavidthermalloy.com) offer a line of products called *extrusions*, which are plate-like pieces of aluminum with many seamlessly attached fins. Because of the large surface area of its fins, an extrusion will very efficiently exchange heat with its surrounding room air, imbuing it with the desired properties of a heat reservoir. I have had good success using 6" lengths of Aavid Thermalloy Model 78780 Extrusion ($30 per 6", available from Newark, http://www.newark.com) as the heat reservoir in the temperature control system of Chapter 12. Tapping two #8–32 screw threads 1.450" apart into the plate-like surface of the extrusion, the TE module with the block atop can be mounted and then secured with two #8–32 (insulating) nylon screws. Again, the use of thermal grease will enhance the conductance of heat between the block and the heat reservoir. A final improvement results by mounting the extrusion on top of a small fan (for example, 120 VAC 4" Cooling Fan, Radio Shack Model 273–238, $20), which forces air flow through the extrusion's array of fins. Legs for the fan and small angle brackets to connect it securely to the extrusion can be fashioned from 1/16" aluminum strips. Attach these strips to the

FIG. A.4   Cooling fan assembly.

fan using the fan's mounting holes. A photograph of the fan with its homemade legs and brackets is shown in Figure A.4.

Photos (from two different angles) of the entirely assembled temperature-control system are provided in Figures A.5 and A.6. A small aluminum bar attached to the plate region of the extrusion acts as a strain relief for the TE module wires.

Alternately, a complete system similar to the one shown in Figure A.6 can be purchased from Laird Technologies (see, for example, Model DA-014–12-02) for about $150.

FIG. A.5   Fully assembled temperature control system hardware.

FIG. A.6   Fully assembled temperature control system hardware.

The bidirectional current driver circuit for the TE module is shown in Figure A.7. For simplicity, the TIP transistors are represented in this diagram as simple bipolar transistors, but are, in actual fact, Darlington transistors. With regard to external connections, a Darlington transistor behaves exactly the same as a single bipolar transistor with a very large current gain β (hence, the acceptability of the simplified representation). The Darlington's large β (on the order of 1000) results from its two-transistor internal construction, in which the collector current of one transistor provides the base current to the other. In its ON state, the Darlington's base-emitter voltage is "two diode-drops" (≈ 1.2 V). Because of the coupling between its two internal transistors, the Darlington is especially susceptible to thermal instabilities, necessitating proper heatsinking of these components for reliable service (see below).

The ± 10 V power sources attached to the TIP transistors' collectors must be capable of providing up to about 2 Amps (the ground of these power sources must be the same ground as the rest of the circuit (i.e., these grounds must be connected by, for example, cabling). The Laboratory DC Power Supplies sold by several electronic equipment manufacturers (for example, Agilent E3610A PS280 or BK Precision 1760A), which contain two identical DC supplies, are ideal instruments for providing both the negative and the positive voltage levels. These Laboratory DC Supplies typically have built-in ammeters that, in the above circuit, can be used to monitor the current flow through the TE device. The 1 μF capacitors in Figure A.7 remove any high-frequency noise present in the power supply voltage levels.

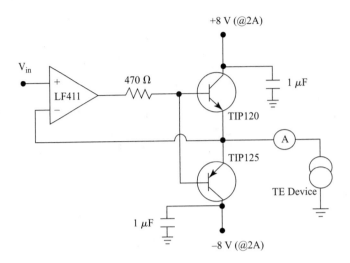

FIG. A.7   Voltage-controlled bidirectional current driver.

When this circuit is in operation, the TIP power transistors generate A LOT of heat. These hot transistors present the following problems: (1) if you touch them, they will hurt your fingers (an avoidable danger, once you're aware of it); (2) if the circuit is set up on a solderless breadboard, they will melt the breadboard's plastic (this is the voice of experience talking); and (3) if the transistors are allowed to get too hot, they will become thermally unstable and possibly destroy themselves (again, the voice of experience). Proper heatsinking of the transistors will avoid these problems. Because this process is somewhat involved, I provide students with a widget that contains the two transistors mounted on a heat sink. The noise-suppressing 1 μF capacitors are included as well. Top and bottom views of the widget (total cost approximately $30) are shown in Figure A.8.

The two TIP transistors have TO-220 packaging; however, I mount them on a TO-3 Heat Sink (Wakefield Engineering Part No. 401K, available from Newark Electronics, $15). The pattern of mounting holes (appropriate for a single TO-3 packaged power transistor) on this finned heat sink allows the two TIP transistors to be affixed side by side and also provides convenient openings through which to pass wiring. A TO-220 heat sink mounting kit (Jameco Part No. 34121, http://www.jameco.com, $4 per five-piece kit) must be used in mounting each TIP transistor to insulate its collector from the heat sink. The heat sink is then attached to a homemade chassis, fashioned from 1/16″ aluminum sheet metal. Additionally, to facilitate necessary electrical connections, banana jacks and binding posts are mounted on the chassis.

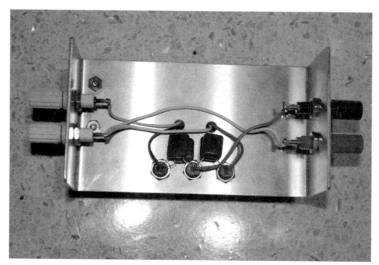

FIG. A.8   Top and bottom views of widget.

The widget is then hardwired according to Figure A.9. All of the parts in this diagram may be purchased from Newark Electronics or from most any other electronics parts distributors.

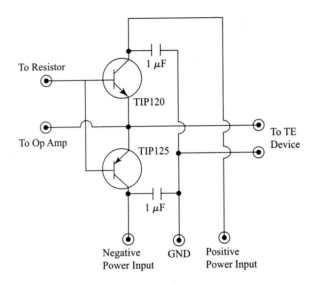

FIG. A.9   Widget circuit.

Program Cross Reference Table

Program [Book section in which created]	Used as Component in this VI [Book section]
Sine Wave Chart (While Loop) [1.1–1.12]	Sine Wave Chart (While Loop with Runtime Options) [7.2]
Waveform Simulator [3.4–3.15]	Waveform Generator (Express) [4.12–4.15], Spreadsheet Storage [5.1–5.7], Spreadsheet Storage (OpenWriteClose) [5.8–5.10]
	FFT of Sinusoids [10.6], FFT (Magnitude Only) [10.8–10.14], Waveform Generator (DAQmx) [11.8]
Spreadsheet Storage [5.1–5.7]	Spreadsheet Read [DIY Chapter 5]
Power Function Simulator [6.4]	Trapezoidal Test [6.6], Power Function Derivative [6.8–6.9], Simpson Test [7.8], Convergence Study (Trap vs. Simp) [7.9]
Trapezoidal Rule [6.6]	Trapezoidal Test [6.6], Power Function Derivative [6.8–6.9], Even Ends [7.7], Convergence Study (Trap vs. Simp) [7.9]
Trapezoidal Test [6.6]	Convergence Study (Trap) [6.7]
Simpson's Rule [7.5–7.8]	Simpson Test [7.8], Convergence Study (Trap vs. Simp) [7.9]
FFT (Magnitude Only) [10.8–10.14]	Spectrum Analyzer [DIY Chapter 10]

Index

Abort Execution button 33
Acquisition mode,
    1 Sample (On Demand) 169
    N Samples 176–177
Activation energy barrier 367, 397
Add icon,
    adding constant to each array element
        383–384
Agilent digital multimeter, commands
    CONFIGURE 528–529
    FETCH? 528–529, 545
    INITIATE 528–529
    SAMPLE 532
    ZERO 560
Agilent digital multimeter, features
    annunciator 564
    autozero 535–536, 560
    beep 564
    date string format 530
    identification string 516–517,
        518, 522
    local/remote mode 564
    measurement types 527–528
    model 34410A vs. 34401A 510, 535–536,
        551–552
    overview 510
    power line cylces (PLC) 535–536
    range and resolution 528, 551–552

Aliasing,
    alias frequency 159
    condition for aliasing 159
    Nyquist frequency 158
    observation of aliasing 159, 204–205,
        443–444, 445, 446
    suppressing with low-pass filter 158,
        204–205, 444, 449–450
Alignment Tool 259
American Standard Code for Information
    Interchange (ASCII),
    backslash codes 237–239
    character set 211
    text file 211–212
Amplitude of sinusoid, defined 414
Analog input (AI),
    Acquisition Mode, 1 Sample (On
        Demand) 169, 455
    Acquisition Mode, N Samples 176–177,
        453, 460
    aliasing 158–160
    channel 155–157
    differential mode 155–157
    nonreferenced singled-ended (NRSE)
        mode 157–159
    range 157
    referenced singled-ended (RSE) mode
        157–159

Analog input (AI) (*cont.*)
   resolution 158
   sampling frequency 158
   voltage resolution 158
Analog output (AO),
   channel 184–185
   Generation Mode, 1 Sample (On Demand)
     185–186, 480
   Generation Mode, Continuous 195–196,
     483–484
   maximum update rate 184
   range 184
   resolution 184
   voltage resolution 184
Analog-to-digital (A/D) conversion, see
     analog input
Append True/False String icon 560–561
Application program 565–567
Array,
   1D indexing 59
   2D indexing 220
   Adding/subtracting constant to each ele-
     ment 383–384
   bracket specifier 66, 73
   building 2D with Build Array 221–222
   empty 304, 331
   initializing 360–362
   inputting entire array into For Loop 319
   inputting from Array Constant 373
   inputting from disk file 239–242, 376–385
   inputting from front-panel control 371–376
   memory allocation 347–348
   multiplying each element by constant
     383–384
   slicing up multi-dimensional 242, 379–385
   storing 1D array in spreadsheet file
     214–219
   storing 2D array in spreadsheet file
     220–224
Array Constant,
   creating 247–248, 373, 553–554
   empty 304, 331

Array Control,
   1D construction 79,371
   2D construction 208
   element display 373
   index display 373
   using to input data array 371–376
Array Indicator,
   construction 79–82
   element display 79
   index display 79, 82
   automatic creation 88–89
   Digits of precision 224
   displaying multiple elements 390–391
   grid cursor 390
Array of clusters,
   creating 116, 326–327
   multiple plots on XY Graph 326–329
Array shell,
   function 79
   index display 79
Array Size icon 271
Array Subset icon 282, 320
Array To Spreadsheet String 232
Artificial data dependency 349–353
ASCII, see American Standard Code for
     Information Interchange
Auto-indexing feature,
   building array with For Loop 65–68
   building array with While Loop 77–78
   For Loop count terminal 268–269, 320
   inputting entire array into For Loop 319
   selecting Disable/Enable Indexing 66–68,
     78
   sequencing input array into For Loop
     268–269, 320
Automatic code generation 466–467
Automatic creation feature,
   Constant, Control, or Indicator 46–48
   subVI 284–289, 548–551
Automatic Tool Selection button 62
Automatic wiring 15
Autoscaling,

Autoscale button 30
Activating from pop-up menu 29
activating from Scale Legend 30
default for Waveform Chart 29
default for Waveform Graph 58
default for XY Graph 112
Scale Lock button 30
AutoSizing, Cluster shell 130

Backslash codes 237–239
Basis functions,
linear combination 370, 405
orthogonal 394
Beep.vi 93, 292
Best fit y-values 386
Beta formula 368–369, 394
Bidirectional current driver circuit 491–492,
579–582
Binary-formatted byte stream files,
comparison with ASCII text file
212–213
reading binary file 246–248
storing binary file 245–246
Block diagram,
placing, moving, and wiring icon 3–20
purpose 1
switching to front panel 2
Bode magnitude plot 206–207
Boolean data type 17
Bracket pair 66, 73
Broken wire,
indicator 17–18
removing 17
Buffered hardware-timed analog input 177
Build Array 116, 221–222, 326–327
Bundle,
creating cluster 71–74
element indexing 131–133, 375–376
resizing 72–73
Bundle By Name,
replacing cluster element 276–277, 327
Byte stream 233

Caption 333, 356–357
Case selector label 296–299
Case Structure,
Add Case for Every Value 306, 474–475,
553
Boolean 296–297, 300–304, 313–317
case selector label 296–299
Duplicate Case 317
enumerated type control 299, 304–309,
474, 552–556
numeric 297–299, 318–321
purpose 296
selector terminal 296, 297, 299
tunnels 316–317, 479
Channel,
analog input (AI) 155–157
analog output (AO) 184–185
physical 167–168, 454–455, 458
virtual 454–455, 489–490
Cloning 63–65
Cluster,
array of clusters 116, 326–327
cluster index 131–133, 375–376
cluster order cursor 133, 376
creating using Bundle 71–74
definition 69–70
Reorder Controls in Cluster 131–133,
375–376
replacing element using Bundle By Name
276–277, 327
shell 129–133
Unbundle By Name 133–134
Cluster index 131–133, 375–376
Cluster order cursor 133, 376
Cluster shell,
AutoSizing 130
using to create front-panel cluster 129–133
Coercion dot 19–20, 126, 173, 261
Coherent gain 436
Column-major order 217
Common commands,
*CLS 540

Common commands (*cont.*)
  *ESE 505, 540, 546–547
  *IDN? 515
  list of mandatory commands 504
  *OPC 505
  *OPC? 540–541
  *SRE 506
Complex-amplitude,
  definition 408
  magnitude 417–421
Complex exponential 405, 427
Complex To Polar icon 419
Complex To Re/Im icon 414
Concatenate Strings icon 234–235
Conditional branching 100, 296
Conditional terminal 9, 51
Confirm button 133, 376
Connector Pane,
  Add/Remove Terminal 143
  assigning terminals 143–145
  changing pattern using Patterns
    142–143
  Flip Horizontal 143
  input/output convention 142
  pattern (default) 142–143
Constant-current circuit 394–396,
  495–496
Context Help Window, see Help Windows
Control algorithms,
  proportional 493–494
  PI 494–495
  PID 495
Control cluster,
  AutoSizing 130
  creating a control cluster 129–130
  element ordering (default) 131
  Reorder Controls In Cluster 131–133
  retrieving using Select a Control 175
  saving using Control Editor 134–136
Control editor,
  activating editor and creating control
    471–473

Control mode 136, 473
  Type Def. and Strict Type Def. mode
    473–474
Control flow 338
Controls Palette,
  activating 22, 57
  changing appearanceusing Customize but-
    ton 22–23
  Modern vs. Silver palettes 23, 56
Count index 32
Count terminal,
  definition 57
  largest value 346
  using auto-indexing to set automatically
    268–269, 320
Convert from Dynamic Data icon 390
Convert to Dynamic Data icon 199, 389
Convolution Theorem 430–434
Create,
  Constant, Control, Indicator 46–48
Customize button 3–4, 22–23
Custom-made VI,
  assigning connector pane terminals
    142–145
  creating icon design 137–141
  placing on block diagram 197–199, 213
Curve fitting, theory
  linear least–squares method 370–371
  residual 386
Curve Fitting Express VI,
  General least square linear mode 385–388
  Non-linear mode 398–400

DAQ Assistant Express VI,
  limitations 201–202, 467–468
  minimizing overhead with stop input
    179–180
  opening dialog window 177
  programming for analog input voltage task
    164–171, 176–177
  programming for analog output voltage
    task 185–186, 195–196

programming for digital output
    task 203
programming for analog input thermistor
    input task 396
programming for analog input tthermo-
    couple input task 205–206
selecting input/output terminal 178
triggering, hardware 182–183
triggering, software 207–210
Use Waveform Timing 195–196, 206
DAQmx, icons
    Clear Task 456
    Create Virtual Channel 453–455
    Physical Channel Constant 457–458
    polymorphic VI selector 452–453
    Read 452, 455–456, 460
    Start Task 452, 455
    Stop Task 452, 455
    Terminal Constant 464–465
    Timing 462–463
    Trigger 464–465
    Write 452, 480–482
DAQmx, tasks
    advantages compared to Express VIs
        152–153, 467–468, 469–480
    automatically starting/stopping task 452,
        455
    creating task using MAX 488–489
    creating virtual channel using MAX
        489–490
    five-step task sequence 451–453
    physical channel 454–455, 458
    polymorphic VI selector 452–453
    programming counter input 485
    programming multiple voltage samples
        input 459–466
    programming single voltage sample input
        453–459
    programming single voltage sample output
        480–481
    programming voltage waveform output
        481–484

task in/out 454, 457
Data acquisition (DAQ) devices
    ELVIS II 154
    list of functions 152
    myDAQ 154
    PCI-6251 153–154
    pinout diagrams 156
    USB-6009 154
Data dependency
    artificial 349–353
    definition 338
    via refnum/error/task/resource name 230,
        349, 457, 502–503
Data flow 338
Data Storage,
    spreadsheet file 213–227
    three-step operation with File I/O VIs
        227–233
Data type,
    Boolean 17
    changing using Representation 43–46
    determining via LabVIEW Help
        Window 44
    dynamic data type 150, 173, 181,
        389
    floating point 19
    format string 225
    integer, signed 19, 45, 64
    integer, unsigned 19, 44–45
    waveform 97–98, 461
Data-type terminal 30–31
DC level 121, 415
Dead time 469
Debugging,
    broken Run button 109
    Error List 109–111
    Highlight Execution 352–353
    Probe Watch Window 83–88
Decrement 272
Default values, determining and defining
    Context Help Window 214, 377
    Make Current Values Default 27

Default values, various programming objects
    conditional terminal (Stop if True) 9
    connector pattern ($4 \times 2 \times 2 \times 4$) 142–143
    enumerated type control (U16) 122
    For Loop auto-indexing (Enabled) 66
    Format string (%.3f) 225
    integer Numeric Constant (I32) 45
    Numeric Control (DBL) 112
    Stop Button (FALSE) 34
    Waveform Chart autoscaling (x OFF, y ON) 29
    Waveform Graph autoscaling (x ON, y ON) 58
    While Loop auto-indexing (Disabled) 78
    XY Graph autoscaling (x ON, y ON) 112
Delimiter,
    multimeter data string 537, 567
    spreadsheet string 211–212, 567
Device name, DAQ 160–161
Device-specific commands 508–509, 527–528
Differential input mode 155–157, 162
Digital Input/Output (DIO)
    line and port 203
Digital oscilloscope,
    analog triggering 182
    properties 175, 182
    digital triggering 182–183, 207–210
Digital signal,
    rising and falling digital edge 182–183, 463–464
    transistor-transistor logic (TTL) 182–183, 464
Digital-to-analog (D/A) conversion, see analog output
Digits of precision 41, 224
Discretely sampled data,
    aliasing 158–159
    description 111, 258–259, 406–407
    Nyquist frequency 158, 406
    sampling frequency 111, 128–129, 158, 406

Displacement 111
Display Format
    Digits and Precision Type 41, 224
    Hide trailing zeros 41, 224
Distribution Tool 259
Divide icon,
    autocreation of Numeric Constant 61–62
    function 40–41
Dynamic data type (DDT)
    automatic labeling for Plot Legend 181
    automatic scaling for Waveform Graph 150, 173, 181
    Convert from Dynamic Data icon 390
    Convert to Dynamic Data icon 199, 389
    Express VI data type 150, 173, 181, 389

Elapsed Time Express VI 363
Element display 79
Element-wise operation
    multiplication and division 120
    power 120, 261
ELVIS II, pins
    analog input 155–156
    analog output 185
    digital output 203–204
    digital triggering 182–183
Empty Array Constant 304, 331
End of Line (EOL),
    platform-dependent definition 237
    icon 234
End of string (EOS) character 522
End or identify (EOI) 522
Enumerated Constant,
    Creating manually 471
    creating with Control Editor 471–474
    obtaining copy with Select a VI 474
    Type Def. and Strict Type Def. 473–474
Enumerated Type Control (Enum),
    Case Structure selector 299, 304–309, 552–556
    data type (default) 122
    digital display 123–124

programming using Edit Items window 122–124

purpose 122, 299

Equal? icon,

  Compare Aggregates vs. Compare Elements 476–477

  function 77

equivalent noise bandwidth (ENBW) 436, 440–442

Error cluster,

  elements 174, 189

  Explain Error 174, 189, 534–535

  wiring directly to Boolean icon 190–191, 458

Error function 371

Error handling 189, 230, 457

Error List 109–111

Error reporting,

  error cluster 174, 190–191, 503

  General Error Handler.vi 230, 457

Expandable Node,

  resizing 171–172

  selecting input/output terminal 178

Explain Error 174, 189, 534–535

Express VIs,

  advantages and limitations 152–153, 201–202, 467–468

  automatic code generation 466–467

  Curve Fitting 385–386

  DAQ Assistant 165–172

  Elapsed Time 363

  Instrument I/O Assistant 573

  Prompt User for Input 363

  Simulate Signal 150, 206, 207, 400–401

  Spectral Measurements 448–449

Fast Fourier Transform (FFT) algorithm 408–409

FFT.vi 413

File,

  ASCII text 211–212

  binary 212–213, 245–248

File I/O VIs 227–233

  including text labels 233–236

  path 226–227

  saving on disk 48–50, 214–224

  spreadsheet 211–212

  storing 1D array 214–219

  storing 2D array 220–224

  using to input data array to VI 376–385

File I/O VIs,

  Close File 229

  Open/Create/Replace File 228

  refnum 228, 230

  three-step file output (or input) process 228–230

  using to create spreadsheet file 227

  Write To Text File 229

Filled Rectangle 138–140

Finite geometric series 427

Flat Top window 436, 440

Floating-point numeric, single- vs. double-precision 19

Floating voltage source 164

For Loop,

  auto-indexing 65–68, 268–269, 320

  count terminal 57, 268–269

  iteration terminal 57

  memory manager 347–348, 360

  purpose 56–57

  resizing 60

Format Value icon 558

Format string 225

Formula Node 100

Fourier component 405

Fourier Transform,

  aliasing 443–444

  complex-amplitude 408

  complex exponentials 405

  continuous 405–406

  continuous inverse 405–406

  DC component 415

  DFT algorithm 409

  discrete 407–408

Fourier Transform (*cont.*)
discrete inverse 408
FFT algorithm 408–409
Fourier component 405
frequency spectrum 408
indexing convention 408, 409–412
magnitude and phase 405, 417–418,
448–449
Fract/Exp String To Number icon 530–531
Frequency doubling 446–447
Front panel,
purpose 1
switching to block diagram 2
Functions Palette,
activating 3, 57
categories 3–4
changing appearance using Customize
button 3–4
placing built-in icon on block diagram
10–13
placing custom-made icon on block diagram 197–199, 213

Gauss, Karl 257–258
General Error Handler.vi 230
Control of Instruments,
common commands 504
device-specific commands 508–509
end of string (EOS) 522
end or identify (EOI) 522
GPIB address 513–514
IEEE 488.2 503–507
interface bus 499, 510
Measurement & Automation Explorer
(MAX) 511–518
message termination 522–523
Standard Commands for Programmable
Instruments (SCPI) 508–509
synchronization methods,
serial poll 538–545
service request 545–551
timeout 523–527

Virtual Instrument Software Architecture
(VISA),
icons 499–503
interface bus independence 520
General Purpose Interface Bus (GPIB), see
Control of Instruments
Generation Mode,
1 Sample (On Demand) 185–186, 480
Continuous 195–196, 483–484
Gensignal command 120–121, 125–127
Geometric series, finite 427
Get Date/Time In Seconds icon 97–98
Gibbs phenomenon 294
Graph Palette, Zoom Tool 422–424
Grid cursor 390

Hanning window 434–435, 436, 438–439
Hardware-Timed Waveform Generation,
buffering 194–196
Help Windows,
aid in wiring 18–19
Context Help 13–14
Context Help vs. LabVIEW Help 22
default inputs 214, 377
Detailed Help 20
LabVIEW Help 20–22
question mark 20–21
required, recommended, optional inputs
214, 377
Show Optional Terminals and Full Path
button 214, 377
Hide Control 356
Hide trailing zeros 41, 224
Hierarchical programming 136, 284
Highlight Execution 352–353
Histogram.vi 91

I32, four-byte signed integer 45, 64
Icon, custom-made
assigning connector terminals 142–145
connector pane pattern (default) 142–143
Context Help Window 146

designing appearance 137–141
pixel size (32 × 32) 139
Icon Editor,
activating 137–138
Filled Rectangle tool 138–140
foreground/background color 139–140
Icon Text tab 137–138, 140–141
list of tools 138–139
Icon pane 137
Icon terminal 30
IEEE 488.2,
common commands 503–504
defined 503–507
end of string (EOS) character 522
end or identify (EOI) 522
message termination 522–523
status reporting 504–507
Index Array icon,
1D array 270–271
disabled index input 380
using to slice multi-dimensional array 242,
379–381
using for a look-up table 554
Index display 79, 82, 373, 391
Indicator cluster,
AutoSizing 130
creating an indicator cluster 129–130
element ordering (default) 131
obtaining copy using Select a Control
175
Reorder Controls In Cluster 131–133
saving using Control Editor 134–136
Initialize Array icon 360–362
Inputting data to VI,
using Array Constant 373
using Array Control 371–376
using disk file 239–242, 376–385
Instrument driver,
creating 551–565
downloading 566–567, 573
using in application program 565–572
Instrument I/O Assistant Express VI 573

Integer,
signed, four-byte (I32) 45, 64
unsigned, four-byte (U32) 44–45
Interface bus 499, 510
Inverse Fourier Transform,
continuous 405–406
discrete 407–408
I/O terminal block 155, 169–170
Iteration terminal 9, 57

Label,
comparison with Caption 333–334
element within Cluster 382, 384, 556
free 60
owned 59–60
including within files 233–236
Labeling Tool 25, 58, 60, 382
LabVIEW Help Window, see Help
Windows
Leakage,
analytic description 427–430
Convolution Theorem description
430–434
observing 422–427
Least-squares method,
linear 370–371
Linramp command 121, 125
Local variable 252
Lock-in amplifier 448
Logical equivalence 125
Look-up table 554
Low-pass filter 158, 204–205, 337, 444,
449–450

Make Current Values Default 27
Mapping Mode 207, 337
Marquee 12
MathScript Interactive Window,
activating 117
executing command in Command Window
117–118
Output Window 117–118

MathScript language, commands
  gensignal (periodic signal) 120–121,
      125–127
  linramp (equally distributed row vector)
      121, 125
  mod (modulo function) 314, 318
  plot (plot data) 118–119
  polar_to_cart (convert polar to Cartesian)
      148
  polyfit (polynomial curve fit) 151, 402–403
  polyval (evaluates polynomial) 404
  quadn_trap (numerical integration) 148
  randnormal (Gaussian random numbers)
      148
  roots (root finder) 148
  sin (sine function) 100–101, 107
  transpose (transpose matrix) 120, 127
MathScript language, syntax
  Array indexing compared to LabVIEW
      411
  classes of functions 101, 120
  creating 1D array 101
  element-wise operation 119–120
  for loop 411–412
  if-else statement 125, 317–318, 411–412
  logical equivalence 125
  matrix multiplication 120
  plotting data 118–119
  row vector 119–120, 126–127
MathScript Node,
  allowed data types 105, 109
  Auto Select output assignment 103–107
  Context Help Window 106–107, 126–127
  creating input and output variables
      103–106, 113–115
  purpose 99–101
  variable name 103
Mean square error (mse) 386
Measurement & Automation Explorer
      (MAX),
  activating 160, 511
  communicating with instrument 511–518

DAQ device pinout 162
device listing 160–161, 511–512
Scan for Instruments 512–513
self-test 161–162
Test Panel,
      DAQ 163–164
      VISA 514–518
using to create task 488–489
using to create virtual channel 499–500
Mechanical Action 35
Memory manager 347–348, 360–362
Menu button 458, 464, 519
Message termination 522–523
Modern Controls palette 23, 56
Modularity 284–289
Modulo function 314, 318
Multiply icon,
  multiply array by constant 383–384
myDAQ, pins
  analog input 155–156
  analog output 185
  digital output 203–204
  digital triggering 182, 207–210
  pinout diagram 156

Natural Logarithm icon 389
NI Device Drivers software 160, 511
NI-DAQmx Global Virtual Channel, creating
      in MAX 499–500
NI-DAQmx Task, creating in MAX 488–489
Node 338
Nonreferenced single-ended (NRSE) mode
      157–159
Number To Boolean Array icon 543–544
Numeric Constant,
  autocreation 46–48
  Display Format 41
Numeric Control,
  representation (default) 112
Numeric Indicator,
  autocreation 88–89
  polymorphic nature 173

Numerical differentiation 278–280
Numerical integration,
    built-in LabVIEW icon 329, 332
    Trapezoidal Rule 264–267
    Simpson's Rule 310–313
    convergence 275–278, 326–329
Nyquist frequency 158, 406

Object Shortcut Menu Tool 28
Open VISA Test Panel 514–518
Operating Tool, purpose 26
Or icon 190–191
Overhead 180, 469

Partial sum 312
Path 226
Path Constant 226–227
PCI–6251, pins
    analog input 155–156
    analog output 185
    digital output 203–204
    digital triggering 182–183
    pinout diagram 156
Peak power 440
Physical channels 167–168, 454–455, 458
Pick Line icon 557–558
Pinout diagram 156, 162
Platform portability 226–227
Plot Legend,
    automatic labeling with DDT 181
    Color 328–329
    Interpolation 38–39
    multiple plots 326–329, 387
    Point Style 39–40
    purpose 25
Polymorphic function 21–22
Polymorphic VI selector 378, 452–453
Pop-up menu,
    popping up 28
    purpose 28
Pop-Up Tool 28
Positioning Tool,

activating 7–8
    movement options 12–13, 221
    purpose 7
    toggling on with spacebar 62
Power function,
    prefactor and power 259
Power line cycle (PLC) 535–536
Probe Watch Window,
    creation 83–86
    purpose 83
    usage 86–88
Probe Tool 84
Program storage 48–50
Prompt User for Input Express VI 363
Property Node,
    BG Color 54, 308
    Caption 333
    Change To Write 54, 303
    creating 301–303
    History Data 54, 301–302
    Mapping Mode 337
    Name Label 337
    Timeout Value 523–527
    Value 354–355
    X-Axis Multiplier 54, 571
Proportional control 493–494
Proportional-Integral (PI) control 494–495
Proportional-Integral-Derivative (PID)
        control 495

Query 499, 515–518, 518–522
Quotient & Remainder icon 314

Random Number (0–1) icon 53, 91, 354
Range 157, 528, 552
Read From Spreadsheet File.vi
    file path 241, 379
    format 379
    function 377–378
    polymorphic VI selector 378
Real-time graphing 360–361, 362
Reciprocal icon 389

Rectangular window 430–434, 434–438
Reference single-ended (RSE) mode
    157–159
Refnum 228–230
Reorder Controls In Cluster,
    cluster index 133
    cluster order cursor 133
    cluster ordering 131–133
Replace Array Subset icon 360–362
Representation 21, 43–46
Residual array 386, 391–394
Resistance, temperature dependent 367–370
Resizing handle 8, 172
Resolution 158, 528, 552
RGB color method 307–308
RGB to Color.vi 307–309
Riemann Zeta function 331
Run button,
    broken 109
    keyboard shortcut 78
    program execution 32
Runtime error 188–191, 533–535

Sample clock 462
Sampling frequency 111, 128–129, 158, 177,
    406
Save 48–50
Save As...
    Plain Text 364
    Text (Tab Delimited) 367
    using to create copy of VI 192
Scale Legend,
    Autoscale button 30
    axes labels 30
    Scale Lock 30
Scaled Time Domain Window.vi 435–436
Scan for Instruments 512–513
Scrollbar 567
Search and Replace String icon 569
Select a Control 175
Select a VI 197–199, 213
Select icon 476

Selector terminal 296, 297, 299
Sequence Structure,
    Flat 339–342
    purpose 338–342
    Remove Sequence 350
    stacked 339
    tunnels 341–342
    use on data dependency diagram 350–351
Serial polling 507, 538, 538–545
Service request (SRQ) 506–507, 538,
    545–551
Set point temperature 491, 493
Shift register,
    creating 252–255
    initializing 256, 268
    multiple registers 253–255, 280
    purpose 252
    uninitialized 289–291
Show Optional Terminals and Full Pathbutton
    214, 377
Sidebands 447
Silver Controls palette 23, 56
Simpson's Rule, theory 310–313
Simulate Signal Express VI 150, 206, 207,
    400–401
Sine icon 10–14
Single-ended input mode,
    referenced vs. nonreferenced 157–159
Slicing multi-dimensional array 242,
    379–385
Software-timed waveform generation
    192–194
Software timing 186
Spacebar, toggling tools 62
Spectral Measurements Express VI 448–449
Spreadsheet file,
    creating with spreadsheet program 367
    creating with VI 213–227
    creating with word processor 364–366
    file format 211–212
    format string for numeric representation
        224–226

including text labels 233–236
opening with spreadsheet program 217
opening with word processor 216–217
reading with VI 239–242, 376–385
Spreadsheet String To Array icon 568
Standard Commands for Programmable
    Instruments (SCPI),
    defined 508–509
    root and subsystem 508
    syntax 508–509, 527–529
Standard Event Status Register (SESR)
    504–507
State machine,
    architecture 334–337, 469–480
    state 335, 470
Status Byte Register (SBR) 505–507
Steinhart-Hart Equation
    definition 370
    linearized 385
Stop Button Boolean switch,
    default value 34
    Mechanical Action 35
Strict Type Def. 473–474
String,
    backslash codes 237–239
    Constant 237
    converting string to numeric 250–251,
        530–531
    defined 211
    Indicator with scrollbar 567
String, icons
    Append True/False String 560–561
    Array To Spreadsheet String 232
    Format Value 558
    Fract/Exp String To Number 530–531
    Pick Line 557–558
    Search and Replace String 569
    Spreadsheet String To Array 568
Subtract icon,
    subtracting constant from each array
        element 390
subVI,

automatic creation 284–289, 548–551
    definition 136–137
    modularity 284–289
Synchronization methods,
    problem defined 532–538
    Serial Poll Method 538–545
    Service Request Method 545–551
Sync output, function generator 183

Tab Constant 234
Task 168, 452
Temperature control system,
    hardware construction 575–582
    defined 495–497
Temperature residual 391–392
Terminal, data-type vs. icon 30–31
Thermistor,
    constant-current circuit 394–396
    properties 367–370
    R-T data 364–366
    using as a thermometer 394–397
Thermocouple,
    calibration equation 397–398
    using as a thermometer 487, 572
Thermoelectric (TE) device,
    bidirectional current driver circuit
        491–492, 579–582
    properties 575–577
Thumbtacking 57
Tick Count (ms) icon,
    function 343
    precision 348
Timeout 523–527, 536
Timestamp 97–98
Tip strip 5
Toolbar 31
Tools Palette,
    Automatic Tool Selection button 62
    activating 7
    selecting tool using spacebar and tab 62
Transistor-transistor logic (TTL) 182–183,
    461

Transpose 2D Array icon 232
Trapezoidal Rule, theory 264–267
Triggering,
   analog level-crossing 182, 463
   digital edge 182–183, 463–464
   slope, falling and rising 182–183, 463–464
   trigger source 182–183, 464–465
Tunnel
Type Def. 473–474

U32, four-byte unsigned integer 44–45
Unbundle By Name,
   output terminal ordering (default) 133
   resizing 133
   selecting output terminal 133
   self-documentation 134
Undo 15
Universal Serial Bus (USB), see Control of
      Instruments
USB-6009, pins
   analog input 155–156
   analog output 185
   digital output 203–204
   digital triggering 182–183
   pinout diagram 156
Use Waveform Timing 195–196, 206

Virtual channel 454–455, 489–490
Virtual instrument (VI), defined 48
VIs, written by users of this book
   Alternating LEDs 203–204
   AM Wave 146–147
   Close (SRQ) 548–551
   Command String 551–562
   Convergence Study (Trap) 275–278
   Convergence Study (Trap vs. Simp)
      326–329
   Data Sampler 565–570
   DC Voltage Source (DAQmx) 480–481
   DC Voltage Source (Express) 185–192
   DC Voltmeter (DAQmx) 457–459
   DC Voltmeter (Express) 164–174

Digital Oscilloscope (DAQmx) 459–466
Digital Oscilloscope (DAQmx State Ma-
      chine) 469–481
Digital Oscilloscope (Express) 175–184
Digital Thermometer 394–397
End-Less Array 285–287
Endpoints 270–272
Estimated Frequency and Amplitude
      440–442
Even Ends 321–323
Extract First Two 281–282
Five Blinking Lights 329–330
Fractorial 289
FFT of Sinusoids 412–414
FFT (Magnitude Only) 418–421, 436–443
Four Flipping Coins 90–92
Frequency Calculator 409–412
Frequency Meter (DAQmx Count Edges)
      485
Frequency Meter (DAQmx Frequency) 485
Get Timeout Value 524–526
Loop Timer (Data Dependency) 349–352
Loop Timer (Sequence) 342–346
Measurement Config 562–564
Numerical Derivative 287–289
Parity Determiner 313–318
PID Temperature Controller 495–497
Power Function Derivative 283–289
Power Function Simulator 259–264
Range and Resolution Decoder 552–557
Reaction Time 354–355
R-T Data Input (Array Control) 371–376
R-T Data Input (Spreadsheet) 376–385
R-T Fit and Plot 387–394
Reformat Data String 567–569
Serial Poll 541–544
Set Timeout Value 526–527
Simple VISA Query 518–522, 530–531
Simple VISA Query (Long Delay)
      532–537
Simpson's Rule 323–324
Simpson's Test 324–325

Sine Wave Chart (While Loop) 2–50
Sine Wave Chart (While Loop with Run-time Options) 299–309
Sine Wave Generator (Software–Timed) 192–194
Sine Wave Graph (For Loop) 57–75
Sine Wave Graph (Mathscript Node) 102–108
Sine Wave Graph (While Loop) 75–90
Spectrum Analyzer 444–445
Spreadsheet Read 239–242
Spreadsheet Storage 213–227
Spreadsheet Storage (OpenWriteClose) 230–239
Status Config (Serial Poll) 538–541
Status Config (SRQ) 545–547
Stirling Test 289
Stopwatch 50
Sum of Partial Sums 318–321
Sum to 100 255–258
Take Data 564–565
Time Evolution of X 571–572
Trapezoidal Rule 267–273
Trapezoidal Test 273–274
VISA Query (Serial Poll) 544–545
VISA Query (SRQ) 547–551
Waveform Generator (DAQmx) 481–484
Waveform Generator (Express) 194–201
Waveform Generator (Modified Express) 201–202
Waveform Simulator 112–146
Virtual Instrument Software Architecture (VISA), icons
Clear 546
Close 502
Disable Event 548
Enable Event 547
interface bus independence 520
Open 500
Read 501
Read STB 507, 543
Resource Name Control 541

Wait for RQS 548
Write 500–501
VISA, session
byte count 501, 520–521, 533
four-step process 499–503
read buffer 501
resource name 500, 513–514, 518, 520, 541
write buffer 500–501
VISA property,
reading with Property Node 523–526
Timeout Value 523–527, 536
writing with Property Node 526–527
VISA resource name,
define 500
syntax,
GPIB 513–514, 520
USB 518, 520
Voltage Resolution 158, 184

Wait (ms) icon,
function 37–38, 173
input representation 43–46
precision 175
Wait Until Next ms Multiple icon 398
Waveform Chart,
autoscaling (default) 29
calibration by multiplicative constant 42–43, 54, 571
Chart Properties window 42–43
changing background color 54–55, 307–309
clearing manually 38
clearing using Property Node 54, 301–304
comparison with Waveform Graph 58–59
count index 32
Label 24–25, 28
Plot Legend 25, 38–40
Property Node 54, 301–309, 571
repositioning and resizing 25–26

Waveform Chart (*cont.*)
  Scale Legend 30
  scaling axes manually 26–27
  strip chart function 23–24, 32
  terminal, icon vs. data–type 30–31
Waveform data type 97–98, 461
Waveform Generation,
  hardware-timed 194–196, 481–484
  software-timed 192–194
Waveform Graph,
  autoscaling (default) 58
  calibration by multiplicative
    constant 69
  calibrating axes using Bundle 69–75
  comparison with Waveform Chart 58–59
  cursor 488
  function 58–59
  label axes with Labeling Tool 58
  multiple plots 70
  Plot Legend 181
  real-time graphing 360–361, 362
  Scale Legend 57–58
  terminal, icon and data-type 59
  using with dynamic data type 181
  using with waveform data type 97–98
While Loop,
  auto-indexing (default) 77–78
  conditional terminal 9
  iteration terminal 9
  iterating *N* times 76–77
  memory manager 347–348, 360–362
  placement on block diagram 6–9
  purpose 9
  Stop if True (default) 9, 18
  stopping gracefully 33–37
  time per iteration 33
Whiskers 19
Windowing,
  coherent gain 436
  commonly used windows 436
  equivalent noise bandwidth (ENBW) 436,
    440–442

  estimating frequency and amplitude
    440–442
  Flat Top window 436, 440
  Hanning window 434–435, 436,
    438–439
  purpose 434–440
  Rectangular window 430–434,
    434–438
  Scaled Time Domain Window.vi
    435–436
Wire clutter 374
Wires,
  1D array 66
  2D array 222
  array of clusters (pink) 327
  Boolean (green) 17
  broken 17–18
  cluster (pink) 74, 116
  dynamic data type (dark blue) 173,
    389
  floating-point (orange) 19
  integer (blue) 19
  single numeric (scalar) 66
  waveform (brown) 461
Wiring,
  automatic 15
  changing direction 17–18
  removing broken wires 17
  tip strips and whiskers 19
  with aid of Help Window 18–19
Wiring Tool,
  wiring procedure 15–18
Write To Text File.vi 229
Write To Spreadsheet File.vi,
  1D array 214–219
  2D array 220–224
  file path 226–227
  format string syntax 224–226
  transpose option 217–219

X-Axis Zoom 422–424
xy cluster, creating 115–116

XY Graph,
   autoscaling (default) 112
   cluster ordering 272–273, 275–276
   icon terminal 115–116
   multiple plots 326–329
   Plot Legend 328–329
   purpose 112

   single plot 115–117
   xy cluster 115–117
   Zoom Tool 422–424

Zoom Tool,
   X-Axis Zoom 422–424
   Zoom to Fit 424